Processing of High Temperature Superconductors

Related titles published by The American Ceramic Society:

Recent Developments in Electronic Materials and Devices
(Ceramic Transactions, Volume 131)
Edited by K.M. Nair, A.S. Bhalla, and S.-I. Hirano,
©2002, ISBN 1-57498-145-5

Dielectric Materials and Devices
Edited by K.M. Nair, Amar S. Bhalla, Tapan K. Gupta, Shin-Ichi Hirano,
    Basavaraj V. Hiremath, Jau-Ho Jean, and Robert Pohanka
©2002, ISBN 1-57498-118-8

The Magic of Ceramics
By David W. Richerson
©2000, ISBN 1-57498-050-5

Electronic Ceramic Materials and Devices (Ceramic Transactions, Volume 106)
Edited by K.M. Nair and A.S. Bhalla
©2000, ISBN 1-57498-098-X

Ceramic Innovations in the 20th Century
Edited by John B. Wachtman Jr.
©1999, ISBN 1-57498-093-9

Impact of Recent Advances in Processing of Ceramic Superconductors
(Ceramic Transactions, Volume 84)
Edited by Winnie Wong-Ng, U. Balachandran, and Amar Bhalla
©1998. ISBN 1-57498-031-9

Dielectric Ceramic Materials (Ceramic Transactions, Volume 100)
Edited by K.M. Nair and A.S. Bhalla
©1997, ISBN 0-57498-066-1

Advances in Dielectric Ceramic Materials (Ceramic Transactions, Volume 88)
Edited by K.M. Nair and A.S. Bhalla
©1996, ISBN 1-57498-033-5

Hybrid Microelectronic Materials (Ceramic Transactions, Volume 68)
Edited by K.M. Nair and V.N. Shukla
©1995, ISBN 1-57498-013-0

Grain Boundaries and Interfacial Phenomena in Electronic Ceramics
(Ceramic Transactions, Volume 41)
Edited by Lionel M. Levinson and Shin-ichi Hirano
©1994, ISBN 0-944904-73-4

Superconductivity and Ceramic Superconductors II
(Ceramic Transactions, Volume 18)
Edited by K.M. Nair, U. Balachandran, Y.-M. Chiang, and A.S. Bhalla
©1991, ISBN 0-944904-38-6

For information on ordering titles published by The American Ceramic Society, or
to request a publications catalog, please contact our Customer Service
Department at 614-794-5890 (phone), 614-794-5892
(fax), <customersrvc@acers.org> (e-mail), or write to Customer Service
Department, 735 Ceramic Place, Westerville, OH 43081, USA.

Visit our on-line book catalog at <www.ceramics.org>.

Ceramic Transactions
Volume 140

Processing of High Temperature Superconductors

Proceedings of the Processing of High Temperature Superconductors symposium held at the 104th Annual Meeting of The American Ceramic Society, April 28–May 1, 2002 in St. Louis, Missouri.

Edited by
Amit Goyal
Oak Ridge National Laboratory

Winnie Wong-Ng
NIST

Masato Murakami
ISTEC Superconductivity Research Laboratory

Judith Driscoll
Imperial College

Published by
The American Ceramic Society
735 Ceramic Place
Westerville, Ohio 43081
www.ceramics.org

Proceedings of the Processing of High Temperature Superconductors symposium held at the 104th Annual Meeting of The American Ceramic Society, April 28–May 1, 2002 in St. Louis, Missouri.

Cover photo: "SEM photographs for the fracture surfaces of melt-quenched samples" is courtesy of S. Nariki, N. Sakai, and M. Murakami, and appears as figure 2(a) in their paper "Processing and Properties of Gd-Ba-Cu-O Bulk Superconductor with High Trapped Magnetic Field," which begins on page 351.

For information on ordering titles published by The American Ceramic Society, or to request a publications catalog, please call 614-794-5890.

Printed in the United States of America.

4 3 2 1–05 04 03 02

ISSN 1042-1122
ISBN 1-57498-155-2

Contents

Bulk Processing

Preface

Major advances have been made during the last decade in the processing of ceramic, high temperature superconductors (HTS). High temperature superconductor materials are layered oxide compounds, which exhibit complex chemistry, including non-stoichiometry and defects (i.e., dislocations, stacking faults, and intergrowths).

For most large-scale, bulk applications of HTS, long lengths of flexible wires carrying large amounts of supercurrents are required. It has been well established that in order to achieve this long range, biaxial or triaxial crystallographic texturing of the superconductor is necessary. Significant efforts are now directed at attempts to fabricate long lengths of near single crystal-like superconducting wires by epitaxial growth on biaxially textured substrates. Typically these substrates comprise either a biaxially textured metal substrate with epitaxial oxide buffer layers, or an untextured metal substrate and biaxially textured oxide buffer layers. Techniques of substrate fabrication, which have received the most interest, include IBAD (ion-beam assisted deposition), RABiTS (rolling assisted biaxially textured substrates) and ISD (inclined substrate deposition). Epitaxial growth of $YBa_2Cu_3O_x$ (YBCO) and oxides on such substrates can be accomplished by a variety of techniques including pulsed laser ablation (PLD), electron beam evaporation, sputtering, chemical combustion vapor deposition (CCVD), jet vapor deposition, ex-situ BaF_2 process, ex-situ sol-gel techniques, and liquid phase epitaxy (LPE).

For other bulk applications, large, single crystal-like pucks of the superconductor are required. Applications requiring large domain levitators include frictionless bearings for flywheels, contact-less transportation, damping, flux-trap magnets, magnetic shields, and current leads. The superconducting material of choice in this case is melt-processed YBCO.

Lastly, basic information about HTS materials concerning phase diagrams, measurement of physical properties, characterization, and effects of various defects including grain boundaries on supercurrent transmission are of great interest and importance for further developments in this field.

This proceedings volume contains papers given at the Processing of High Temperature Superconductors symposium held during the 104th Annual Meeting of The American Ceramic Society (ACerS), April 28–May 1, 2002, in St. Louis, Missouri. The symposium focused on the above-mentioned issues pertaining to HTS materials as well as the following areas:

Materials Processing for Conductors
Biaxially textured substrates - RABiTS, IBAD, ISD
Other novel approaches to form biaxial texture
Physical vapor deposition techniques for deposition of oxide buffer
    layers and superconductors
Chemical vapor deposition techniques for deposition of oxide buffer
    layers and superconductor Novel synthetic methods
Physical properties (transport, flux pinning, field trapping)
Microstructure-property correlations
Issues related to scale-up
Application demonstrations

Materials Processing for Levitators
Fabrication of large area YBCO levitators
Magnetic characterization of levitators
Mechanical characterization of levitators
Issues related to scale-up
Application demonstrations

Basic Issues
Thermodynamics and phase equilibria
Grain boundary doping effects
Fundamental growth studies
Dopants, impurities and stability
Defects and microstructures
Non-stoichiometry
New materials such as MgB_2

The contents of this transaction volume comprise the proceedings of the focused session. A total of 34 scientific papers are featured in this volume. These contributions are divided into two parts: second generation wires and bulk processing. The order in which the papers appear here and the division into which they are organized may be different from that of their presentation at the meeting. It is hoped that this comprehensive volume will be a good summary of the latest developments in high-temperature superconductor research as well as good source material for researchers and managers working in this field.

We acknowledge the service provided by the session chairs and appreciate the valuable assistance from ACerS programming coordinators. We are also in debt to Ms. Mary Cassells and Ms. Sarah Godby for their involvement in editing and producing this book. Special thanks are due to the speakers, authors, manuscript reviewers, and ACerS officials for their contributions.

Amit Goyal

Winnie Wong-Ng

Masato Murakami

Judith Driscoll

Second Generation Wires

SURFACE-OXIDATION EPITAXY METHOD FOR CRITICAL CURRENT CONTROL OF YBa$_2$Cu$_3$O$_{7-\delta}$ COATED CONDUCTORS

Kaname Matsumoto
Department of Materials Science and
Engineering, Kyoto University
Yoshida-honmachi, Sakyo-ku
Kyoto 606-8501, Japan

Izumi Hirabayashi
Superconductivity Research
Laboratory, Div. V, ISTEC
2-4-3, Mutsuno, Atsuta-ku
Nagoya 456-8587, Japan

Kozo Osamura
Department of Materials Science and
Engineering, Kyoto University
Yoshida-honmachi, Sakyo-ku
Kyoto 606-8501, Japan

ABSTRACT

The surface quality of NiO buffer, produced by surface-oxidation epitaxy (SOE) method, was improved by a polishing and a cap layer coating techniques. Critical current density (J_c) of the YBa$_2$Cu$_3$O$_x$ (YBCO) film directly deposited on the polished SOE-NiO buffer by pulsed laser deposition (PLD) method was increased up to 0.17MA/cm^2 (77K, 0T) from 0.03-0.05MA/cm^2 (77K, 0T, on the non-polished buffer).

In addition, a cap layer of perovskite oxide (BaSnO$_3$) was prepared for promoting the epitaxial growth of YBCO films on SOE-NiO buffer. As a result, J_c of the YBCO film formed on BaSnO$_3$/NiO/Ni substrate reached 0.45MA/cm^2 (77K, 0T). This is because the number of superconductive weak coupling in the YBCO film was reduced, by the flattening of the NiO surface and the coating of the perovskite oxide cap layer.

INTRODUCTION

Simple and low cost production methods to form biaxially textured $YBa_2Cu_3O_{7-\delta}$ (YBCO) films on long-length metallic substrates are necessary; in order to obtain high critical current density (J_c) YBCO coated conductors. We have proposed a method to produce YBCO coated conductors by using surface oxidized NiO layer grown on long-length nickel tapes. The biaxial orientation of NiO can be realized by considering the epitaxial growth of NiO on the nickel surface. We call this technique surface-oxidation epitaxy (SOE) method.[1, 2] The formation of NiO on cube-textured Ni, or Ni-based alloys has been widely investigated recently.[3,4,5]

Biaxially textured NiO layer functions as a barrier layer of chemical reaction between YBCO and nickel, and as a template for the epitaxial growth of YBCO film. However, according to the research so far, the J_c of YBCO film directly deposited on NiO was only 0.03-0.05MA/cm^2 (77K, 0T).[6] Although the NiO crystal made by SOE is highly in-plane oriented, the surface roughness induced by grooves in the NiO grain boundaries remains large. Such a rough surface may generate the tilt grain boundaries in YBCO films deposited on NiO, and bring about the J_c decrease when the tilt angle is high. Furthermore, a Ni contamination may also bring about the degradation of critical temperature (T_c) of YBCO, since Ni atom from the underlying Ni substrate is easy to diffuse in the grain boundary.

An insertion of MgO cap layer between NiO and YBCO film was very effective in order to solve these problems, and J_c of 0.3 MA/cm^2 (77K, 0T) was achieved on SOE-NiO.[2] Nevertheless, the possibility that the local degradation of J_c happens, in the rough surface area, has still remained because the roughness of the NiO surface has not been essentially improved. We have tried to polish SOE-NiO surface mechanically, and also insert a perovskite oxide cap layer between NiO and YBCO film for promoting the epitaxial growth of YBCO films on SOE-NiO buffer. Consequently, high J_c value of 0.45 MA/cm^2 (77K, 0T) was successfully achieved on SOE-NiO.[7] In this paper, we describe J_c control of the YBCO films on the Ni tapes through SOE process.

ORIENTATION CONTROL OF NiO LAYER BY SOE METHOD

An oxide of underlying metal is formed on the metal surface by a thermal oxidation. In an initial stage of the formation of the oxide, the fixed relation exists between crystal orientation of the grown oxide and that of the underlying metal surface. The typical preferred orientations of NiO on Ni(100) face are NiO(100) and NiO(111).[8] Crystal orientation relationships between thermally oxidized NiO and Ni surfaces, which were observed at 500°C in oxygen atmosphere, are summarized in Table I.

Table I.   Orientation relationships between oxidized NiO on several Ni surfaces

Metal	Oxide	Oxidation condition	Crystal plane		Crystal orientation	
			Ni	NiO	Ni	NiO
Ni (fcc) a=3.53 Å	NiO (cubic) a=4.17 Å	1atm, O₂ 500℃ 10~120min	{100}//{111} {100}//{100} {110}//{110} {110}//{114} {111}//{111} {113}//{110}		[110]//[110] [110]//[110] [110]//[110] [110]//[110] [110]//[110] [110]//[110]	

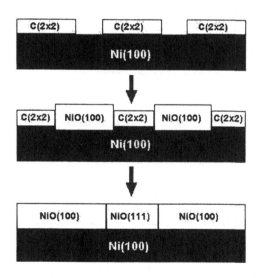

Fig.1. Schematic view of the initial stages of oxidation of Ni(100) surface.

The oxidation of metal begins by adsorbing oxygen molecules on the metallic surface. In the oxidation of (100) face of nickel single crystal, an adsorption of oxygen atoms by a two-dimensional arrangement is generated. A schematic view of the initial stages of Ni oxidation is presented in Fig.1. The layer of the simple cubic lattice in the period equal to nickel is gradually formed, when the adsorption advances, and finally, the isolated NiO nuclei is formed.[9] The NiO nuclei grow and coalesce, and the Ni surface would be covered with NiO. Then, a fixed crystal orientation relation occurs between NiO and Ni by the regulation of crystal orientation on the surface of underlying metal. Generally, (100)- and (111)-oriented NiO species coexists on the Ni(100) face because lattice misfit between NiO and Ni is large (misfit=18.1%).

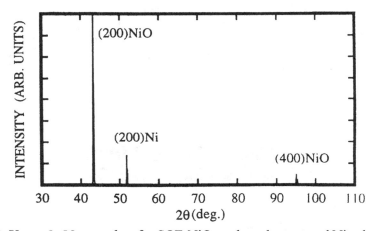

Fig.2. X-ray θ–2θ scans data for SOE-NiO on the cube-textured Ni substrate.

We have discovered, however, the NiO(100) face became dominant at the elevated temperature above 1000°C in air. In particular, the (100)-oriented NiO grains grew predominantly on the cube-textured {100}<001> nickel tape when it was oxidized under appropriate oxidation condition. Texture degree defined by I(200)/{I(200) +I(111)} have reached 0.99 at 1200°C, where I(200) and I(111) mean the X-ray θ–2θ intensities for NiO(200) and NiO(111) peaks. Figure 2 shows the typical X-ray θ–2θ scan data for the SOE-treated NiO/Ni substrate.

High Temperature Superconductors

Strong NiO(200) and (400) peaks are observed. This indicates the outermost layer of NiO is almost composed of the (100)-oriented NiO grains. SOE method utilizes these phenomena.

Cube-textured Ni tapes, prepared by the combination of rolling and heat treatment, are used for the SOE processing. The recrystallization heat treatment at 700-800°C is conducted in order to achieve both the flatness and the in-plane orientation of the Ni surface. The SOE heat treatment is carried out subsequently at around 1200°C to form the biaxially textured NiO layer on the Ni surface. Then, crystal orientation relation between NiO and Ni becomes NiO{001}//Ni{001} and NiO<110>//Ni<110>. Highly biaxial texture of SOE-grown NiO layer has been confirmed by the X-ray pole figure measurement. A typical FWHM of in-plane texture, $\Delta\phi$, of NiO(111) peak is 10-14 degrees and FWHM of out of plane texture, $\Delta\theta$, of NiO(200) peak become 6-8 degrees.

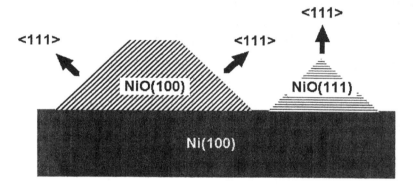

Fig.3. Orientation mechanism of biaxially textured SOE-NiO on Ni substrate.

At the initial stage of SOE process, (100)- and (111)-oriented NiO grains coexist. Why does NiO(100) face dominates on the cube-textured Ni tape after high temperature thermal oxidation? The reason would be speculated as follows: the growth rate of (111) face of cube crystal is larger than that of (100) face because (111) face is the non-singular face. Therefore, the NiO(111) grains grow only in the perpendicular direction, while it is permitted that the NiO(100) grains grow perpendicularly and horizontally, as illustrated in Fig.3. This means that the

NiO(100) grains laterally and quickly grow faster than the NiO(111) grains.

The (100)-oriented NiO grains coalesce and bury (111)-oriented NiO grains, and eventually covers all NiO surfaces. The NiO(100) grains grow by keeping epitaxial relation with Ni(100) surface, so that the biaxial orientation of NiO layer realizes. However, NiO surface becomes rough due to the coalescence of NiO(100) grains. The typical surface roughness of the SOE-NiO is approximately 30-150nm. The defects on the oxide buffer will induce the degradation of J_c of the YBCO films; hence the improvement of surface quality of SOE-NiO is strongly desired.

IMPROVEMENT OF J_c IN YBCO FILM ON SOE-NiO

YBCO films, prepared on SOE-NiO by pulsed laser deposition (PLD) with KrF excimer laser, show cube-on-cube epitaxy. A typical film formation condition is as follows: 700°C-substrate temperature, 200mTorr oxygen pressure and 2 J/cm^2 energy density. In Fig.4, X-ray pole figures of NiO(111) and YBCO(103) peaks for YBCO/NiO are shown. $\Delta\phi$ values of NiO(111) and YBCO(103) are 12 and 11 degrees, respectively.

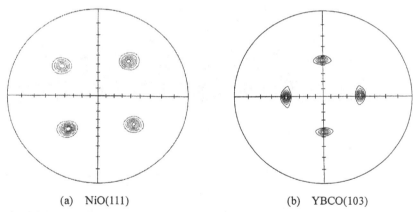

(a) NiO(111)   (b) YBCO(103)

Fig.4. X-ray pole figures of NiO(111) and YBCO(103) peaks for YBCO/NiO.

However J_c of YBCO films deposited on the bare SOE-NiO was only 0.03-0.05 MA/cm^2 (77K, 0T). This is one order of magnitude lower than that expected from in-plane texture of YBCO. Recently, the YBCO films with high J_c of 2-3MA/cm^2 (77K, 0T) have been successfully deposited on NiO(100) single crystal by PLD.[10] Considering the potentiality of NiO(100) surface as a template for epitaxial growth of YBCO, surface roughness of NiO and/or contamination of YBCO film by Ni element through NiO grain boundaries seem to be causes of J_c degradation in YBCO/SOE-NiO system.

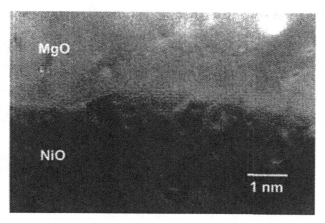

Fig.5. TEM cross-section of the interface between MgO layer and SOE-NiO.

It is effective to form a thin oxide cap layer on NiO in order to solve this problem. Improvement of surface quality of NiO has been attained by using thin MgO cap layer, and J_c of 0.3MA/cm^2 (77K, 0T) was obtained. The surface roughness of the buffer might be reduced with an excellent wetting property of MgO on NiO. TEM photograph of the interface between NiO and MgO is shown in Fig.5, where MgO was epitaxially and continuously grown on the SOE-NiO surface. MgO and NiO have a same crystal structure (rock-salt type) and they are completely soluble each other in solid state at all compositions. As a result the rough surface of NiO was nicely masked with MgO cap layer. MgO was also very effective in diffusion prevention of the Ni element. According to local EDX analysis, Ni did not diffuse in the MgO cap layer so that the Ni contamination was perfectly suppressed.

Even though MgO grows epitaxially on NiO, a careful control of the exact YBCO film deposition condition is necessary in order to form high J_c YBCO films on MgO. This is because the YBCO grains with cube-on-cube relation and with 45-degrees rotated orientation are easy to coexist on MgO.[11] Thus, the perovskite-type materials, which have better chemical compatibility to YBCO than that of MgO, were examined as a new cap layer in the following work.

MECHANICAL POLISHING OF SOE-NiO SURFACE

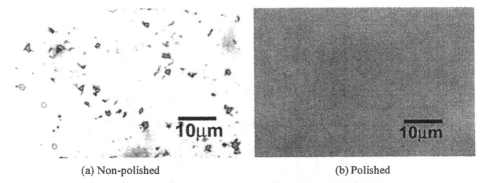

(a) Non-polished   (b) Polished

Fig.6. Photographs of Non-polished and polished SOE-NiO surfaces

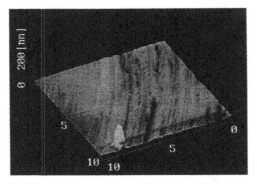

Fig.7. AFM picture of the polished SOE-NiO surface

In order to solve the rough-surface problem, the polishing of the SOE-NiO was investigated by using both 1 μm and 0.25 μm diamond pastes. Photographs of the non-polished and polished NiO surfaces are shown in Fig.6. The highly

High Temperature Superconductors

smooth and dense NiO surface was realized by this technique. The thickness of polished NiO layer was about 3 μm. No cracking in NiO occurred during polishing. We observed the surface morphology of polished NiO by AFM. Fig.7 shows the AFM image of the surface morphology of smooth NiO. Scan area of NiO was set to 10×10 μm^2. From the measurement of AFM image, the average surface roughness was 1.1 nm for smooth area and 5.5 nm for whole area.

The YBCO film deposited on the polished NiO surface by PLD showed strong c-axis orientation and cube-on-cube epitaxy. In-plane texture $\Delta\phi$ that determined from YBCO(103) peak was also 11 degrees. This value was similar to those of the other texturing technique. We have measured critical current (I_c) of the film at 77K and 0T. The bridge width and length formed in the YBCO film were 1.8 mm and 2.0 mm, and also, YBCO film thickness on NiO was 0.5 μm. Four terminals were attached by silver paste to the YBCO and the specimen was immersed into the liquid nitrogen. I_c of the YBCO film reached 1.52 A; namely, this corresponded to J_c of 0.17 MA/cm^2 (77K, 0T). These results indicate that the smooth NiO surface is crucial in order to produce the high J_c YBCO films on SOE-NiO.

PEROVSKITE CAP LAYER COATING

On the polished SOE-NiO, the perovskite oxide cap layer and YBCO films were grown by PLD. The perovskite materials, such as BaTiO$_3$, BaSnO$_3$ and BaZrO$_3$ were considered as a candidate since they included BaO layer and the excellent chemical compatibility with YBCO film was expected. As a first step, we chose BaSnO$_3$ material as a new cap layer because the crystal lattice constant is 0.411 nm.[12] The value is approximate for the lattice constant of NiO (0.416 nm) and misfit between them is -1.2%. Figure 8 shows the new configuration of YBCO /cap layer/SOE-NiO.

Deposition of BaSnO$_3$ and YBCO on the polished SOE-NiO was carried out by PLD at an energy density of 2J/cm^2 and repetition rate of 10-25 Hz. The substrate temperature was varied in the temperature range of 700-780°C and oxygen pressure was changed from 30 to 200mTorr to determine the optimal deposition condition. The bridges of 1.0 mm width and 2.0 mm length were patterned in YBCO films for I_c measurement.

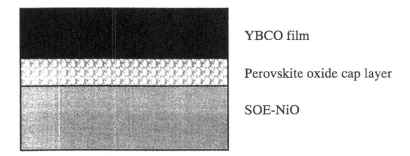

YBCO film

Perovskite oxide cap layer

SOE-NiO

Fig.8. Multi-layer structure of YBCO/cap layer/SOE-NiO.

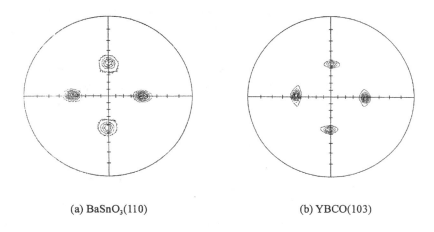

(a) BaSnO₃(110)   (b) YBCO(103)

Fig.9. X-ray pole figures of BaSnO₃(110) and YBCO(103) peaks
for YBCO/BaSnO₃/SOE-NiO configuration.

The BaSnO$_3$ films deposited on the NiO were c-axis oriented and in-plane textured and their thickness were 0.3-1.0 μm. Subsequently, YBCO films were deposited on the BaSnO$_3$ layer by PLD. The thickness of YBCO films on BaSnO$_3$/NiO was arranged to 0.5 μm. Figure 9 shows the pole figures of BaSnO$_3$(110) and YBCO(103) planes for YBCO/BaSnO$_3$/SOE-NiO configuration.

High Temperature Superconductors

It is proven that $BaSnO_3$ and YBCO films epitaxially grow on the polished NiO(100) surface with cube on cube relation. In-plane texture, $\Delta\phi$ of YBCO(103) X-ray peak, was 10.4 degrees and out of plane texture, $\Delta\theta$, for YBCO(005) X-ray peak was 2.5 degrees. This means that YBCO film on the $BaSnO_3$/NiO substrate had the advanced biaxial texture.

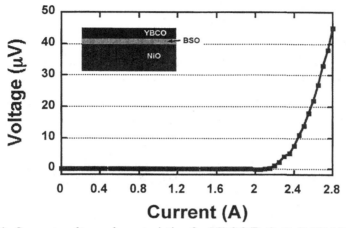

Fig.10. Current-voltage characteristics for YBCO/$BaSnO_3$/SOE-NiO system.

Current-voltage curve at 77K and 0T for YBCO/$BaSnO_3$/SOE-NiO specimen is presented in Fig.10. I_c of the patterned specimen was 2.25 A, which corresponded to J_c of 0.45 MA/cm^2 (77K, 0T). The improvement of J_c is due to the flattening of the NiO surface by the polishing and the good chemical compatibility between YBCO and $BaSnO_3$. It is considered that the perovskite material, such as $BaSnO_3$, is one of the candidates of cap layer on the NiO surface to obtain high J_c properties in the YBCO coated conductors by SOE method.

CONCLUSIONS

The surface roughness of NiO buffer by surface-oxidation epitaxy was improved by polishing technique. The surface roughness of NiO was reduced in this manner to 1-5nm, and J_c of 0.17MA/cm^2 (77K, 0T) was achieved for even YBCO films deposited on the bare NiO surface. In addition, the coating of the perovskite oxide cap layer was also examined instead of the MgO cap layer in order to promote the epitaxial growth of YBCO. We have obtained J_c of 0.45AM/cm^2 (77K, 0T) in YBCO/BaSnO$_3$/NiO/Ni system. The improvement of J_c was due to both the flattening of NiO surface and good epitaxy between YBCO and BaSnO$_3$. It is considered that the perovskite material, such as BaSnO$_3$, is one of the candidates of cap layer on the NiO surface to obtain high J$_c$ properties in YBCO coated conductors by SOE method.

ACKNOWLEDGMENTS

The author would like to thank T. Watanabe of Furukawa Electric Co., for valuable discussion. This work was supported by the New Energy and Industrial Technology Development Organization (NEDO) as Collaborative Research and Development of Fundamental Technologies for Superconductivity Applications.

REFERENCES

[1] K. Matsumoto, Y. Niiori, I. Hirabayashi, N. Koshizuka, T. Watanabe, Y. Tanaka, and M. Ikeda, "New Fabrication Method of High-J_c YBa$_2$Cu$_3$O$_7$ Superconducting Films on Flexible Metallic Substrates", pp.611 in *Adv. Superconductivity X*, Springer, Tokyo, 1998.

[2] K. Matsumoto, S. B. Kim, I. Hirabayashi, T. Watanabe, N. Uno, and M. Ikeda, "High Critical Current Density YBa$_2$Cu$_3$O$_7$ Tapes Prepared by the Surface-Oxidation Epitaxy Method", *Physica C* **330** 150(2000).

[3] T. Watanabe, K. Matsumoto, T. Tanigawa, T. Maeda, and I. Hirabayashi, "Surface-Oxidation epitaxy of Ni-Clad Ni-20wt%Cr and Ni-Clad Austenitic Stainless Steel tapes for Y-Ba-Cu-O Coated Conductors", *IEEE Trans. Appl. Supercond.*, **11** 3134(2001).

[4] Z.Lockman, X. Qi, A. Berenov, R. Nast, W. Goldacker, J. MacManus-Doriscoll, "Study of Thermal oxidation of NiO buffers on Ni-based Tapes for Superconducting Substrates", *PhysicaC* **351** 34(2001).

[5] T. Petrisor, V. Boffa, G. Celentarno, L. Ciontea, F. Fabbri, V. Galluzzi, U. Gambardella, A. Mancini, A. Rufoloni, and E. Varesi, "Epitaxial Oxidation of Ni-V Biaxially Textured Tapes", *PhysicaC* (2002), in press.

[6] K. Matsumoto, S. B. Kim, J. G. Wen, I. Hirabayashi, T. Watanabe, N. Uno, and M. Ikeda, "Fabrication of In-Plane Aligned YBCO Tapes Using NiO Buffer

Layers Made by Surface-Oxidation Epitaxy", *IEEE Trans. Appl. Supercond.* **9** 1539(1999).

[7]K. Matsumoto, I. Hirabayashi, and K. Osamura, "Surface-Oxidation Epitaxy Method to Control Critical Current of $YBa_2Cu_3O_7$ Coated Conductors", *Physica* C (2002), in press.

[8]N.N. Khoi, W. W. Smeltzer, and J. D. Embury, "Growth and Structure of Nickel Oxide on Nickel Crystal Faces", *J. of Electrochem. Soc.*, **122** 1495(1975).

[9]R.S. Sakai, A. P. Kaduwela, m. Sagurton, J. Osterwalder, D. J. Friedman, and C. S. Fadley, "X-ray Photoelectron Diffraction and Low-Energy Electron Diffraction Study of the Interaction of Oxygen with the Ni(100) Surface: c(2x2) to Saturated Oxide", *Surface Science* **282** 33(1993).

[10]T. Maeda, H. Iwai, T. Watanabe, and I. Hirabayashi, unpublished.

[11]D. M. Hwang, T. S. Ravi, R. Ramesh, S. W. Chan, C. Y. Chen, L. Nazar, "Application of a Near coincident Site Lattice Theory to the Orientations of $YBa_2Cu_3O_7$ Grains on (001)MgO Substrates", *Appl. Phys. Lett.* **57** 1690(1990).

[12]S. Miyazawa and M. Mukaida, "Transmission electron Microscope Observation of Interfaces in YBa2Cu3Ox/BaSno3/YBa2Cu3Ox Trilayers", *Jpn. J. Appl. Phys.*, **37** L949(1998).

CONTROL OF THE SULFUR C (2 × 2) SUPERSTRUCTURE ON {100}<100>-TEXTURED METALS FOR RABITS APPLICATIONS.

C. Cantoni, D. K. Christen, L. Heatherly, F. A. List, A. Goyal, G. W. Ownby, and D. M. Zehner

Oak Ridge National Laboratory, 1 Bethel Valley Rd., Oak Ridge, TN 37831

ABSTRACT

We investigate the influence of a chemisorbed S template with centered (2 × 2) structure on the epitaxial growth of commonly used oxide buffer layers on {100}<100> metals for RABiTS fabrication. Our study involves growth of CeO_2, Y-stabilized Zr_2O_3, Gd_2O_3, $LaMnO_3$ and $SrTiO_3$ seed layers on biaxially textured Ni and Ni-alloys. We also discuss the effect of an incomplete c (2 × 2) surface coverage on the seed layer texture and the use of H_2S in a pre-deposition anneal as a mean to control the superstructure coverage and optimize the seed layer texture.

INTRODUCTION

Recently we found that sulfur impurity atoms present on the Ni (001) surface are determinant for the epitaxial growth of certain oxide films used as seed buffer layers in RABiTS conductors (e.g.: CeO_2 or Y_2O_3-stabilized ZrO_2) [1,2]. Our results showed that chemical and structural properties of the textured metal surface have to be considered in order to obtain the desired texture for the buffer layers. Using reflection high-energy electron spectroscopy (RHEED), and Auger electron spectroscopy (AES), our experiments revealed the existence of a sulfur superstructure on the textured Ni surface that forms after surface segregation of sulfur contained as a common impurity in the metal bulk. During the high-temperature texturing anneal S atoms diffuse to the surface of the tape and arrange in a centered (2×2) superstructure. However, depending on the initial S concentration and/or specific annealing conditions, the superstructure layer can exhibit different coverage.

Here we analyze the relationship between S surface concentration and seed layer texture, and present a method that controls and optimizes the superstructure coverage on the metal, providing reproducible quality of the seed layer texture. In addition, we extend our study of the influence of the S superstructure to the case of biaxially textured Ni alloys, such as Ni-3%W, and Ni-13%Cr.

SEED LAYERS ON {100}<100> Ni BY PULSED LASER DEPOSITION

CeO_2 seed layers are successfully deposited by vapor deposition techniques like e-beam evaporation, reactive sputtering, and pulsed laser deposition (PLD). All these techniques use a background H_2O pressure of about 1×10^{-5} Torr and a substrate temperature in the range of 600 to 750 °C. Water dissociation provides enough oxygen in the background to form and stabilize the CeO_2 phase without oxidizing the Ni [3]. The stability of the Ni surface in the presence of water with a partial pressure in the range of 1×10^{-5} Torr was investigated by *in situ* RHEED performed inside the laser molecular beam epitaxy (MBE) chamber used to fabricate films for this study. It was found that such H_2O partial pressure does not cause any structural modification of the sample surface that can be related to growth of NiO islands or adsorbed O. The documented first stage of Ni oxidation consists of a chemisorbed O layer with p (2×2) structure up to a coverage of 0.25 ML, and with subsequent c (2×2) structure up to a coverage of 0.5 ML [4] [1 ML $\equiv$ (number of surface adsorbate atoms) / (number of surface substrate atoms)]. None of these structures was observed to form as a consequence of water exposure on a clean {100}<100> Ni surface at a typical deposition temperature of 600 °C. The clean Ni surface was obtained by depositing a Ni film *in situ* on the textured Ni substrate in ultrahigh vacuum conditions at 600 °C by PLD. RHEED

High Temperature Superconductors

and AES studies revealed that the as-grown films were free of impurities at typical buffer-layer deposition temperatures.

Figure 1a shows a RHEED pattern acquired during the nucleation of CeO_2 on a {100}<100> Ni substrate having the c (2 × 2)-S superstructure shown in Fig. 1b. The electron beam is directed along the Ni <100>. The CeO_2 film was grown at 600 °C in the presence of a water partial pressure of 1.5×10^{-5} Torr. The pattern shows diffracted spots that correspond to a (002) biaxially textured CeO_2 film viewed along the <110> crystal direction. This is consistent with the general observation that CeO_2 grows with a 45° in-plane rotation with respect to the {100}<100> Ni substrate. The presence of spots in the RHEED pattern indicates that the nucleation is occurring by formation of islands, which are tall enough to produce a transmission diffraction pattern. The spacing between the RHEED reflections in Fig.1a provides a rough estimate of the in-plane lattice parameter of the film, which is 5.43 ± 0.05 Å.

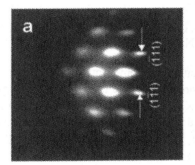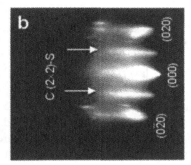

Figure 1. RHEED pattern obtained at 600 °C during the nucleation of a CeO_2 seed layer on a biaxially textured Ni substrate with a c (2 × 2)-S surface structure. The electron beam is directed along the Ni <100> and the spacing indicated by the arrows corresponds to the interatomic distance $a_{CeO_2}\sqrt{2}/2$, (a). Substrate RHEED pattern showing the extra reflections indicating the sulfur c (2 × 2) superstructure, (b).

Figure 2 shows an XRD φ–scan of the (111) peak for the completed CeO_2 film, approximately 400 Å thick. We found that the deposition of CeO_2 on the c (2 × 2)-S template consistently produced highly textured (200) films, while deposition on a clean Ni surface always gave rise to (111) oriented films.

Among other oxides that have been successfully used as seed layers, those grown directly on a textured metal by PLD for this study are: $SrTiO_3$ (STO) and

LaMnO$_3$ (LMO), with perovskite and pseudo-perovskite structure, respectively; Y-stabilized Zr$_2$O$_3$ (YSZ), with fluorite structure; and Gd$_2$O$_3$, with RE$_2$O$_3$ structure. For all these oxides, RHEED and X-ray measurements have shown that the S superstructure is needed to obtain (001) epitaxial growth.

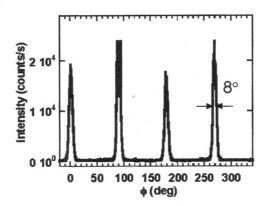

Figure 2. XRD ϕ-scan of the (111) peak for a CeO$_2$ seed layer deposited on the c (2 × 2) sulfur template chemisorbed on {100}<100> Ni.

YSZ seed layers were deposited at substrate temperatures ranging between 600 °C and 800 °C using the following procedure. After an initial ~100-Å-thick layer was deposited in vacuum ($P_{base} \leq 5 \times 10^{-8}$ Torr), the O$_2$ partial pressure was increased to the value 1×10^{-5} Torr and a final 1200-Å-thick film was grown at the same chosen deposition temperature. YSZ films grown on the c (2 × 2) surface showed single (002) orientation, with a (111) pole figure indicating the same degree of grain alignment as the substrate. YSZ films grown on the superstructure-free Ni overlayer showed only the (111) peak in the θ-2θ scan. In this case, the pole figure of the (200) reflection showed 4 different in-plane domains rotated 30° with respect to each other. This epitaxial relation is expected for the nucleation of a threefold symmetric lattice on a square symmetric lattice. The quality of the YSZ seed layers on c (2 × 2)/Ni was tested by growing a 0.3-μm-thick YBCO film by the *ex situ* BaF$_2$ method [5] on some of the samples. A 20-nm-thick CeO$_2$ cap layer enabled compatibility of the precursor layer with the YSZ. The resulting YBCO critical current density was 1.15 MA/cm^2 in self-field at 77 K, indicating that a 120-nm-thick YSZ film is a good Ni diffusion barrier.

High Temperature Superconductors

Although STO is thermodynamically less stable than bulk NiO, it can be easily grown on Ni at oxygen pressures as low as 1×10^{-9} Torr, in conditions at which surface NiO will not form[*]. The ease of growth of STO on Ni can be attributed in part to the stability of the STO crystal structure in presence of oxygen vacancies [6].

Figure 3a shows the RHEED pattern for a (200) epitaxial STO film grown on the textured Ni substrate with S superstructure. The electron beam is directed along the <100> direction and the epitaxial relationship between STO and Ni is cube-on-cube. The corresponding STO (111) pole figure is shown in Figure 3b. The STO films were grown at a substrate temperature of 700 °C. The oxygen pressure was kept in the range $1 \times 10^{-9} - 1 \times 10^{-8}$ Torr during the deposition of a 20-to-100 Å-thick nucleation layer and then increased to a value of about 1.0×10^{-5} Torr. On a clean (1×1) Ni surface, the STO seed layer grows multi-domain, with the cube axis remaining parallel to the substrate normal, but with two additional in-plane orientations besides the one with STO <100> // Ni <100>.

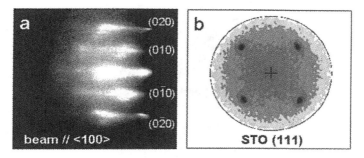

Figure 3. RHEED pattern of an STO film grown epitaxially on a {100}<100> Ni substrate showing the c (2×2) sulfur superstructure, (a). Corresponding (111) logarithmic pole figure of the STO film indicating cube on cube epitaxial relationship with the Ni substrate, (b).

---

[*] According to the thermodynamics for the bulk chemical reaction $2Ni + O_2 = 2NiO$, an oxygen pressure in the range of 10^{-9} Torr is sufficient to form NiO at a typical deposition temperature of 600 °C. However, surface contributions to the free energy of formation of NiO appear to be significant. Our RHEED experiments have shown that the surface of {100}<100> Ni is in fact stable in an oxygen partial pressure as high as 5×10^{-7} Torr at a temperature of 600 °C.

Recently La$_{0.7}$Sr$_{0.3}$MnO$_3$ (LSMO) and LMO were evaluated as promising single buffer layers for coated conductors. YBCO films deposited by PLD on sputtered LSMO or LMO buffered Ni RABiTS have shown high critical currents [7]. The LMO phase could be nucleated on Ni, by PLD, only at oxygen partial pressures higher than those used for CeO$_2$. We found, for example, that, for a substrate temperature of 600 to 700 °C, growth in a water partial pressure of 1×10^{-5} Torr gave rise to amorphous films, while a water background of 6 mTorr was sufficient to obtained a crystalline (001) textured film. Figure 4 shows the RHEED patterns of an LMO film, approximately 80 nm thick deposited on c (2 × 2)/Ni at a temperature of 600 °C and a H$_2$O partial pressure of 6 mTorr. The RHEED pattern is consistent with cube on cube texture of the LMO on Ni. The spot-like reflections are indicative of a rough surface morphology.

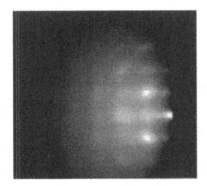

Figure 4. RHEED pattern of an 80 nm-thick LMO film deposited on Ni by PLD.

It is evident that the c (2 × 2)-S superstructure plays a very important role in the nucleation of an oxide buffer layer on Ni. We propose that the effect of the S superstructure can be partially explained on the basis of structural and chemical considerations. The S layer behaves like a template that well matches and mimics the arrangement of the oxygen atoms in particular (001) sub-lattice planes of all the seed layer considered in this study, as shown in Figure 5. Sulfur belongs to the VI group and is chemically very similar to oxygen, often exhibiting the same electronic valence. Therefore, it is plausible that during the seed layer deposition the cations easily bond to the S atoms already present on the substrate surface, giving rise to the (001) epitaxial growth of the film, which otherwise would not take place. Ironically, the desired structural and chemical properties of RABiT substrates made in the past few years can be attributed, partially, to the involuntary presence of a sufficient S content in the rolled Ni metal. In cases for which the bulk S content of particular batches of Ni was much lower than 30 wt. p.p.m., the seed layer deposition process produced films with degraded texture or even partially (111) orientation. In those cases, as a recent Auger electron spectroscopy (AES) analysis revealed, S was depleted in the near-surface layer during a high temperature anneal (~1100 °C). This may be the result of the formation and subsequent desorption of SO_2 in the presence of a sufficient partial pressure of oxygen. Sulfur depletion impeded formation of a continuous $c(2 \times 2)$ layer across the entire Ni surfaces, drastically modifying the oxide film nucleation.

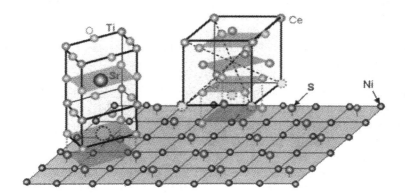

Figure 5. Schematic model for the nucleation of CeO_2 (or YSZ) and STO on a (001) Ni surface with chemisorbed c (2×2) sulfur superstructure. The dashed lines indicate correspondence between oxygen sites in (001) planes of the seed layer and sulfur sites on the Ni surface. The seed layer cations impinging on the Ni surface bond easily to the sulfur atoms present on the metal surface, promoting the (001) orientation of the growing film. In the STO case, there is a 1:1 correspondence between oxygen atoms in the SrO plane and sulfur atoms on the Ni surface. Therefore, it is plausible that Ti ions initially bond to the S surface atoms to form the first TiO_2 plane of the STO structure. In the CeO_2 case, two of four oxygen ions per unit cell match the sulfur atoms of the c (2×2). During nucleation of CeO_2, oxygen atoms may fill in the empty fourfold Ni hollows and the Ce cations subsequently bond to the template formed by S and O.

QUANTIFICATION AND CONTROL OF THE S-SUPERSTRUCTURE

The above-mentioned considerations illustrate a connection between the quality of the seed layer texture and the degree of coverage of the c (2×2)

superstructure on Ni. To investigate this aspect, we conducted combined AES and RHEED experiments aimed at quantifying the coverage of the c (2 × 2) structure on different Ni samples. During Auger spectroscopy experiments, depending on the material under test and the beam energy, Auger electrons escape and are detected from a layer 5 to 10 Å deep below the atomistic surface. Therefore, the S concentration as deduced from AES can differ from the actual atomic concentration of the top monolayer of the surface 1), because of the uncertainty related to the measurement itself; and 2), because, after segregation, S could exhibit a concentration gradient in the layer measured by the Auger technique. We solved this problem by comparing the AES results obtained on typical samples with those obtained on initially superstructure-free samples on which the c (2 × 2) was intentionally grown through S adsorption. In these experiments, a clean Ni surface was first produced by depositing a Ni layer *in situ*. Subsequently, a small amount of H_2S with a partial pressure of $5 \times 10^{-7} - 1 \times 10^{-6}$ Torr was introduced in the vacuum chamber at a substrate temperature of 700-800 °C for few minutes and then pumped away. As known from several surface studies, the H_2S molecules dissociate at the Ni surface and S atoms chemisorb forming a c (2 × 2) superstructure with a coverage that saturates at 0.5 ML, corresponding to one complete atomic layer of the c (2 × 2)-S [8-10]. Exposures to H_2S as low as a few L [1L (langmuir) $\equiv 10^{-6}$ Torr-s] produced very strong c (2 × 2) reflections in the RHEED patterns as shown in Fig. 6a. The sulfur superstructure was stable at 800 °C after the H_2S was removed. CeO_2 and YSZ seed layers deposited *in situ* after S adsorption were highly oriented with a percentage of cube texture very close to 100%, as indicated by the (111) pole figure of Fig. 6b. Longer exposure of the Ni surface to H_2S (≥ 1800 L) did not produce any degradation of the RHEED pattern that could be attributed to adsorption of surface S in excess of 0.5 ML, or formation of a Ni sulfide phase. The as-grown superstructure was found to be stable also after exposure to air at room temperature and consequent re-heating of the sample in vacuum, which reproduced the same initial RHEED pattern. AES experiments performed *in situ* after S adsorption yielded a sulfur signal of about 25% at saturation. The same value was obtained on samples where the c (2 × 2)-S had formed consequent to segregation, and that showed RHEED patterns identical to the ones acquired in the S adsorption experiments. From these observations we associated an AES value of ~25% to a full layer of c (2 × 2)-S on the Ni surface. Consequently, we were able to quantify the S surface content in cases in which the superstructure did not cover the entire Ni surface and the c (2 × 2) reflections in the RHEED pattern were less intense than in fully covered samples.

We noticed that, in cases for which the S coverage was equal or less than 0.25 ML after the texturing anneal, S appeared to have still formed a c (2 × 2) structure and not a p (2 × 2) as typically observed in S adsorption experiments by others

[11]. In fact, no p (2×2) reflections could be identified in RHEED patterns with the electron beam parallel to the Ni <110> direction in these cases.

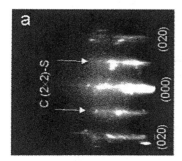 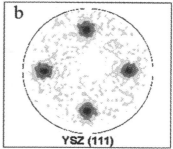

Figure 6. RHEED pattern showing strong c (2×2) reflections obtained after exposing a clean {100}<100> Ni surface to few L of H_2S at 800 °C, (a). Logarithmic (111) pole figure of a YSZ seed layer grown *in situ* on the c (2×2)-S shown in (a), (b). For this sample the x-ray background signal was eliminated by subtracting a second pole figure obtained after changing the Bragg angle by 1 degree. The calculated percentage of cube texture for this film was 99.96%.

Figure 7 is a comparison of ϕ scans of the (111) peak for 3 different STO films grown on Ni samples showing 1), a complete c (2×2) layer; 2), 40% of the full $c(2 \times 2)$ coverage; and 3), no superstructure, respectively. We notice that an incomplete $c(2 \times 2)$ coverage is not sufficient to obtain full cube texture of the seed layer, and, in the case of 40% coverage, additional peaks indicating crystal domains grown with orientations other than cube on cube, are still present in the ϕ scan.

Figure 8 shows a similar comparison for two CeO_2 films deposited on two different Ni samples cut from the same tape but subjected to two different texturing anneals. In the first case the sample was annealed in high vacuum at 1100 °C for 2 h and exhibited a complete S superstructure. In the second case the anneal was much longer (15 h) and caused partial evaporation of the surface S resulting in a weaker $c(2 \times 2)$ superstructure with about 70 % coverage (see insets of Fig. 7). In both cases the Ni substrate showed a FWHM of the (111) peak of 8.2° ± 0.4°. The CeO_2 films were cube textured and showed single (002) orientation on both the Ni substrates. However, it is evident that the degree of grain alignment for the CeO_2 films correlates with the coverage of the c (2×2)

superstructure on the substrate, and the incomplete S template results in a broadening of the seed layer texture. We found that the texture of the seed layer replicates that of the substrate only when the $c(2 \times 2)$ has a full 0.5 ML coverage.

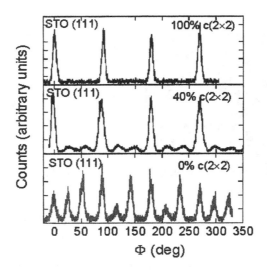

Figure 7. Comparison between ϕ scans of the (111) peak for 3 different STO films grown on {100}<100> Ni substrates having different sulfur coverage: 100% of a c (2×2) layer (or $\theta = 0.5$), top; 40%, middle; 0%, bottom.

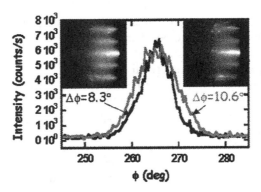

Figure 8. ϕ scans of the (111) reflection of a CeO_2 seed layer grown on a Ni substrate annealed for 2 h (black line), and on a Ni substrate annealed for 15 h (gray line).

Sulfur adsorption experiments provide a method for quantifying the S coverage on the Ni surface after the texturing anneal. But, more importantly, H_2S exposure provides a rapid and efficient way to obtain a full coverage of the c (2 × 2)-S on textured Ni, and, consequently, optimize the seed layer texture without being restricted by the amount of S actually present in the bulk and the slow, less efficient segregation process. H_2S exposure has been successfully introduced in a continuous process for texturing and CeO_2 coating of meter-long Ni tapes, yielding highly oriented and reproducible long lengths of RABiTS [12]. YBCO films deposited on short sections of these RABiTS tapes by PLD and *ex situ* BaF_2 method have consistently shown critical currents larger than 1 MA/cm^2.

RESULTS OF SEED LAYER GROWTH ON Ni-3%W AND Ni-13%Cr

Because of the magnetic properties of Ni, coated conductors fabricated using Ni tape show losses in the presence of AC electromagnetic fields, which represents a serious problem for certain applications [13]. Alloying Ni with small quantities of metals like W and Cr results in lower or zero magnetization (the latter being the case of the Ni-13%Cr alloy). In addition, Cr and W add strength to the tape, making handling of lengths easier.

Knowing the importance of the S superstructure for epitaxial seed layer growth on Ni, it is natural to wonder whether the superstructure plays the same role in textured Ni alloys. Figure 9 shows Auger data and a RHEED pattern for a textured Ni-3%W substrate heated from 500 °C to 900 °C in high vacuum. It is evident in this case that, just as for Ni, S surface segregation during texturing has resulted in a c (2 × 2) superstructure with complete coverage. The oxygen and carbon signals in the Auger plot derive from air exposure after the texturing anneal, and desorb easily by mild heating. As expected from these observations, high quality buffer layers were deposited on such Ni-W surfaces.

Figure 10 shows the RHEED pattern of a Gd_2O_3 grown by PLD on a textured Ni-3%W substrate at a temperature of 600 °C and a water pressure of 1×10^{-5} Torr. The figure indicates that the seed layer has grown epitaxially with (004) orientation. The weaker streaks marked by the arrows are superlattice reflections originated by the periodicity of the large cubic cell (a = 10.813) of Gd_2O_3. [3]. Fully textured CeO_2, YSZ and Y_2O_3 seed layers have also been deposited on c (2 × 2)-S/Ni-3%W, by PLD or other deposition techniques, using the same growth conditions established for deposition on Ni. We would like to point out that, because of the higher reactivity with O_2 of W, as compared to Ni, careful control of background oxygen is important during buffer layer deposition on Ni-W alloys. RHEED observations showed that while the Ni-3%W surface is stable in the presence of a water pressure in the range of 10^{-5} Torr, H_2O pressures in the range of 10^{-4} or higher cause changes in the diffraction patterns that can be interpreted as due to oxide formation. The surface of pure Ni remained stable against oxidation at water pressures as high as tenths of mTorr.

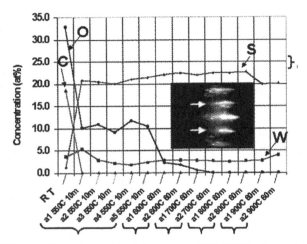

Figure 9. AES data collected on a biaxially textured Ni-3%W alloy substrate after heating at different temperatures and holding for 10 or 60 minutes. The inset shows the RHEED pattern obtained after heating to 550 °C.

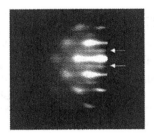

Figure 10. RHEED pattern for a 100 nm-thick G_2O_3 film on Ni-3%W

The reactivity with oxygen is substantially higher for Ni-13%Cr substrates. In fact, attempts to deposit usual seed layers like CeO_2 or Y_2O_3 by several vapor deposition techniques on Ni-13%Cr have failed. A combination of RHEED and Auger measurements on these substrates has shown that C and O species are strongly bound to the Ni-Cr surface after air exposure, and high temperature anneals (700-750 °C) in high vacuum, or 200 mTorr of forming gas, before

High Temperature Superconductors

deposition are necessary to clean the surface and expose the underlying S superstructure. Once cleaned, the Ni-13%Cr surfaces with the top c (2 × 2)-S superstructure are stable in background oxygen pressures lower that 10^{-8} Torr. Above this value oxidation of the substrate occurs rapidly, disrupting seed layer growth. As mentioned above, the deposition conditions for most of the seed layers used for RABiTS involve a water pressure of 10^{-5} Torr or higher. Such a water pressure in the deposition chamber gives rise to an equilibrium oxygen pressure in the range of 10^{-7} Torr, as indicated by a quadrupole mass spectrometer installed on our vacuum system, and therefore is sufficient to oxidize the Ni-13%Cr surface. The only successful attempt to oxide seed layer growth on the bare c (2 × 2)/Ni-Cr surface by PLD was obtained using YSZ. Unlike other seed layers, YSZ can be deposited from a ceramic target using oxygen partial pressures as low as 10^{-10} Torr. Cube textured YSZ films were grown on Ni-13%Cr at a temperature of 600 °C by ablation of a YSZ target in the base pressure of the vacuum chamber ($0.5-1 \times 10^{-8}$ Torr, corresponding to an O_2 partial pressure of $0.5-1 \times 10^{-10}$ Torr). In addition to the (002) reflections, these films showed a small (111) component with no preferential in-plane orientation. The (111) component was not observed when the initial part of the YSZ deposition was conducted in a partial pressure of H_2S equal to 2×10^{-6} Torr. Figure 11 shows the logarithmic YSZ (111) pole figure for a 170 nm thick film that was grown using H_2S for the initial layer of 70 nm in thickness. Although the background x-ray signal was subtracted in order to bring out (highlight) possible secondary orientations, the pole figure is clean and the percentage of cube texture is 93%.

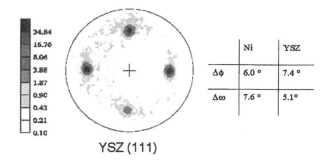

YSZ (111)

	Ni	YSZ
$\Delta\phi$	6.0 °	7.4 °
$\Delta\omega$	7.6 °	5.1°

Figure 11. Logarithmic (111) pole figure of a YSZ film deposited on c (2 × 2)/ Ni-13%Cr

Despite the low background oxygen pressure, activated oxygen species present in the ablation plume can lead to local formation of Cr-oxide, and consequently partial nucleation of YSZ with (111) orientation. It is useful to recall that Cr-oxide is thermodynamically more stable than YSZ, and therefore its formation is expected for sufficiently high oxygen impingement rates. We think that the presence of H₂S during the first stage of film growth reduces the number of oxygen species in the plume by formation of SO₂ molecules. Consequently, the substrate oxidation rate decreases, allowing a well-textured YSZ film to be deposited. After the YSZ deposition, an 80-nm-thick CeO_2 cap layer was grown in situ at 660 °C in an oxygen partial pressure of $5 \ 10^{-4}$ Torr. A 200-nm-thick YBCO film grown on this sample by PLD showed a J_c of 0.44 MA/cm^2 at 77 K and in self magnetic field.

In conclusion, the c (2 × 2)-S superstructure is not only present on {100}<100> Ni but forms also on the surface of biaxially textured Ni-alloys such as Ni-3%W and Ni-13%Cr. In all cases, the superstructure acts as a template that enables the epitaxial growth of many oxide seed layers on the metal. In addition, a complete coverage of this chemically stable layer is necessary to replicate the substrate texture in the buffer layer. Annealing Ni substrates at 800 °C in the presence of a small amount of H₂S is sufficient to produce a stable c(2 × 2)-S with a coverage of 0.5ML. This simple step allows creating a complete S template independently of the coverage obtained through segregation. Implementing this step in a continuous RABiTS fabrication process allows the initial Ni texturing anneal to be carried out at conditions that yield the best texture (and often involve high temperatures and consequent S evaporation) without being limited by the S segregation process.

Research sponsored by the U.S. Department of Energy under contract DE-AC05-00OR22725 with the Oak Ridge National Laboratory, managed by UT-Battelle, LLC.

High Temperature Superconductors

REFERENCES

1. C. Cantoni, D. K. Christen, R. Feenstra, D. P. Norton, A. Goyal G. W. Ownby, and D. M. Zehener, *Appl Phys Lett* **79** 3077 (2001)
2. C. Cantoni, D. K. Christen, L. Heatherly, A. Goyal, G. W. Ownby, and D. M. Zehner, *Journal of Materials Research* submitted
3. M. Paranthaman, D. F. Lee, A. Goyal, E. D. Specht, P. M. Martin, X. Cui, J. E. Mathis, R. Feenstra, D. K. Christen, D. M. Kroeger, *Supercond Sci Tech* **12** (5): 319 (1999)
4. P. H. Holloway, *J Vac Sci Technol*, **18**(2) 653 (1981)
5. D. F. Mitchell, P. B. Sewell, and M. Cohen, *Surf. Sci.* **61**, 355 (1976)
6. P. C. McIntyre, *J Appl Phys* **89** 8074 (2001)
7. T. Aytug, M. Paranthaman, B. W. Kang, S. Sathyamurthy, A. Goyal, and D. K. Christen, *Appl. Phys. Lett.* **79** 2205 (2001)
8. M. Perdereau and J, Oudar, *Surf Sci.* **20** 80 (1970)
9. S. Andersson and J. B. Pendry, *J Phys C* **9** 2721 (1976)
10. C. A. Papageorgopoulos, and M. Kamaratos, *Surf Sci* **338** 77 (1995)
11. A. Patrige, G. J. Tatlock, F. M. Leibsle, and C. F. J. Flipse, *Phys Rev B* **48** 8267 (1993)
12. M. M. Kowalewski, F. A. List, L. Heatherly, and D. M. Kroeger, unpublished
13. J. R. Thompson, A. Goyal, D. K. Christen, and D. M. Kroeger, *Physica C* **370** (3) 169 (2002), also cond-mat/0104239.

INCLINED SUBSTRATE DEPOSITION OF MAGNESIUM OXIDE FOR YBCO-COATED CONDUCTORS

B. Ma,[1] M. Li, [1] B. L. Fisher, [1] R. E. Koritala, [1] S. E. Dorris, [1] V. A. Maroni, [2] and U. Balachandran[1]
[1]Energy Technology Division
[2]Chemical Technology Division
Argonne National Laboratory
Argonne, IL 60439

ABSTRACT
Thin films of $YBa_2Cu_3O_{7-x}$ (YBCO) were grown on MgO buffered metallic substrates by pulsed laser deposition (PLD). The MgO buffer films, which provide the initial biaxial texture, had been grown on polished Hastelloy C276 (HC) tapes using inclined substrate deposition (ISD). The ISD process is promising for the fabrication of coated superconductor wires because it produces biaxially textured template films on nontextured substrate at high deposition rates. Biaxially aligned MgO films were deposited at deposition rates of 20 to 100 Å/sec. The buffer films were deposited on these template films before ablation of the YBCO films by PLD. The microstructure was studied by scanning electron microscopy and atomic force microscopy. X-ray pole figure analysis and ϕ- and ω-scans were used for texture characterization. Good in- and out-of-plane textures were observed on the ISD MgO films (≈ 1.5 μm thick). The full width at half maximums were 9.2° for the MgO (002) ϕ-scan and 5.4° for the ω-scan. Cube-on-cube epitaxial growth of yttria-stabilized zirconia (YSZ) and ceria (CeO_2) films on the ISD MgO films was also achieved by PLD. A superconducting critical temperature of 90 K, with a sharp transition, and transport critical current density of $>2.5 \times 10^5$ A/cm^2 were obtained on a 0.5-μm-thick, 0.5-cm-wide, and 1-cm-long YBCO film with MgO buffer layer at 77 K in self-field.

INTRODUCTION

Thin-film superconductors and coated conductor wires have many applications, including high-power transmission cables, high-field magnets, generators, fault-current limiters, magnetic shields, and large-scale microwave devices [1-3]. $YBa_2Cu_3O_{7-X}$ (YBCO) can readily be deposited on single-crystal substrates to form biaxially textured (e.g., with both c-axis and in-plane alignment) thin films that carry high critical current density (J_c). However, for coated-conductor applications, YBCO films must be coated onto polycrystalline, nontextured, and flexible metal substrates and also be able to carry high J_c. Biaxially textured template films are necessary for successful deposition of textured YBCO films on metallic substrates and thus to achieve high J_c [4-6]. Research efforts in the past few years have accelerated the processing, fabrication, and manufacturing of high-temperature coated conductors to meet the needs of the U.S. electric power industry [1-3,7]. Several techniques, including ion-beam-assisted deposition (IBAD) [8-10], rolling-assisted biaxially textured substrates (RABiTS) [11,12], and inclined-substrate deposition (ISD) [13-15], were developed. Compared to the first two processes, ISD produces textured films at high deposition rates (20-100 Å/sec) and is independent of the recrystallization properties of the metallic substrates. It is also simpler and easier to accomplish, without the need of an assisting ion source or complicated heat treatment.

We grew biaxially textured MgO thin films on mechanically polished Hastelloy C276 (HC) substrates by ISD using an e-beam evaporation system. To decrease the surface roughness of the as-deposited ISD MgO films, an additional thin layer of MgO was deposited at an elevated temperature and a zero inclination angle. YSZ and CeO_2 buffer layers were epitaxially deposited on these ISD-MgO-buffered substrates with an excimer laser system before YBCO ablation. The surface morphology of the films was investigated by scanning electron microscopy (SEM), and atomic force microscopy (AFM) was used to determine surface roughness. Raman spectroscopy was measured on selected samples to study the phase integrity of the YBCO films. X-ray pole figure analysis and ϕ- and ω-scans were conducted to analyze texture. The superconducting transition temperature (T_c) was determined inductively, and J_c was measured by the four-point method at 77 K in self-field. In this paper, we discuss the growth mechanism, crystalline texture, microstructure, and superconducting properties of YBCO deposited on biaxially textured ISD MgO buffer layers fabricated on polished HC substrates.

EXPERIMENTAL PROCEDURE

Mechanically polished HC coupons measuring $\approx$0.1 mm thick, $\approx$5 mm wide, and 1 cm long were used as substrates for the deposition of ISD MgO template films, buffer layers, and YBCO films. The ISD MgO films were grown from a MgO source evaporated by an electron beam. Fused lumps of MgO (Alfa Aesar, 99.95% metal basis, 3-12 mm pieces) were used as target material. The HC substrate was attached using silver paste to a tiltable sample stage that was above the e-beam evaporator, as shown in Fig. 1. The angle between substrate normal and the MgO vapor direction, the inclination angle, was varied between 10 and 70°. Base pressure of the evaporation chamber was $\approx$1 x 10^{-7} torr. High-purity oxygen flow was introduced into the system at $\approx$3 sccm to maintain an operating pressure of $\approx$1 x 10^{-5} torr. A crystal monitor was mounted besides the sample stage to monitor and control the deposition rate. High deposition rates of 20 to 100 Å/sec were used, and the substrate temperature was maintained between room temperature and 50°C during deposition. After deposition of the ISD films, a thin layer of MgO was deposited at a zero degree inclination angle at elevated temperatures to reduce the surface roughness.

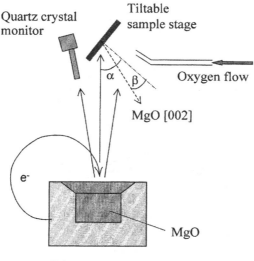

Figure 1. Schematic illustration of experimental setup for ISD MgO.

Buffer films and YBCO films were deposited by pulsed laser deposition (PLD) with a Lambda Physik LPX 210i excimer laser using a $Kr-F_2$ gas premixture as the lasing medium. Figure 2 shows a schematic illustration of the PLD system, which includes an optical beam raster to produce films with better uniformity over broader area. The laser beam was focused at the target through a quartz lens (1000-mm focal length) coated with anti-reflective coating and was reflected by a mirror that was mounted as part of the beam raster. The reflected beam hits the target at a 45° incident angle. The rotating target carrousel carries four targets to accommodate the needs for multiple layer ablation without breaking the vacuum. Commercial targets (Superconductive Components, all better than 99.99% pure), 45 mm in diameter and 6 mm in thickness, were used. The substrates were attached to a heatable sample stage with silver paste and heated to high temperature (700-800°C) during deposition. The base pressure of the chamber was $\approx 1 \times 10^{-5}$ torr. The desired operating pressure (100-300 mtorr) was maintained by flowing ultra-high-purity oxygen at ≈ 10 sccm and pumping the chamber with a molecular turbo pump. The size of the laser spot focused at the rotating target was ≈ 12 mm^2, which resulted in an energy density of ≈ 1.5 J/cm^2. The distance between the target and the substrates was ≈ 7 cm.

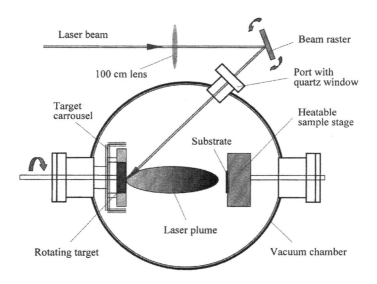

Figure 2. Schematic illustration of PLD system.

T_c and J_c were determined by an inductive method, and transport J_c was measured by the four-point transport method at 77 K in liquid nitrogen using an 1 $\mu V/mm$ criterion. The inductive method is a standard characterization tool used to measure the superconducting properties of YBCO films. Thin-film superconductor samples were placed between a pair of primary and secondary coils with inner diameters of ≈1 mm and outer diameters of ≈5 mm. Alternating current of 1 kHz was introduced to the primary coil and detected from the secondary coil by a lock-in amplifier (Stanford Research Systems SR830 DSP). Samples used for transport measurements were first coated with 2-μm-thick silver by e-beam evaporation and then annealed in flowing high-purity oxygen at 400°C for 2 h.

The film texture was characterized by X-ray diffraction pole figure analysis with Cu-K_α radiation. For the ISD MgO films and subsequently deposited buffer films, in-plane texture was characterized by the full-width at half maximum (FWHM) of the ϕ-scan for (002) reflection, and the out-of-plane texture was characterized by the FWHM of the ω-scan at the [001] pole for the same reflection. As for the YBCO films, in-plane texture was measured by the FWHM of the YBCO (103) ϕ-scan, and out-of-plane texture was measured by the FWHM of the YBCO (005) ω-scan. Plan-view and fracture cross-sectional SEM (Hitachi S-4700-II) were conducted to study the morphology of MgO films. Surface roughness was measured by taping-mode AFM with a Digital Instruments Dimension 3100 SPM system.

RESULTS AND DISCUSSION

Typical X-ray pole figures of an ISD MgO film deposited at an inclination angle $\alpha = 55°$, with a thickness of 1.5 μm, are shown in Fig. 3. Unlike the YSZ films prepared by inclined-substrate PLD [13], where the (001) planes are nearly parallel to the substrate surface, the [001] axis of the ISD MgO buffer layer is tilted away from the substrate normal. The asymmetric distribution of the pole peaks reveals that the MgO (001) planes have a tilt angle β toward the deposition direction. These ISD MgO films exhibit good texture; distinct in-plane alignment can be seen by the well-defined poles for not only the [001] axis but also the [010] and [100] axes in Fig. 3a. Out-of-plane alignment was characterized by the ω-scan; data were taken at the [001] pole. The tilt angle, as determined from the chi angle value of the [001] reflection in the MgO (002) pole figure, was ≈32°.

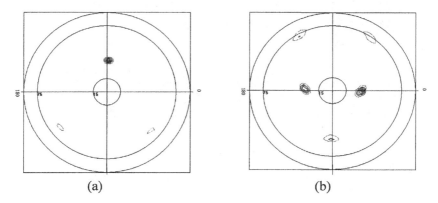

(a)   (b)

Figure 3. (a) MgO (002) and (b) MgO (220) pole figures for ≈1.5-μm-thick ISD MgO film deposited at room temperature with $\alpha = 55°$.

Plan-view SEM revealed (Fig. 4a) a roof-tile structure for a MgO film deposited at room temperature with $\alpha = 55°$. Columnar grains nearly perpendicular to the substrate surface were observed on the cross-sectional fracture surface (Fig. 4b). The MgO grain size increased when the film grew for the first 0.25-μm thickness; it then became stabilized at ≈0.1 μm, without noticeable change in size when the film grew thicker. To reduce surface roughness, an additional thin layer of MgO was deposited on the ISD MgO films at elevated temperatures (600-800°C) by e-beam evaporation at $\alpha = 0$. Figures 4c and 4d are SEM images of the top plan-view and fracture cross-sectional view of a 0.5-μm-thick homoepitaxial MgO layer deposited at 700°C on 1.5-μm-thick ISD MgO. Surface smoothness of the film was improved. Plate-shaped grains were formed during the homoepitaxial deposition at 700°C, in contrast to columnar grains during ISD deposition at room temperature, as shown in Fig. 4d. Figure 5 shows the AFM images of ISD MgO films with and without a homoepitaxial MgO layer. AFM analysis revealed that the root-mean-square (RMS) surface roughness of the MgO films improved from ≈28 nm to ≈10 nm after deposition of the homoepitaxial layer.

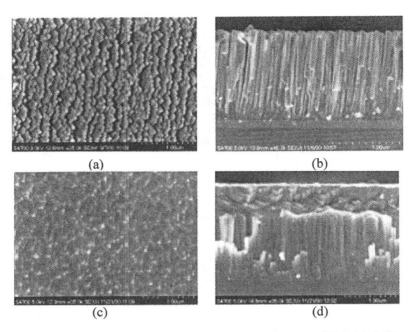

Figure 4. (a) Plan view and (b) cross-sectional SEM images of ISD MgO film deposited at room temperature with $\alpha = 55°$; (c) Plan view and (d) cross-sectional SEM images of MgO film after depositing additional layer of MgO by e-beam evaporation at 700°C with $\alpha = 0°$.

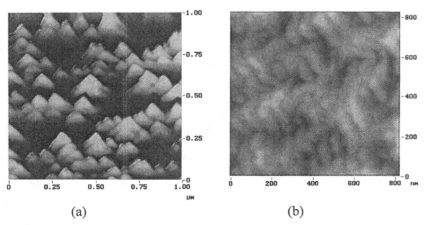

Figure 5. AFM images of (a) ISD and (b) homoepitaxial MgO films.

Figure 6 shows a TEM image and selected area diffraction (SAD) pattern of MgO columnar grain in a film deposited with α = 55°. The top facet of MgO grains is a (002) plane. Film morphology and texture evolution in the ISD MgO films can be understood from the self-shadowing effect. It has been demonstrated [16] that in-plane texture during polycrystalline film growth develops through a combination of fast growth along a certain crystallographic direction and the self-shadowing that occurs when deposition is at an inclined angle. In the case of MgO, the fast growth plane is {200} [17]. Because maximizing the (002) faces can decrease the surface free energy, the {200} plane is also the equilibrium crystal habit, as confirmed by the cubic morphology exhibited in the MgO film [18]. With deposition at an inclined angle, the {200} plane rotates toward the vapor source, so the (002) surface grows faster than other crystalline faces.

[200]

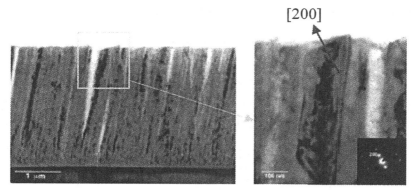

Figure 6. TEM cross-sectional view and selected area diffraction demonstrating that columnar grain terminates with a (200) crystal face in an ISD MgO film.

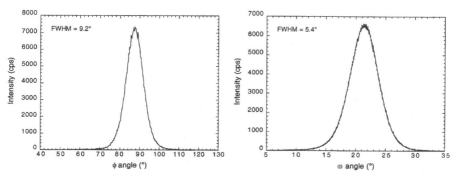

Figure 7. MgO (002) (a) φ-scan and (b) ω-scan patterns after homoepitaxial growth of 0.5-μm-thick MgO layer on ISD MgO film at elevated temperature.

Figure 7 shows typical φ-scan and ω-scan patterns for MgO (002) after homoepitaxially growing a 0.5-μm-thick MgO layer on ISD MgO film at elevated temperature. Both in- and out-of-plane textures were improved after deposition of the homoepitaxial MgO layer. FWHMs were reduced from 12.2 to 9.2 and from 6.3 to 5.4°, respectively, in the MgO (002) φ-scan and ω-scan. To study the effect of film thickness on film texture, we deposited ISD MgO films of various thickness using the same inclination angle (55°) and deposition rate (120 nm/min.). Following the deposition of the ISD MgO layer, a homoepitaxial layer of ≈0.5 μm was deposited. The samples with different ISD layers and the same thickness of homoepitaxial layer of MgO were used for texture analysis. Figure 8 shows the FWHM of MgO (002) φ-scan as a function of ISD layer thickness. We found that the film thickness of ISD layer has little effect on the texture of a final for samples with an initial ISD layer thickness of 0.25 μm or thicker. The value of 0.25 μm appears to be the critical thickness for growing high-quality ISD MgO films.

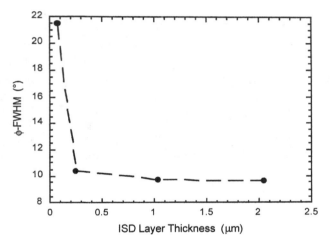

Figure 8. In-plane texture dependence on ISD layer thickness.

The YBCO films were deposited on ISD MgO-buffered HC substrates using the conditions listed in Table 1. To improve lattice mismatch and therefore enhance the superconducting properties of YBCO films, the YSZ and CeO_2 buffer layers were deposited by PLD. Details of the buffer layer architecture will be reported elsewhere [19]. Figure 9 shows a Raman spectrum of the YBCO deposited on ISD MgO-buffered HC substrate. Only a sharp peak at 340 cm^{-1} and a weak peak at 500 cm^{-1} were observed in the Raman shift region from 200 to 800 cm^{-1}. Both the 340 and the 500 cm^{-1} peaks are associated with YBCO-123 phase [20]. Raman data revealed that the YBCO film was of good quality, lacking secondary phases or cation disorder.

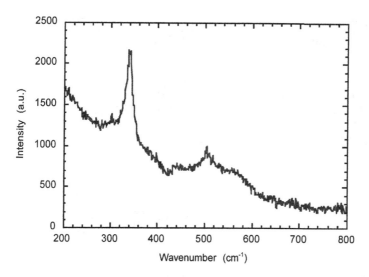

Figure 9. Raman spectrum of a YBCO film deposited on ISD MgO-buffered HC substrate.

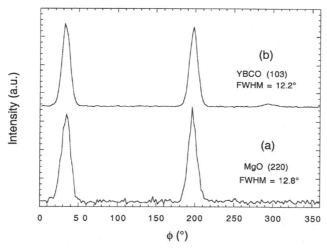

Figure 10. φ-scan patterns of MgO (220) and YBCO (103) showing epitaxial growth.

TABLE 1. Conditions used for epitaxial growth of YBCO by PLD

Laser wavelength	248 nm (Kr-F)
Repetition rate	8 - 70 Hz
Pulse width	25 ns
Energy density	1-3 J/cm^2
Substrate temperature	700-800°C
Operating pressure	100-300 mtorr
Oxygen flow rate	10 sccm
Target-to-substrate distance	4-8 cm

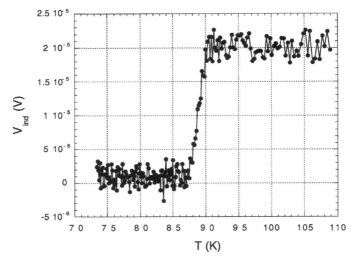

Figure 11. Critical-temperature transition curve for YBCO film deposited on ISD-MgO-buffered HC substrate.

Figure 10 shows φ-scans for the MgO (220) and YBCO (103) grown on the MgO-buffered HC substrate; it reveals epitaxial growth, with cubic-on-cubic biaxial alignment: YBCO [001] // MgO [001] and YBCO [110] // MgO [100] (or MgO [010]). The FWHMs of the YBCO films were generally 1-2° smaller than those of the underlying MgO films; the FWHMs of the MgO buffer layers on which the YBCO films had been deposited were also slightly smaller. We obtained a T_c of 90 K for a 0.5-μm-thick YBCO film deposited on an ISD MgO-buffered HC substrate. From the inductive measurements, Fig. 11 shows that the superconducting transition was complete at 88 K. The measured transport J_c was >2.5 x 10^5 A/cm^2 at 77 K in self-field on a sample that was 0.5 μm thick, 0.5 cm wide, and 1 cm long.

High Temperature Superconductors

CONCLUSIONS

Biaxially textured MgO films were successfully grown by the ISD method, which is much more time-efficient for fabrication of buffer layers than is the IBAD YSZ process. The MgO films grown by the ISD process contained columnar grains that were terminated at the surface by (002) planes. Plan-view SEM revealed a roof-tile-shaped structure. The surface roughness and biaxial texture of the ISD MgO thin films were significantly improved by deposition of an additional thin layer of MgO at elevated temperature. Surface roughness of the MgO films reduced from ≈ 28 nm to ≈ 10 nm as measured by AFM, and FWHMs were reduced from 12.2 to 9.2 and from 6.3 to 5.4° in the MgO (002) ϕ-scan and ω-scan, respectively, after deposition of a ≈ 5-μm-thick homoepitaxial layer at 700°C. The value of 0.25 μm appears to be a critical thickness for growing high quality ISD MgO films. Raman data revealed high-quality and phase integrity for the YBCO films grown on ISD-MgO-buffered HC substrates. A T_c of 90 K with a sharp superconducting transition, and a $J_c > 2.5 \times 10^5$ A/cm^2 at 77 K in self-field were measured on YBCO films fabricated using ISD architecture. These results demonstrate that biaxially textured ISD MgO buffer layers deposited on metal substrates are excellent candidates for fabrication of high-quality YBCO-coated conductors.

ACKNOWLEDGMENTS

SEM/TEM analysis was performed in the Electron Microscopy Center for Materials Research at Argonne National Laboratory. This work was supported by the U.S. Department of Energy (DOE), Energy Efficiency and Renewable Energy, as part of a DOE program to develop electric power technology, under Contract W-31-109-Eng-38.

REFERENCES

1. D. K. Finnemore, K. E. Gray, M. P. Maley, D. O. Welch, D. K. Christen, and D. M. Kroeger, *Physica C*, 320, 1-8 (1999).
2. Y. Iijima and K. Matsumoto, *Supercond. Sci. Technol.*, 13, 68-81 (2000).
3. J. O. Willis, P. N. Arendt, S. R. Foltyn, Q. X. Jia, J. R. Groves, R. F. DePaula, P. C. Dowden, E. J. Peterson, T. G. Holesinger, J. Y. Coulter, M. Ma, M. P. Maley, and D. E. Peterson, *Physica C*, 335, 73-77 (2000).
4. D. Dimos, P. Chaudhari, J. Mannhart, and F. K. Legouges, *Phys. Rev. Lett.*, 61, 219-222 (1988).
5. D. Dimos, P. Chaudhari, and J. Mannhart, *Phys. Rev. B*, 41, 4038-4049 (1990).
6. X. D. Wu, R. E. Muenchausen, S. Foltyn, R. C. Estler, R. C. Dye, C. Flamme, N. S. Nogar, A. R. Garcia, J. Martin, and J. Tesmer, *Appl. Phys. Lett.*, 56, 1481-1483 (1990).
7. "Coated Conductor Technology Development Roadmap," U.S. Department of Energy, Superconductivity for Electric System Program (Aug. 2001).
8. Y. Iijima, N. Tanabe, O. Kohno, and Y. Okeno, *Appl. Phys. Lett.*, 60, 769-771 (1992).
9. R. P. Reade, P. Berdahl, R. E. Russo, and S. M. Garrison, *Appl. Phys. Lett.*, 61, 2231-2233 (1992).

10. C. P. Wang, K. B. Do, M. R. Beasley, T. H. Geballe, and R. H. Hammond, *Appl. Phys. Lett.*, 71, 2955-2957 (1997).
11. D. P. Norton, A. Goyal, J. D. Budai, D. K. Christen, D. M. Kroger, E. D. Specht, Q. He, B. Saffain, M. Paranthaman, C. E. Klabunde, D. F. Lee, B. C. Sales, and F. A. List, *Science*, 274, 755 (1996).
12. M. Schindl, J.-Y. Genoud, H. L. Suo, M. Dhalle, E. Walker, and R. Flukiger, *IEEE Trans. Appl. Supercond.*, 11, 3313-3316 (2001).
13. K. Hasegawa, K. Fujino, H. Mukai, M. Konishi, K. Hayashi, K. Sato, S. Honjo, Y. Sato, H. Ishii, and Y. Iwata, *Appl. Superconductivity*, 4, 487-493 (1996).
14. M. Bauer, R. Semerad, and H. Kinder, *IEEE Trans. Appl. Supercon.*, 9, 1502-1505 (1999).
15. B. Ma, M. Li, Y. A. Jee, B. L. Fisher, and U. Balachandran, *Physica C*, 366, 270-276 (2002).
16. O. P. Karpenko, J. C. Bilello, and S. M. Yalisove, *J. Appl. Phys.*, 82, 1397-1403 (1997).
17. A. F. Moodie and C. E. Warble, *J. Crystal Growth*, 10, 26-38 (1971).
18. R. E. Koritala, M. P. Chudzik, Z. Luo, D. J. Miller, C. R. Kannewurf, and U. Balachandran, *IEEE Trans. Appl. Supercond.*, 11, 3473-3476 (2001).
19. U. Balachandran, B. Ma, M. Li, B. L. Fisher, R. E. Koritala, R. Erck, and S. E. Dorris, to be published in proceedings of Materials Research Society Fall 2001 Meeting, Boston, Nov. 25-29, 2001.
20. M. N. Iliev, P.X. Zhang, H. U. Habermeier, and M. Cardona, *J. Alloys Compounds*, 251, 99-102 (1997).

ION-BEAM-ASSISTED DEPOSITION OF MAGNESIUM OXIDE FILMS FOR COATED CONDUCTORS

T. P. Weber, B. Ma, U. Balachandran
Energy Technology Division
Argonne National Laboratory
Argonne, Il, 60439

M. McNallan
Department of Materials Science and Engineering
University of Illinois at Chicago
Chicago, IL 60612

ABSTRACT

The development of high critical-temperature thin-film superconductors and coated conducting wires is important for electric power applications. To achieve high transport current density, template films are necessary for the successful deposition of biaxially aligned $YBa_2Cu_3O_{7-x}$ (YBCO) on flexible metal substrates. We grew biaxially aligned magnesium oxide (MgO) template films by ion-beam-assisted deposition with electron-beam evaporation. MgO films of ≈ 100 Å thickness were deposited on Si_3N_4-coated Si substrates at a deposition rate of ≈ 1.5 Å/sec with an ion flux of ≈ 110 $\mu A/cm^2$ bombarding the substrate at a 45° angle. To study crystalline structure by X-ray diffraction, we deposited an additional layer of MgO. Good in- and out-of-plane alignment was observed, with (111) ϕ-scan full-width half-maximum (FWHM) of 6.2° and (002) ω-scan FWHM of 2.2°.

INTRODUCTION

High-temperature superconducting wires and tapes have a variety of applications, including transmission wires and microwave devices [1,2]. Fabrication of tapes that can carry the necessary current has been the focus of a great deal of research. Production of high temperature superconducting

$YBa_2Cu_3O_{7-x}$ (YBCO) tapes on nonaligned metallic substrates requires high-quality biaxially textured template layers. Yttria-stabilized zirconia (YSZ) has been used successfully as a template layer when deposited by the ion-beam-assisted deposition (IBAD) process. While superconducting films made with YSZ buffer layers have been successful, with transport current densities $\approx 10^6 A/cm^2$, a limitation to the production process is the time required for YSZ deposition. Due to the competitive growth process of YSZ, films with thickness of $\approx 8,000$ to 10,000 Å are necessary in order to achieve the desired in-plane alignment of $\approx 13°$ [3,4]. The excessive time required to deposit such films has driven the search for new buffer-layer materials.

MgO, on the other hand, exhibits a much more rapid growth mechanism. Because texturing of MgO begins at the nucleation stage, its films need only be ≈ 100 Å thick in order to obtain quality texture [5-7]. Processing times for these films are thus reduced by at least one order of magnitude from those of YSZ films [8]. Using MgO buffer layers is a significant step toward producing high critical-temperature superconductors for practical applications.

We grew biaxially textured MgO films on Si_3Ni_4 substrates with the IBAD process and an e-beam evaporation system. An additional homoepitaxial layer was deposited in order to facilitate film characterization via X-ray analysis; ϕ-scans and ω-scans were used to analyze film texture.

EXPERIMENTAL PROCEDURE

IBAD MgO films were deposited on as-received Si_3N_4 substrates measuring $\approx 1 \times 0.5$cm and mounted on a heatable stage with silver paste. A 3 cm Kaufman-type ion source was used to accelerate argon ions at 750 eV. All substrates were presputtered for 5 minutes before deposition. The angle of incidence, α, for the ion beam was maintained at $45°$, the channeling direction for MgO. A Faraday cup was used to measure the beam current density of 110

$\mu A/cm^2$. A 10:1 argon-to-oxygen ratio provided a background pressure of 8×10^{-5} torr during IBAD. Vapor flux was provided by e-beam evaporation and monitored with a quartz crystal monitor at 1.5 Å/s for an ion-to-atom ratio of ≈0.9. The IBAD films were deposited to a thickness of 100 Å. The experimental setup is illustrated in Fig. 1 and deposition conditions are given in Table 1.

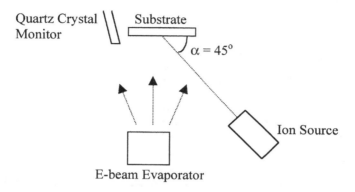

Fig. 1. Experimental setup for IBAD process.

After IBAD MgO deposition, the substrates were heated to $\approx300^\circ$C before addition of a homoepitaxial layer of MgO deposited without ion fluence. The homoepitaxial layer was necessary in order to characterize the films by x-ray diffraction and is ≈1000 Å thick. Oxygen flow was introduced into the system at a rate of ≈0.35 sccm during deposition. The deposition rate for the homoepitaxial layer was the same as that for IBAD (1.5 Å/s).

Table 1. Deposition conditions for biaxially textured IBAD MgO films.

Operating Pressure	8×10^{-5} torr
Deposition Rate	1.5 Å/s
Ion beam incident angle	45°
Beam energy	750 eV
Ion fluence	110 $\mu A/cm^2$
Gas flow (Ar + O_2)	3.35 sccm

RESULTS AND DISCUSSION

Biaxial texture of the IBAD MgO films was characterized by X-ray φ-scans and ω-scans. In-plane texture determined by the full-width half-maximum (FWHM) of (111) φ-scans is illustrated in Fig. 2. The average FWHM is 6.2°, which compares favorably with the obtainable in-plane texture of YSZ films.

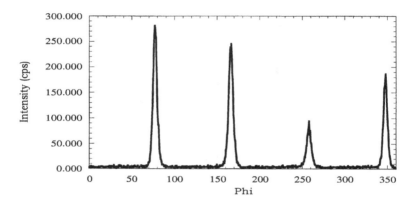

Fig. 2. X-ray diffraction phi-scan of MgO film.

Out-of-plane texture as characterized by the ω-scan of Fig. 3, shows a FWHM of 2.2°. The appearance of high-quality texture in films of only ≈100 Å thickness illustrates the nucleation growth mechanism of IBAD MgO films. Our experience has shown that texture deteriorates rapidly if film thickness deviates more than ≈20 Å from this thickness. The relationship between film texture and thickness of the film was analyzed and described in reference 9; our experiments support these findings. Figure 4 shows diffraction patterns from a series of 2-theta scans performed on films of varying thickness. All films were deposited on Si_3N_4 substrates under the growth conditions mentioned previously. The figure

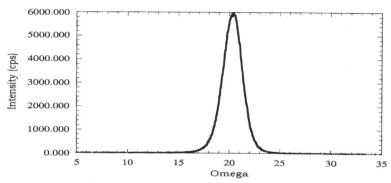

Fig. 3. X-ray diffraction omega-scan of MgO film.

illustrates the presence of the MgO <200> peak at ≈75 Å, followed by a much stronger peak in films of ≈100 Å thickness. In films of ≈125 Å, the <200> peak becomes much less intense and the presence of the <220> peak is evident.

Phi-scans of our films of various thickness also exhibit the broadening of FWHM values to ≈8° for films of ≈80 Å and ≈15° for films of ≈120 Å. The results of our studies of these films agree with the critical thickness for optimal texture of IBAD MgO films at 100 Å thickness.

We also performed studies on deposition temperature for the homoepitaxial layer. Immediately after deposition of the IBAD layer, the films were heated to 300, 400, and 500°C for the subsequent application of the 1000 Å homoepitaxial layer. Our preliminary findings indicate no observable difference in obtainable texture among these deposition temperatures and hence no apparent advantage to depositing at the higher temperatures of 400 and 500°C.

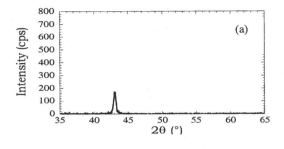

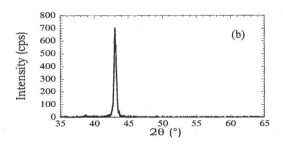

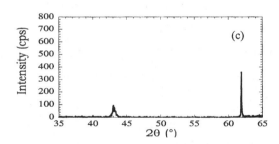

Fig. 4. X-ray 2-theta scans for IBAD MgO films of (a) 75 Å, (b) 100 Å, and (c) 125 Å thickness.

CONCLUSIONS

Biaxially textured MgO films were successfully grown by using the IBAD MgO process. MgO films grown with this approach require much less time than IBAD YSZ films and exhibit favorable texture. We examined texture variation with respect to film thickness and found the in- and out-of-plane texture to be optimal in films of ≈100 Å thickness. Additionally, our research indicates that these films can be fabricated at a homoepitaxial deposition temperature of only 300°C. IBAD MgO films having in-plane texture of 6.2° and out-of-plane texture of 2.2° were observed. This work has demonstrated that high quality textured MgO films can be grown with the IBAD process in very thin films (≈100 Å) in a very time-efficient manner. Such films are excellent candidates for buffer layers in the fabrication of YBCO coated conductors.

ACKNOWLEDGMENT

This work was supported by the U.S. Department of Energy (DOE), Office of Energy Efficiency and Renewable Energy, as part of a DOE program to develop electric power technology, under Contract W-31-109-Eng-38.

REFERENCES

1. J.O. Willis, P.N. Arendt, S.R. Foltyn, Q.X. Jia, J.R. Groves, R.F. DePaula, P.C. Dowden, E.J. Peterson, T.G. Holesinger, J.Y. Coulter, M. Ma, M.P. Maley, and D.E. Peterson, "Advances in YBCO coated conductor technology," *Physica C*, **335**, 73-77 (2000).
2. Y. Iijima, M. Kimura, T. Saitoh, and K. Takeda, "Development of Y-123 coated conductors by IBAD process", *Physica C*, **335**, 15-19, (2000).
3. K.G. Ressler, N. Sonnenberg, and M.J. Cima, "The development of biaxial alignment in YSZ films fabricated by IBAD," *Journal of Electronic Materials*, **25** [1] 35-42 (1996).

4. B. Ma, M. Li, B.L.Fisher, and U. Balachandran, "Ion-beam-assisted deposition of biaxially aligned yttria-stabilized zirconia template films on metallic substrates for YBCO0-coated conductors," *Superconductor Science and Technology*, (March 2002).

5. J.R. Groves, P.N. Arendt, H. Kung, S.R. Foltyn, R.F. DePaula, and L.A. Emmert, "Texture development in IBAD MgO films as a function of deposition thickness & rate," *IEEE Transactions on Applied Superconductivity*, **11** [1] 2822-2825 (2001).

6. J.R. Groves, P. N. Arendt, S. R. Foltyn, R. F. DePaula, E. J. Peterson, T. G. Holesinger, J. Y. Coulter, and Raymond W. Springer, "Ion-beam assisted deposition of biaxially aligned MgO template films for YBCO coated conductors," *IEEE Transactions on Applied Superconductivity*, **9** [2] 1964-1966 (1999).

7. J.R. Groves, P. N. Arendt, S. R. Foltyn, Q. Jia, T. G. Holesionger, H. Kung, R. F. DePaula, P. C. Dowden, E. J. Peterson, L. Stan, and L. A. Emmert, "Development of the IBAD MgO process for HTS coated conductors", *International Workshop on Superconductivity*, 43-46 (2001).

8. C.P. Wang, K.B. Do, M.R. Beasley, T.H. Geballe, and R.H. Hammond, "Deposition of in-plane textured MgO on amorphous Si_3N_4 substrates by IBAD & comparisons with IBAD YSZ," *Applied Physics Letters*, **71** [20] 2955-2957 (1997).

9. J.R. Groves, P. N. Arendt, H. Kung, S. R. Foltyn, R. F. DePaula, L. A. Emmert, and J. G. Storer, "Texture development in IBAD MgO films as a function of deposition thickness and rate," *IEEE Transactions on Applied Superconductivity*, **11** [11] 2822-2825 (2001).

SCALING-UP OF HIGH-Tc TAPES BY MOCVD, SPRAY PYROLYSIS AND MOD PROCESSES

Sandrine Beauquis, Sébastien Donet, François Weiss
Laboratoire des Materiaux et du Genie Physique-ENSPG-INPG,CNRS-UMR5628, Bp46, F-38402 S. Martin d'Hères, France

ABSTRACT

Fabrication of coated conductors is a key aspect of the development of High Temperature Superconductors (HTS). Chemical deposition processes (MOCVD, spray pyrolysis and MOD) become actually the more promising techniques for the production of HTS tapes, due to the good relation between sample quality, mass production and low capital costs of these processes.

We compare the advantages and drawbacks of each process and specify the criteria which may be considered to select the more appropriated one.

Several compounds have been synthesized using the aforementioned techniques, including buffer layers (CeO_2, YSZ, $BaZrO_3$, Y_2O_3...) and the different HTS phases (Y-123, Hg-1223, Tl-1223). Properties of these films and of various architectures will be reported.

INTRODUCTION

The Coated conductor technology requires flexible and mechanically strong textured substrate tapes, with a good chemical compatibility with the deposited superconducting oxide film. The Substrate thickness is generally over 50μm (for a good mechanical behaviour) and the superconducting layer has to be as thick as possible (thickness if possible over 1 μm) in order to increase the overall engineer current flowing in the conductor. The matching between metallic substrates and HTS layers can only be realised by adding adequate buffer layers (CeO_2 and YSZ have actually been identified as good materials for this purpose) [1,2].

We discuss in the present paper the chemical engineering of different Chemical Deposition Processes used for the synthesis of thin or thick superconducting oxide films on RABiTS or IBAD based substrates. The different deposition techniques considered are : i) MOCVD, ii) spray pyrolysis and iii) Metal-Organic Decomposition (MOD). These Chemical routes are very promising for coated conductor production, since they are considered as low cost techniques that can be extrapolated to a large-scale production.

In the present review, the basic discussion, concerning film preparation and film characterisation, is concentrated on YBCO based conductors. The other families of superconducting oxides (Tl and Hg based superconductors) have also been investigated and the main characteristics of these films will also be presented.

For the complex architecture of coated conductors different compounds have been synthesised by the Chemical deposition processes described, from buffer layers (CeO_2, YSZ, Y_2O_3...) to the superconducting phases (Y-123, Hg-1223, Tl-1223).

We present for each of these techniques the basic principle involved, the main results obtained today for the fabrication of HTS thin and thick films and their development for the realisation of coated conductors in reel to reel systems.

CHEMICAL DEPOSITION: BASIC PRINCIPLES

Chemical Deposition processes, used for the formation of HTS films, are based on chemical reactions between inorganic or organic metal containing species, leading to the formation of a thin layer on an appropriate substrate.

These processes can be classified in two main groups :

1.One step processes, (MOCVD) using volatile chemical precursors, where the superconducting oxide phase is formed directly after chemical reactions from the gas phase on the surface of a determined substrate

2.Two step processes, (MOD, sol-gel, spray pyrolysis..) which are based on the deposition of an unreacted precursor film, followed by a recrystallisation step, where solid state reactions induce the formation of the desired superconducting phase.

All these processes are in general simple and less expensive than physical deposition processes. They are commonly used in the production of various coatings and thin films with the main advantage that they allow an easy scaling up to large size and a conformal coverage of complex shapes.

METALORGANIC CHEMICAL VAPOR DEPOSITION (MOCVD) OF HTS

In a MOCVD process thin films are formed on a surface from the thermal decomposition of gaseous precursor molecules. The total pressure used in the gas phase is generally a low pressure, in the range of a few Torrs. This pressure range is commonly attained with classical one stage vacuum pumps and does not need ultra high vacuum equipments, which increase considerably the equipment and maintainance costs in a deposition system.

Precursors for MOCVD should in general meet the following properties:
1) The precursor should have an evaporation rate which is large enough, stable and constant with the time.
2) The precursor molecule should be chemically and thermally stable during the transport through the gas phase to the surface,
3) The precursor should be relatively easily to synthesize
4) The precursor should not be dangerous and should not produce dangerous side products.
5) The precursor must be soluble and stable in a suitable solvent without formation of precipitates (In the case of a droplet derived deposition process)

For MOCVD of HTS materials, chelate compounds are the main candidates, basically β-diketonates compounds are used (they are also named tmhd). They are free of fluorine and have vapor pressures which are high enough for MOCVD experiments.

Since a few years, new liquid delivery systems have been introduced to overcome the problem of thermal stability of Ba complexes (Pulsed Injection (PIMOCVD) [3, 4, 5] Band Evaporation (BEMOCVD) [6]). The main advantages of these liquid delivery systems are :

- A single precursor mixture can be used for multicomponent systems
- Very high growth rates can be obtained

- In PIMOCVD, the solution is injected directly by a micro-valve. The frequency, the opening time and the solution concentration are the main parameters controlling the growth rate.

- In BEMOCVD, a band system is wet with the precursor solution. An evaporator separates the solvent molecules from the precursor molecules and dry precursors are evaporated to the deposition zone. As solvent mostly diglyme or monoglyme are used.

Actually, on the basis of these two new liquid delivery systems, high quality YBCO films have been obtained on static substrate wafers [7, 8, 9].

Fabrication of long Tapes: In order to deposit HTS thin or thick films on moving tapes a special reel-to-reel deposition system has been developed (Fig.1).

In this tape casting unit, the tape is wound on two reels, before and after deposition, the reels remaining under vacuum during the whole process. The precursors are dissolved in monoglyme and injected (with a Pulsed Injection system) into an evaporator heated at 250°C. The solution injected is flash volatilized and carried into the deposition chamber by a strong flux of Ar/O$_2$. The transfer time, between the evaporator and the deposition zone, is very short (1second).

A post annealing furnace is set-up on line in order to control, after deposition, the tape cooling and the oxygen loading of the films.

Figure 1 : Reel-to-reel MOCVD deposition system The substrate wind on a supply reel passes through the deposition chamber to be wound on a winding reel.
A : annealing chamber , B : deposition chamber, C : winding reel, D : supply reel.

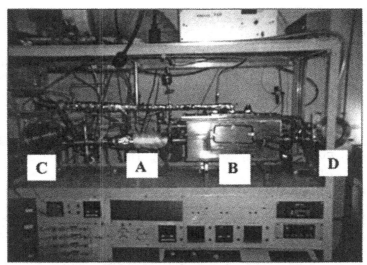

Using this system, buffer layer deposition (YSZ, GdSZ, CeO_2, Y_2O_3, Gd_2O_3...) on 3meters tapes, YBCO (ie: SmBCO) deposition on 2m tapes, as well a combination of complex architectures (CeO_2/ YBCO for example) have been realised.
The experiences were focused on:
- reducing the weak link behaviour of films, high angle grain boundaries
- increasing film thickness and film deposition rates.

High quality YBCO films have been obtained on Ni RABiTS and on SS//YSZ (IBAD) tapes (Fig 3, 4) [10, 11]. Special buffer layer architectures have been developed in order to obtain high-critical current densities (Jc >1MA/cm^2, 77K, OT) as well as a high degree of epitaxial texture.

On RABiTS tapes a successful stacking was realised with Ni//NiO/YSZ/Y_2O_3/YBCO. NiO was prepared by a Substrate Oxidation Epitaxy technique [12], and all the following buffer layers were prepared in-situ by

MOCVD with a final YBCO layer of 750nm. The Jc value reaches $0.9 \ 10^6$ A/cm^2 in this stacking.

On Hastelloy substrates, covered with an YSZ IBAD layer, the best results have been obtained for the following architecture : SS//YSZ/CeO$_2$/YBCO (Fig. 5). Critical current densities , measured on short samples are of Jc= $1.3 \ 10^6$ A/cm^2 for a 300nm thick CeO$_2$ and 750 nm thick YBCO layer.

Figure 3 : Epitaxial growth of YSZ/CeO$_2$/YBCO   Figure 4 : Tc of YBCO on
on SS/YSZ (IBAD) substrates   IBAD and RABITS substrates

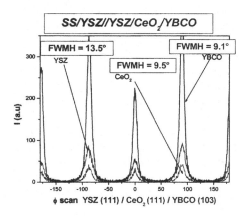
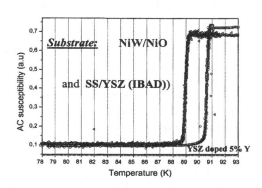

Thicker 2µm HTS layers have also been deposited, based on a multilayer stacking of Y123/Sm123/Y123 (Fig.6) with a Jc value of $1.2 \ 10^6$ A/cm^2.

Figure 5 : SEM picture of a   Figure 6 : SEM picture of a 2µm
SS/YSZ//YSZ/CeO$_2$/YBCO films   thick Y123/Sm123/Y123 layer

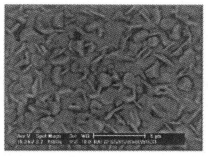
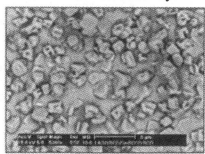

High Temperature Superconductors

SPRAY PYROLYSIS DEPOSITION

Aerosol routes offer a variety of approaches for film generation. These processes have been used to deposit many materials including ceramic superconductors, simple metal oxides, non-oxide ceramics... at high rates. Experimental procedure involves the pyrolysis on a heated substrate of droplets produced by an aerosol of non-volatile liquid source precursors. This procedure allows in some extend to have a reasonable control over film thickness and over the homogeneity.

In general the standard deposition route for the preparation of HTS materials involves a deposition stage of a non completely reacted film (precursor film), followed by one or several annealing treatments allowing control of the film recrystallisation and a final oxidation step to improve the superconducting properties.

For the spray pyrolysis of HTS materials (YBCO, Tl and Hg based superconductors) aqueous nitrate solutions have been used in most of the cases but sometimes also carboxylates. Nitrate solutions are generally preferred, specially in the case of Ba based superconductors, because they allow to circumvent problems of carbon contamination in the layer.

Aqueous nitrate solutions are prepared either by dissolving $Y(NO_3)_3 \cdot 3H_2O$, $Ba(NO_3)$, and $Cu(NO_3) \cdot 6H_2O$, or $Ca(NO_3)_2 \cdot 4H_2O$ in highly distilled and deionized water to give the suitable cationic ratio for deposition, or from single oxides or carbonates by dissolution in nitric acid. This last procedure gives a better control of stoichiometry because the water content of nitrates is not always exactly known.

YBCO films: YBCO films have been sprayed over single crystalline substrates, MgO, SrTiO$_3$ and on C fibers and wires [13]. After the deposition, films are in an amorphous state and therefore an annealing step at elevated temperature is necessary to build the proper structure. This annealing is performed either with an inert atmosphere or in a mixture of an inert atmosphere with O$_2$, with a final post-annealing under O$_2$ atmosphere.

Recent studies have shown that biaxial texturing of YBCO can be reached using nitrate deposition at elevated temperatures on single crystalline substrates and on biaxially textured Ag substrates. Jc in the range 10^4-10^5 A/cm^{-2} were reported [14,15,16].

In the present study, YBCO thick films have been synthesized using the spray pyrolysis method, with the aim to develop an in-situ deposition process. We used an all nitrate precursor approach to grow epitaxial HTS oxide film on

SrTiO$_3$(001) substrate at a high deposition rate and at high temperature (over 800°C). We performed 1μm thick YBCO films [17, 18] in situ at 850°C in a very short deposition time (4 minutes) corresponding to a very high local deposition rate of 12 μm/hr. The final sintering process is performed at the same temperature for 2 hours under oxygen atmosphere before low temperature oxygen loading (Figure 7). X-ray diffraction (XRD) results (θ-2θ scans and φ scans) show that the films do not have only a strong c-axis texture but also a sharp in-plane biaxial orientation. FWHM values for YBCO (102) and YBCO (005) diffraction lines can be as low as ~1.4° (in-plane epitaxy, Δφ) and ~0.4° (out-of-plane epitaxy, Δω), respectively. Furthermore, the critical temperature (T$_c$) and the critical current density (**J$_c$**) of this film are 91K and 1.4 x 10^{6}A/cm^2 (@ 77K, 0T) (Figure 8).

YBCO thick films have actually been grown on different single crystal substrates with high current performance (**J$_c$** >1 MA/cm^2) and on Ag single crystals or biaxially textured tapes. The deposition conditions are actually adapted to buffered coated conductor substrates.

The all-nitrate spray pyrolysis technique, using a very low cost equipment and "green" raw materials for the production of YBCO films at high deposition rate (12μm/hr.), encourages very much to attempt the scaling up of this process for producing long length-coated conductors in the future.

A Reel-to-reel equipment is actually under development for this purpose.

Figure 7: SEM photomicrograph of YBCO film on SrTiO$_3$

Figure 8 : Tc, Jc and critical current profile of YBCO film on SrTiO$_3$

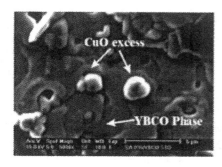

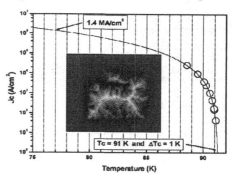

Tl(Hg)-BCCO films: Tl (and Hg) based oxide superconductors have also been synthesized using spray pyrolysis techniques. The synthesis procedure is splitted in two distinct steps :

- deposition of a Ba-Ca-Cu nitrate precursor layers on single crystalline or Ag substrates.
- Tl or Hg introduction into the pre-annealed precursor layer, using a Tl or Hg source from a Tl(Hg)BCCO pellet or from a Tl_2O_3 ($HgCaO_2$) powder.

The second step annealing is carried out at high temperatures (typically 800-860°C) with a careful control of Tl (Hg) partial pressures.

Tl based films have be prepared with very good superconducting properties. The most interesting Tl compound, Tl-1223, can be obtained relatively as a pure phase [19]. The best superconducting properties obtained so far are $T_c = 120$ K, $J_c = 10^5$ Acm^{-2} (77 K, 0 T).

Hg-1223 films present actually the most promising superconducting properties. Films with a thickness of 1-2 μm and T_c~130K have been epitaxially grown by spray pyrolysis on $SrTiO_3$ substrates (Figure 9). X ray diffraction data show a good c-axis orientation, the $CaHgO_2$ phase should disappear with a longer high temperature treatment.. In these films, the critical current density can be as high as: j_c~4.4·10^5 Acm^{-2} (77 K, 0 T) [20,21].

Figure 9: SEM micrograph of a Hg-1223 thin films.	Fig 10: XRD diagram of a Hg-1223 thin films

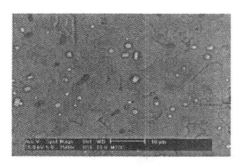 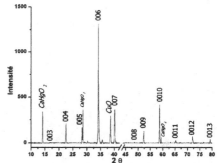

METALLORGANIC DECOMPOSITION (MOD)

Metal-Organic Decomposition (MOD) methods appeared recently as very appealing alternatives to grow biaxially textured YBCO films in combination with textured metallic substrates [22]. It has first been shown that the MOD technique has a very high flexibility in preparing simple oxides buffer layers (CeO_2, YSZ) and binary compounds such as $BaZrO_3$ and $LaNiO_3$. Moreover, the successful deposition of high Jc YBCO at MIT by using Tri-fluoro-acetate (TFA) precursors

High Temperature Superconductors

demonstrates that the chemical processing of new generation of coated conductors can be processed by MOD.

The MOD method consists in combining precursors of the desired metals in the corresponding ratio followed by their thermal decomposition. The decomposition then yields an atomic level mixture of the oxides/carbonates. After thermal treatment the desired mixed metal oxide material is obtained. MOD uses as precursors a metal linked to organic groups, most commonly metal carboxylate and modified metal alkoxides. The MOD ligands are separated from the precursor by thermolysis; therefore, compounds that sublime, evaporate and are not suitable as MOD precursors. Usually, the thermal decomposition leads to the formation of metal carbonates and the corresponding ketone, both of which decompose further following their own thermodynamic characteristics. In this system, relatively stable $BaCO_3$ forms as intermediate compound during the Y-123 synthesis. In order to avoid this problem, trifluoacetate are used as a metal carboxylate. They are forming oxofluorides which are converted to oxides by reacting with the moist atmosphere.

Critical current densities of $Jc=7x10^6A/cm^2$ and $Jc=1,7x10^6A/cm^2$ (77k,0T) have been obtained for very thin films (200nm) on single crystalline $LaAlO_3$ substrates and on IBAD substrates with a CeO_2 cap layer respectively [23, 24, 25].

As an alternative MOD process, and following the same idea, which is to prevent the formation of intermediate $BaCO_3$, we have developed an all-iodide deposition route [26]. The solution is spun at 6000rpm for 40sec on *(100)* $LaAlO_3$ single crystals. The thermal treatment applied, consists in a heating ramp at 200°C/hr up to 600°C under pure O_2 (1 bar) after what, the gas is switched to 100ppm O_2 at 800°C for 1hr. The samples are later cooled down at a rate of 200°C/hr in pure O_2 up to 500°C where the sample is maintained for 1h following a furnace cooling to room temperature.

Fig.10 : XRD pattern of Y-123 obtained by the iodine route.   Fig.11 : Rocking curve of (005) peak of Y-123 obtained by the iodine route.

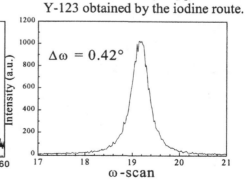

Epitaxial YBCO layers have been obtained (Fig. 10) with a good out of plane orientation of $\Delta\omega = 0.42°$ (Fig.11). These films have a critical temperature of 90K and critical current densities over 10^5 A/cm^2. These first values can certainly be optimised. Nevertheless, the solution handling of the iodide mixture is difficult and raw materials (mainly YI$_3$) are expensive. Alternative combinations of iodides and carboxylates are actually studied in order to economically scale up this process.

PROCESS RELATED TECHNICAL ISSUES

All the Chemical Deposition Methods presented in this study have shown that they are able to produce high quality HTS films either on single-crystalline (LAO,STO...) substrates or on technical metallic substrates with Jc over 1MA/cm^2 in most of the cases.

The major issues for the economical manufacturing of long length YBCO coated conductors pertain to the capability of uniform, epitaxial deposition at high rates over a large area with a continuous process. Such capability has to meet the other important requirements of proper stoichiometric composition and epitaxial structure. The volumic growth rate Vr (growth speed * covered surface aera A) is a major parameter splitting the different deposition techniques. Vr will fix the whole production time of a given coated conductor as well as its average effective cost.

Table I sums-up the performance of various deposition techniques, extrapolated to the largest deposition zones actually covered, and gives a comparison between Physical deposition techniques and Chemical deposition techniques.

Table I : Deposition technique and deposition rate

	r (μm/hr)	Vr (μm.cm^2/hr)	A (covered surface area cm^2)
Pulse Laser Deposition	3 - 7	25 - 60	10-20
Magnetron Sputtering	0.3	25 - 100	80 (4")
Thermal Evaporation	1.2	400	320 (8")
PLD - IBAD for buffer		36	
MOCVD	5 -10	250 - 500	50
MOCVD (extrapolated)	40	15000	400
MOD, spray pyrolysis	>40	> 15 000 !!!	> 400 !!

In Chemical Processes, the covered surface area can normally be extended to large sizes by simple chemical engineering techniques and a proper gas phase distribution. The local growth rate can easily be a factor 10 higher than in physical deposition processes. Chemical Processes are therefore much faster than Physical processes and are certainly better adapted for mass production of long length coated conductors. As an example, a $400cm^2$ MOCVD reactor (as shown in Table I) will be able to produce a 1cm wide coated conductor tape transporting 100A (1µm YBCO with $Jc=10^6A/cm^2$) at a speed of 150m/hr.

For each of the processing option presented, it is finally very important to assess clearly the cost of chemicals, the mass transfer (reactor yield) and reaction kinetic data, the need and cost of diagnostic and control elements and the environmental issues, in order to select the best process for coated conductor production.

Concerning the cost of chemicals, it appears in Table II that the nitrate route is certainly the cheapest one. Equipment (no vacuum), production and maintenance costs of this "green" process strengthen the assess that, among the Chemical deposition techniques, it is actually the only one able to reach the 10EUROS/KA.m criterion fixed for coated conductors.

Table II : Cost comparison of chemicals (raw estimations) for three 1µm thick layers(YSZ, CeO_2 and YBaCuO) on a 1cm wide and 1 meter long tape, with a reactor yield of 50%.

COSTS in EUROS	Nitrate spray pyrolysis	MOCVD (tmhd)	MOD (TFA)
Research cost	0.052	31.3	75
Production cost	0.018	3.1	7.5
Cost / KA.m	0.18	31	75

CONCLUSION

Chemical Deposition Processes have been shown to be, in the present state of the art, good alternatives for the technology development of coated conductors. If good quality YBCO based short samples of conductors can actually be obtained with all of these techniques, strong efforts have always to be made by improving the fabrication pathways, on the quality of substrates and on simplified buffer layer architectures. Easy scaled-up processes and fast deposition rates are the main advantages of Chemical Deposition techniques.These parameters are crucial for the mass production of long lengths coated conductors.

AKNOWLEDGEMENTS.

This work has been supported by the European Community under Brite Euram contracts READY (BRPR-CT-98-0676), SUPERTEXT (BRPR-97-556). Parts of the basic process research were financially supported by the Region Rhône-Alpes.

[1]X.D. Wu, S.R. Foltyn, P.N. Arendt *et al.*, "Properties of $YBa_2Cu_3O_{7-\delta}$ thick films on flexible buffered metallic substrates", *Appl.Phys.Lett.*, **67** [16] 2397-9 (1995).

[2] J.O. Willis, P.N. Arendt, S.R. Foltyn, Q.X. Jia, J.R. Groves, R.F. DePaula, P.C. Dowden, E.J. Peterson, T.G. Holesinger, J.Y. Coulter, M. Ma, M.P. Maley, D.E. Peterson, "Advances in YBCO-coated conductor technology". *Physica C* **335** 73|-7 (2000).

[3]J.P. Sénateur, R. Madar, F. Weiss, O. Thomas, A. Abrutis, 93/08838, France, (1993), extended : FR94/0000858, Europe/USA, (1994).

[4]F. Felten, J.P. Senateur, F. Weiss, R. Madar, A. Abrutis, "Deposition of Oxide Layers by computer Controlled "Injection LPCVD ", *J. de Physique IV*, C5 1079 (1995).

[5]J.P. Senateur, F. Felten, S. Pignard, F. Weiss, , A. Abrutis, V. Bigelyte, A. Teiserskis, Z. Saltyte, B. Vengalis, "Synthesis and characterization of YBCO thin films grown by Injection-MOCVD", *Journal of Alloys and Compounds*, **251** 288 (1997).

[6]O. Stadel, J. Schmidt, G. Wahl, F. Weiss, D. Selbmann, J. Eickemeyer, O. Yu. Gorbenko, A. R. Kaul, C. Jimenez, "Scale up of a single source MOCVD system for deposition of YBCO on travelling tapes", *Journal de Physique IV*. **11**, Pr11-233 (2001).

[7]F. Weiss, U. Schmatz, A. Pisch, F. Felten, S. Pignard, J.P. Senateur, A. Abrutis, K. Fröhlich, D. Selbmann, L. Klippe, "HTS thin films by innovative MOCVD processes" *Journal of Alloys and Compounds*, **251** 264 (1997).

[8]J. Lindner, F. Weiss, J.P. Senateur, A. Abrutis, "YBa2Cu3O 7-x /SrTiO3//LaAlO heterostructures obtained by injection MOCVD, "Integrated Ferroelectrics", 3([1-4] 301-8 (2000).

[9]Jimenez-C; Weiss-F; Senateur-JP; Abrutis-A; Krellmann-M; Selbmann-D Eickemeyer-J; Stadel-O; Wahl-G, "YBaCuO deposition by MOCVD on metalli(substrates: a comparative study on buffer layers", *IEEE Trans. App Supercond.*, 1] [1] Part 3 2905-8 (2001).

High Temperature Superconductors

[10]O. Stadel, J. Schmidt, G. Wahl, C. Jimenez, F. Weiss, M. Krellmann, D. Selbmann, N.V. Markov, S.V. Samoylenkov, O.Y. Gorbenko, A.R. Kaul, "Continuous YBCO deposition onto moved tapes in liquid single source MOCVD systems", *Physica C*, **341** Part 4 2477-2478 (2000).

[11]S. Donet, F. Weiss, J.P. Senateur, P. Chaudouet, A. Abrutis, A. Teiserskis, Z. Saltyte, D. Selbmann, J. Eickemeyer, O. Stadel, G. Wahl, C. Jimenez and U. Miller "YBCO coated nickel-based tapes with various buffer layers", *Journal de Physique IV,* **11**, Pr11-319-23 (2001).

[12] D. Selbmann, J. Eickemeyer, H. Wendrock, C. Jimenez, S. Donet, F. Weiss, U. Miller, O. Stadel, "NiO layer on Ni RABiTS for epitaxial buffer deposition by LS MOCVD", *J. de Physique IV*, 11 Pr11 239-245 (2001).

[13]M. Jergel, S. Chromik, V. Strbik, V. Smatko, F. Hanic, G. Plesch, S. Buchta, S. Valtyniova, ," Thin YBCO films prepared by low-temperature spray pyrolysis", *Supercond. Sci. Technol.*, **5** 225-230 (1992).

[14]A. Ferreri, J.J. Wells and J.L. MacManus-Driscoll, "Fabrication and post-annealing studies of ultrasonically spray pyrolysed Y-Ba-Cu-O films", *IEEE Trans. App Supercond.*, **11** [1] Part 3 2742-5 (2001).

[15]G.B. Blanchet, C.R. Fincher Jr., " High temperature deposition of HTSC thin films by spray pyrolysis" *Supercond. Sci. Technol.*, **4** [2] 69-72 (1991).

[16]J.L. MacManus-Driscoll, A. Ferreri, J.J. Wells, J.G.A. Nelstrop, "In-plane aligned YBCO thick films grown in situ by high temperature ultrasonic spray pyrolysis" *Supercond. Sci. Technol*, **14** [2] 96-102 (2001).

[17]A. Sin, Z. Supardi, A. Sulpice, P. Odier, F. Weiss, L. Ortega, M. Nunez-Regueiro, "Synthesis by aerosol process of superconductor films and buffer layer materials", *IEEE Trans. Appl. Supercond*, **11** [1] Part 3 2877-2880 (2001).

[18]P. Odier, F. Weiss, Z. Supardi, patent hanging

[19]S. Phok, A. Sin, Z. Supardi, P. Galez, J. L. Jorda and F. Weiss, "Preparation of Tl-1223 and Tl-2223 superconducting films by spray pyrolysis", *Journal de physique IV*, **11** Pr11-157 (2001).

[20]A. Sin, Z. Supardi, P. Odier, F. Weiss, L. Ortega, A. Sulpice, M. Nunez-Regueiro, "Synthesis of $Hg_{0.75}Re_{0.25}Ba_2Ca_2Cu_3O_{8+delta}$ bi-axially textured thin film by the aerosol process", *Thin solid film*, **388** [1-2] 251-5 (2001).

[21]A. Sin, F. Weiss, P. Odier, Z. Supardi, M. Nunez-Regueiro, "Synthesis under in situ pressure control of (Hg,Re)-1223 thick films by aerosol technique", *Physica C*, **341** Part 1 399-402 (2000).

[22] M. W Rupich. et al., "Low cost Y-Ba-Cu-O coated conductors", *IEEE Trans. Appl. Supercond.*, **11**, 2927–30 (2001).

[23] T. Araki, Y. Takahashi, K. Yamagiwa, Y. Iijima, K. Takeda, Y. Yamada, J. Shibata, T. Hirayama, I. Hirabayashi, "Firing conditions for entire reactions of fluorides with water in metalorganic deposition method using triflouroacetate", *Physica C,* **357-360** 991-4 (2001).

[24] Y. Takahashi, T. Araki, K. Yamagiwa, Y. Yamada, S.B. Kim, Y. Iijima, K.Takeda, T. Hirayama, I. Hirabayashi, "Preparation of YBCO films on CeO$_2$ buffered metallic substrates by the TFA-MOD method", *Physica C*, **357-360** 1003-6 (2001).

[25] Y. Yamada, S.B. Kim, T. Araki, Y. Takahashi, T. Yuasa, H. Kurosaki, I. Hirabayashi, Y. Iijima, K. Takeda, "Critical current density and related microstructures of TFA-MOD coated conductors" , *Physica C*, **357-360** 1007-10 (2001).

[26] I. Matsubara, M. Paranthaman, A. Singhal, C. Vallet, D.F. Lee, P.M. Martin, R.D. Hunt, R. Feenstra, C.Y. Yang, S.E. Babcock, "Preparation of textured YBCO films using all-iodide precursors", *Physica C*, **319**, 127-32 (1999).

TOWARDS AN ALL CHEMICAL SOLUTION COATED CONDUCTOR

O. Castaño, A. Cavallaro, A. Palau, J.C. González, M. Rossell, T. Puig, F. Sandiumenge, N. Mestres, S.Piñol, X.Obradors
Institut de Ciència de Materials de Barcelona, CSIC, 08193 Bellaterra, Spain

ABSTRACT
    Chemical solution techniques have been investigated for the growth of oxide buffer layers suitable for coated conductors and for the growth of $YBa_2Cu_3O_7$ thin films. The growth conditions of CeO_2 and $BaZrO_3$ have been investigated and a high quality epitaxy has been demonstrated on single crystalline substrates. Trifluoroacetate precursors have been used for growing $YBa_2Cu_3O_7$ films (thickness of 250nm) on single crystal substrates. The kinetic hindrances for the formation of single phases have been investigated by means of Raman spectroscopy and Fluorine analysis. After optimisation of the deposition and growth conditions very high critical currents have been demonstrated (J_c^{ab}=3.2 10^6 A/cm^2 at 77K and 2,7 10^7A/cm^2 at 5K).

INTRODUCTION
    Chemical solution growth of thin films has arisen as a new very exciting opportunity for the development of advanced functional ceramic materials. It has been particularly shown that epitaxial thin films can be grown on single crystalline substrates thus rising new opportunities for many applications where the material anisotropy or the granular character needs to be controlled. This includes ferroelectric-based devices, magnetoelectronic oxides, superconducting materials, etc. [1].
    A very outstanding challenge, from the materials preparation point of view, is the preparation of superconducting coated conductors where a biaxial texture of the superconducting oxide needs to be reached on metallic substrates. These metallic substrates can be either textured themselves through metallurgical processes, such as the RABiT technique [2], or they can have an oxide buffer oxide template textured by the IBAD technique, or the so-called Inclined substrate deposition (ISD) method where the substrate is directed at a certain angle of the deposition beam generated by pulsed laser deposition (PLD). Over the past years many experimental techniques have been investigated to obtain $YBa_2Cu_3O_7$ thin films, with a thickness as high possible, on metallic substrates. This includes PLD, sputtering, thermal co-evaporation, CVD, Liquid Phase Epitaxy, etc. Among these

techniques those based on a chemical solution growth process appear as very promising in view of the preparation of low cost conductors [4,5]. It´s however a complex issue to reach a full control of all the microstructural factors which may influence the superconducting properties of these conductors. It appears then very appealing to investigate first the influence of the processing parameters on the growth mechanisms and the microstructural development in single crystalline substrates. In this work we present a study of the growth of oxide buffer layers, particularly $BaZrO_3$ and CeO_2, which have a good crystallographic and chemical compatibility with the high T_c oxide superconducting materials, and the growth of $YBa_2Cu_3O_7$ by means of the so called trifluoroacetate (TFA) route [4-7].

EXPERIMENTAL DETAILS

Oxide buffer layers with composition CeO_2 and $BaZrO_3$ have been grown in (100) YSZ single crystalline substrate, in the first case, and in (100) $SrTiO_3$ (STO) and (100) $LaAlO_3$ (LAO) substrates in the second case. The precursor solutions for Ce and Zr were 2,4- pentadionate dissolved in glacial acetic acid and Barium acetate for Ba. The precursor solutions were prepared in a magnetic stirrer and the concentration was modified in order to obtain a spinnable solution with controlled thickness. The spin coater speed was maintained in the 3.000-6.000 r.p.m. range with a constant acceleration. After a drying process at 150-200°C during 30-60 minutes, a high temperature treatment under $Ar/5\%H_2$ atmosphere allowed to grow the layers. The temperature range where single phases were observed was 650°C-900°C.

Growth of $YBa_2Cu_3O_7$ was performed by using trifluoroacetate (TFA) precursor solutions which were prepared by dissolving stoichiometric mixtures of the Y, Ba and Cu acetates in an aqueous solution of 25% of trifluoroacetic acid. The mixture was continuously submitted to a reflux at 75°C for 4 hours. The resulting solution was dried getting a blue glassy residue, which is redissolved in methanol to make a solution of low viscosity. The thin film samples were prepared by coating the substrates with the precursor solution spinned at room temperature at angular speeds up to 6000 r.p.m.. The final film thickness was altered by adjusting the concentration of the mixed-metal precursor solution. In the present case the thickness of the films was $t\approx250nm$, as determined by transverse SEM observations and interferometric methods.

The wet films were calcined in three stages. First, the organic material was pyrolyzed at 400°C in a moist O_2 atmosphere. The films were then crystallized at high temperatures (750-830°C) in a humid atmosphere ($PH_2O=7.3\%$) and $PO_2=20$ $ppmO_2$ .The elapsed time depends on temperature and film thickness. Typically 3 hours allowed to complete the reaction for films of 250 nm. In the last stage the films were oxidized at 450°C in a dry oxygen atmosphere during 1 hour to convert the tetragonal phase in the superconducting orthorhombic phase.

High Temperature Superconductors

X-ray diffraction (θ-2θ, ω-scan, ϕ-scan and pole figures) was used to determine the phase purity and the crystallographic texture of the films. Micro-Raman spectroscopy allowed to further analyse the kinetics of the formation reaction and to detect the misoriented grains in the films. Fluorine analysis of the exhaust gases in the furnace performed with a selective electrode allowed to determine the optimum duration of the high temperature reaction transforming BaF_2 to $YBa_2Cu_3O_7$ through the formation of HF. SEM and AFM were used to study the microstrutural modifications of the films and the surface roughness. Finally, the superconducting properties were studied through 4-points electrical resistivity measurements and low field dc susceptibility and inductive critical currents measured with a SQUID magnetometer. The critical currents were calculated from the irreversible magnetization using the critical state model which states that $J_c^{ab} = 30\Delta M/2R$, where ΔM is the irreversible magnetization and R the radius of the sample.

RESULTS AND DISCUSSION
Oxide buffer layers

A typical θ-2θ X-ray diffraction pattern of CeO_2 oxide grown on (100)YSZ single crystal is shown in Figure 1, while Figure 2 shows the excellent in-plane epitaxy achieved (FWHM $\Delta\phi$=0.27°). The quality of the out-of plane epitaxy was determined by measuring the ω-scan of the (400) reflection (FWHM $\Delta\omega$=0.23°) (inset of Fig.2) and the uniqueness of the cube on cube texture was verified by measuring pole figures. Similar excellent epitaxy was achieved for films having thickness in the range 20 nm to 100 nm.

The epitaxy achieved in the case of $BaZrO_3$ films grown on STO and LAO single crystals was also of a high quality (Figure 3) (FWHM $\Delta\omega$=0.5° in ω-scan and FWHM $\Delta\phi$=1.17° in ϕ-scan), in agreement with the similar lattice matching in both cases (5.3 % for CeO_2/YSZ vs 6.9 % for BZO/STO). It´s straightforward to note, however, that no residual stresses due this lattice mismatch were detected through X-ray diffraction $\sin^2\chi$ plots performed in both thin film materials, thus we may conclude that these films relax the strong lattice misfit through some kind of microstructural accommodation.

The evolution of the microstructural modifications induced by changes in the growth temperature was followed through AFM observations. Figure 4 display a typical image showing a nanometric grain size ($\Phi\approx$30nm) and a very low porosity and surface roughness (rms$\approx$1 nm). Similar values were achieved in the case of $BaZrO_3$ layers. These microstructural features of the buffer layers could be widely modified through the high temperature thermal treatments thus we may conclude that chemical solution growth techniques allow to prepare oxide layers with excellent performances for the deposition of superconducting layers.

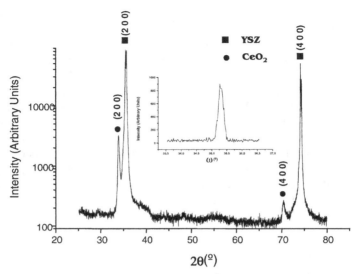

Figure 1.- X-ray diffraction θ-2θ pattern of CeO₂ grown on a YSZ single crystal.
Inset: ω-scan of the (400) reflection

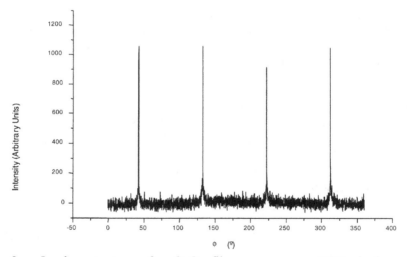

Figure 2 .- In-plane texture of a CeO₂ film grown on a YSZ single crystal
measured with the (111) reflection.

High Temperature Superconductors

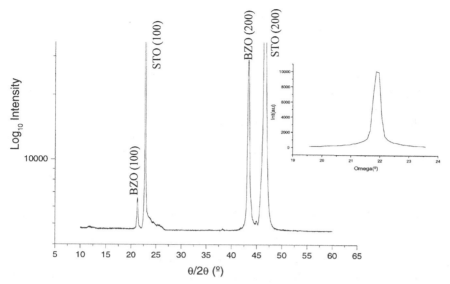

Figure 3.- X-ray θ-2θ diffraction pattern and (inset) out-of-plane texture of BaZrO₃ layers deposited on (100) SrTiO₃ single crystals.

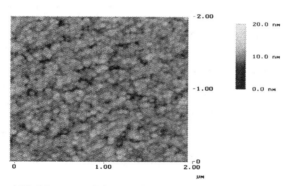

Figure 4.- AFM image of the surface of a CeO₂ thin film grown on a YSZ single crystal.

Superconducting layers

The conversion of the TFA salts into $YBa_2Cu_3O_7$ thin films is a complex process which involves, first, a low temperature decomposition of the metalorganic precursors into nanometric crystals of oxides (CuO, Y_2O_3) and fluoride (BaF_2) [4,5]. This is the pyrolisis process that should be performed in such a way that the metallic ions segregation is minimized and that the film shrinkage does not generate cracks or striations which would degradate the film quality after the growth process. The second reaction step occurs at high temperatures where the decomposition of BaF_2 and its reaction with the oxides leads to the formation of the $YBa_2Cu_3O_7$ phase. The reaction kinetics involves diffusion of the the atomic species diffusion and the elimination of HF, which is controlled through the wet atmosphere within the furnace.

There are multiple processing parameters which require optimisation for the preparation of high quality $YBa_2Cu_3O_7$ thin films and probably not a single combination of them do exist. In our present films we optimised the solution deposition parameters (viscosity, spinning rate) and we performed a slow pyrolisis process ($\approx$20h) to ensure that no degradation of the nanocristalline precursor films occurs. In Figure 5 we show, for instance, a SEM image displaying the formation of striations in the film after the growth process, which are generated when the spinning rate in the coating process is too high. As it can be observed the epitaxy is strongly degraded within these striations where an enhanced porosity, secondary phases and a-axis grains is observed thus decreasing the critical currents of the films.

The formation of the $YBa_2Cu_3O_7$ phase at temperatures within the range 700°C-820°C was monitored by θ-2θ X-ray diffraction, Fluorine analysis in the gas exhaust line and μ-Raman spectroscopy. The processing parameters during this phase conversion step are also multiple. They have a strong influence on reaction kinetics, nucleation energy and grain growth rate and hence they should be widely investigated to control the microstructure. We should mention particularly the temperature, water pressure, gas flow, oxygen pressure and reaction time. In this work we have investigated the relationship among temperature and reaction time to monitor microstructural changes during the growth of films.

Fluorine analysis indicated that the reaction time was doubled when the temperature was reduced from 830°C to 700°C, if the remaining parameters remain fixed. μ-Raman spectroscopy proved very useful to detect small amounts of unreacted phases [8]. Figure 6 show two typical spectra corresponding to an uncomplete reacted sample and to a sample where only the modes of the $YBa_2Cu_3O_7$ phase are observed. In the first case (Fig.6(a)), additionally to the BaF_2 phase,

Figure 5.- SEM image of a $YBa_2Cu_3O_7$ film deposited in a STO single crystal by spin coating at 6.000 r.p.m. The films were reacted at 790°C (a) Low magnification image where the formation of striations is observed, (b) Magnification of the microstructure observed within the striations where the film has an enhanced porosity, secondary phases and a-axis oriented grains.

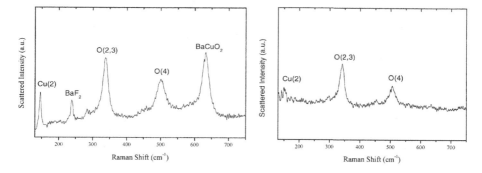

Figure 6.- Raman spectroscopy of $YBa_2Cu_3O_7$ films, (a) sample which has been uncompletely reacted. Note that BaF_2 and $BaCuO_2$ peaks are identified in the spectra, (b) fully reacted sample where only the modes corresponding to $YBa_2Cu_3O_7$ are observed.

we can observe a peak corresponding to $BaCuO_2$. This an impurity that is hardly observed in the X-ray diffraction spectra and which is probably associated to a kinetic hindrance effect promoted by any kind of macrosegregation generated during the deposition or the pyrolisis step. High temperature reaction has been found to promote the disappearance of this impurity.

Figure 7 displays a typical X-ray diffraction spectra corresponding to a pure sample having a high quality texture. ω-scan and φ-scan of these samples lead to FWHM values of Δω=0.4° for the (005) reflection and Δφ=0.5° for the (115) reflection indicating that good out-of-plane and in-plane biaxial texture have been reached during the growth process.

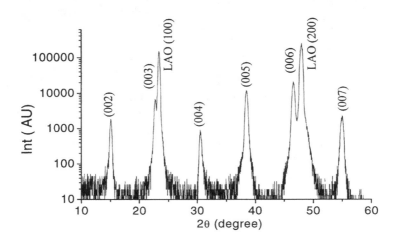

Figure 7.- θ-2θ X-ray diffraction pattern of a YBa$_2$Cu$_3$O$_7$ film deposited on a LaAlO$_3$ single crystalline substrate

The increase of reaction temperature during the growth process has proved to have a tremendous influence on the final microstructure and the superconducting properties of the samples. We have found particularly that the porosity of the samples is strongly dependent on this parameter. As a typical example we show in Figure 8 two SEM images of the sample surface of two samples fired at 700°C and at 830°C during the required time for the reaction completion. The differences in the pore concentration in both samples are easily detected. A quantitative indirect determination of this effect was reached through measurements of the electrical resistivity in the normal state. As it may be observed in Figure 9 (a), the slope of the temperature dependence of the resistivity and the residual resistivity are enhanced in samples with a high porosity, thus indicating that the percolative behavior of current is hindered in such disordered films. The resistivity observed at 300K in the samples having a low porosity, ρ(300K) ≈ 300 μΩcm, together with the high resistivity ratio

High Temperature Superconductors

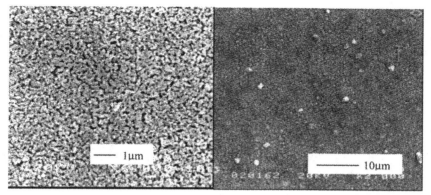

Figure 8.- SEM images of the surface of $YBa_2Cu_3O_7$ films prepared at different temperatures: (a) 700°C, a high porosity is observed, (b) no residual porosity is observed

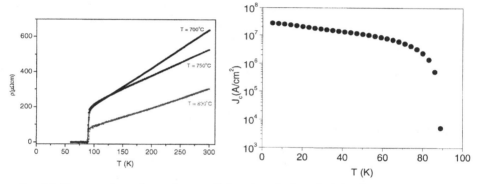

Figure 9.- (a) Temperature dependence of the electrical resistivity in $YBa_2Cu_3O_7$ thin films grown at different temperatures and displaying a very different degree of porosity, (b) Temperature dependence of the critical currents $J_c^{ab}(T)$ in a high quality $YBa_2Cu_3O_7$ thin film

$\rho(300K)/\rho(100K) \approx 3,3$ are typical features of high quality thin films. In this Figure we can also observe that the transition temperature $T_c \approx 90K$ was not modified by the annealing temperature.

Finally, the critical currents of the $YBa_2Cu_3O_7$ thin films were measured inductively by using a SQUID magnetometer. The critical currents were found to be very sensitive to the residual porosity of the films. Very high J_c^{ab} values were measured, instead, in samples showing no residual porosity. As it can be seen in Figure 9(b), at zero field these samples displayed very high critical currents ($J_c^{ab}=3,2 \ 10^6$ A/cm^2 at 77K and $2,7 \ 10^7$A/cm^2 at 5K). To obtain such high performances in thin films usually require the use of vacuum processes and hence our work further support the view that chemical solution growth techniques have the potential to become competitive, from the performance point of view, with these techniques. We note also that in the present work any special refinement process of the metalorganic precursors was applied, as it was previously required [6].

In conclusion, the present work has advanced in the knowledge of the growth mechanisms on single crystalline substrates of oxide buffer layers and superconducting coatings deposited from chemical solutions. The control of the influence of the processing parameters will allow to extend this low cost technique to obtain chemically based coated conductors.

ACKNOWLEDGEMENTS

This work has been supported by the Spanish Science Ministry (MAT 2001-1698 and 2FD97-1628), by the Catalan Pla de Recerca (00206) and the EU-Marie Curie Program (HPMT-CT-2000-00106). A.P. and M.R. acknowledge to C.S.I.C. and J.C.G. to A.E.C.I. for a doctoral fellowship.

REFERENCES

[1] F.F.Lange, "Chemical solution routes to single-crystal thin films", *Science*, **273**, 903 (1996).

[2] D.P.Norton et al, "Epitaxial YBCO on biaxially textured Nickel (00l): An approach to superconducting tapes with high critical current density", *Science*, **274**, 755 (1996).

[3] Y.Ijima et al, "In-plane aligned YBCO thin films deposited on polycrystalline metallic substrates", *Appl.Phys.Lett.*, **60**, 769-771 (1992).

[4] P.C. McIntyre and M.J. Cima, "Heteroepitaxial growth of chemically derived ex situ YBa$_2$Cu$_3$O$_{7-x}$ thin films", *J. Mater. Res.*, **9**, 2219-2230 (1994).

[5] J. A. Smith et al.,"High critical current density thick MOD-derived YBCO films", *IEEE Trans. Appl. Supercond.*, **9**, 1531 (1999)

[6] T.Araki et al, "Coating processes for YBa$_2$Cu$_3$O$_{7-x}$ superconductor by metaorganic deposition method using trifluoroacetates", *Supercond. Sci. Technol.* **14**, 783-786 (2001)

[7] Y-A. Jee et al, "Texture development and superconducting properties of YBa$_2$Cu$_3$O$_x$ thin films prepared by a solution process in low oxygen partial pressure", *Supercond. Sci. Technol.* **14**, 285-291 (2001)

[8] T. Puig et al., "Texture analysis of coated conductors by micro-Raman and synchrotron X-ray diffraction", *Mat. Res. Soc. Symp.* **659**, II5.6.1.-6.6 (2001)

AN ECONOMICAL ROUTE FOR PRODUCTION OF HIGH-QUALITY YSZ BUFFER LAYERS USING THE ECONO™ PROCESS (INVITED)

M. A. Zurbuchen, S. Sambasivan,
I. Kim,[*] J. Rechner,
J. Wessling, and J. Ji
Applied Thin Films, Inc.
Evanston, IL 60201

S. Barnett
Materials Science and Engineering
Northwestern University
Evanston, IL 60201

B. F. Kang and A. Goyal
Metals and Ceramics Division
Oak Ridge National Laboratory

P. A. Barnes and C. E. Oberly
Power Propulsion Directorate
U.S. Air Force Research Lab., OH

ABSTRACT

A new scalable and economical process has been developed to produce high-quality YSZ buffer layers on metal and alloy RABiTS. Current densities of 1 MA/cm^2 were achieved reproducibly on 0.3 μm-thick $YBa_2Cu_3O_{7-\delta}$ layers deposited on the YSZ buffer using PLD with an intermediate ~20 nm ceria layer. The buffer process essentially involves depositing a ~200 nm-thick epitaxial yttrium zirconium nitride directly on RABiTS, with subsequent conversion to a ~300 nm-thick epitaxial yttria stabilized zirconia film via a simple oxidation step. The new process is termed ECONO™, for Epitaxial Conversion to Oxide via Nitride Oxidation, and is also potentially useful for developing oxide epitaxial templates on non-oxide substrates (e.g., silicon or GaAs) for a broad range of device applications. The key advantage of this approach is that it avoids interference from substrate oxidation during epitaxial growth. For coated conductors, it eliminates both the need for an oxide seed layer and the need for a sulfur superstructure. In addition, the deposition of nitrides by reactive sputtering can be performed at much higher rates relative to oxides and the resulting YSZ films are dense and relatively defect-free. A description of the process and the associated conversion mechanism will be discussed.

---

[*] Now at Functional Coating Technology, LLC.

INTRODUCTION

High-temperature superconductor (HTS)-coated conductors represent the next generation of HTS wire technology, with primary application in power generation, power transmission, and compact motors for military and commercial use. HTS tapes, formed by deposition onto biaxially textured metal substrate tapes, are intended to replace copper as a conductor in these applications. The current approach (e.g., RABiTS) being pursued by several research groups worldwide involves deposition of a 10-20 nm-thick a ceria/yttria layer, followed by 200-500 nm-thick YSZ layer, capped with a 10-20 nm-thick ceria layer prior to deposition of the HTS layer.

Initial demonstrations of the proof-of-concept at Oak Ridge National Laboratory (ORNL), now duplicated by other research groups worldwide, utilized a buffer layer configuration of CeO_2 (0.01 μm) / YSZ (0.5 μm) / CeO_2 (0.01 μm) to provide a surface for $YBa_2Cu_3O_{7-\delta}$ (YBCO) epitaxy and to act as a Ni diffusion barrier. The purpose of the first buffer layer is to provide a good epitaxial oxide layer on the reactive, biaxially textured Ni substrate without the formation of undesirable NiO. CeO_2 is special among oxides in its ability to readily form single orientation cube-on-cube epitaxy with Ni. Deposition of CeO_2 was done using a range of deposition techniques under a background of forming gas with small amounts of water vapor (4% H_2O - 96% Ar). Under such conditions, the formation of CeO_2 is thermodynamically preferred over NiO. It is not possible to deposit YSZ under such conditions without undesirable orientations, but in the case of CeO_2, one can readily obtain a single epitaxial-orientation texture. A CeO_2 layer that is thick enough to serve as a chemical diffusion barrier to Ni readily forms microcracks, so a YSZ layer is required. YSZ provides an excellent chemical barrier to Ni diffusion and does not crack when grown thick because there is a significant lattice mismatch between YSZ and YBCO (~5%), leading to a second, 45°-rotated orientation that sometimes nucleates and functions to relieve misfit stresses. A thin CeO_2 layer, which has an excellent lattice match with YBCO (~0.1%), is deposited epitaxially on the YSZ layer to provide an epitaxial template.

The drawbacks of this buffer layer structure are that the deposition of the first CeO_2 layer is quite difficult. Strict control of deposition conditions, in particular the O_2 partial pressure, is required to avoid formation of undesirable NiO. (NiO typically nucleates in mixed orientations and is very weak.) Furthermore, CeO_2 can possess a wide range of oxygen stoichiometry, and is brittle and electrically non-conducting. It will be a challenging engineering task to develop a large-scale continuous process for producing thick (>0.5 μm) crack- and porosity-free oxide films based on a vapor phase process. For example, in a continuous process

High Temperature Superconductors

involving reactive electron beam evaporation of Ce to form CeO_2, the formation of an oxide on the target complicates matters relating to rate of deposition as well as stability of the melt pool. Any change of conditions during deposition is known to have profound affects on the film microstructure. Moreover, oxidation of the biaxially textured metal, even after the successful deposition of CeO_2, can induce undesirable interfacial stresses leading to spallation or further cracking, thus deteriorating the material properties. Microcracks in the oxide buffer layer will adversely affect the epitaxial quality of the growing YBCO film and create weak-links, besides serving as diffusion paths for Ni.

Despite these difficulties, excellent results have been reported by a number of researchers for the current-carrying capacity of HTS films on short samples, the industry is still in search of a low-cost scalable approach that will help produce HTS films in long lengths without compromising on quality and consistency. This has been a significant technical and engineering challenge and has led to substantial R&D investments over the past 5 years. The critical need for second-generation wire technology is a low-cost, scalable, and high-quality biaxially textured yttria stabilized zirconia (YSZ) buffer layer on metal tapes.

A primary concern with the current architecture of the multilayer stack is the precise control needed to deposit the initial ceria/zirconia seed layer. Any oxidation of the underlying substrate during growth of the seed layer results in poor-quality texture. In the case of pure nickel substrates, NiO forms quite readily and interferes with the epitaxial growth of the ceria layer. With non-magnetic substrates (e.g., Ni-Cr or Ni-W), the concern for substrate oxidation is exaggerated by presence of Cr or W and hence requires even more stringent control for seed layer growth. In addition, it is well established that for direct growth of oxide seed layers, a monolayer of 2x2 sulfur superstructure is necessary to promote epitaxy. This necessitates a well controlled annealing procedure in sulfur-containing atmosphere to create the superstructure. For transitioning the current approach to an industrially scalable process, the aforementioned issues present significant technical and engineering challenges, and more importantly, impose a high cost which impedes its deployment as a competitive product.

The performance criterion for superconducting tapes is defined by the texture quality (both in and out of plane) and the microstructural characteristics of the film (YSZ film density and surface morphology). The surface morphology characteristics are especially critical for YBCO films produced using the TFA process,[1] where a surface roughness below 10 nm RMS is desired.

THE ECONO™ PROCESS

The problems encountered with the current buffer layer approach are circumvented by using an innovative and novel process developed at Applied Thin Films, Inc. (ATFI), termed "ECONO" for **E**pitaxial **C**onversion to **O**xide via **N**itride **O**xidation. The ECONO™ process represents a technological breakthrough that has significant implications for the intended application of HTS coated conductors and beyond. This new process provides a distinct cost and performance advantage for fabrication of buffer layers needed for the HTS coated conductor tapes.

Essentially, the process allows the growth of high-quality oxide epitaxial layers on non-oxide substrates. A precursor *epitaxial* nitride layer of composition $Y_{0.2}Zr_{0.8}N$ (YZN) is deposited on the substrate and subsequently converted to an epitaxial yttria stabilized zirconia (YSZ) layer by a simple oxidation step. The process is shown schematically in Fig. 1. The oxidation can be conducted either *in-situ* after nitride deposition or *ex-situ* in a conventional furnace with controlled oxygen atmosphere. As described in more detail below, the conversion from nitride to oxide is very fast and proceeds from the top of the nitride layer which then serves as the template for subsequent oxide conversion downward. The oxidation proceeds in a highly cooperative manner whereby the epitaxy both in- and out- of plane is maintained. The key advantages of the process include: (a) deposition of a high-quality nitride layer which is dense with smooth morphology and relatively free of defects (which are prevalent in oxide deposition by reactive or RF sputtering), (b) does not require a seed layer or a sulfur superstructure, as is the case with the current approach, (c) the nitride can be deposited at relatively high rates compared to oxides, and (d) YSZ, having a larger molar volume than YZN, translates into ~50% thicker films after conversion which reduces the effective deposition time.

It is important to note that even though the above approach can yield a compositionally complex nitride stack, the deposition of all nitride layers can be performed in a single step, in the same chamber, without compromising the process efficiency as compared to YZN single layer deposition.

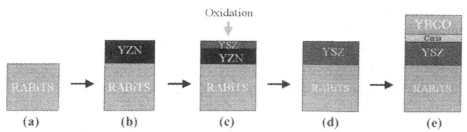

Figure 1. Schematic of YBCO growth on a RABiTS using the ECONO™ process. (a) A bare RABiTS is prepared, (b) $Y_{0.2}Zr_{0.8}N$ is deposited(c) The nitride is then converted to the oxide (yttria stabilized zirconia) by a thermal oxidation step to form an oxide film on the alloy RABiTS, (d). CeO_2 and YBCO are then deposited by the conventional. technique.

To meet the cost target of the final YBCO conductor, costs associated with fabrication of each component must be minimized without compromising film quality. A key parameter affecting the cost of buffer layer production is the deposition rate. Direct deposition of oxide layers by reactive sputtering is relatively slow compared to nitrides. One of the key competitive advantages offered by the ECONO process is the ability to deposit high quality nitrides (dense and uniform with high-quality texture) at high rates.

Broader significance of the ECONO process beyond the use for coated conductors or HTS devices is also apparent and deserves much attention. The ECONO process should be considered as a "generic epitaxial template technology" suitable for a broad range of commercial applications. In addition to RABiTS, the process has been demonstrated for depositing high-quality epitaxial YSZ layers on silicon and sapphire. Reactive sputtering is a well established thin film deposition process routinely used in the semiconductor industry, and can be easily integrated into a manufacturing process. YSZ, with cubic symmetry, is an excellent template layer for subsequent growth of oxide device layers and has been used for deposition of $Pb(Zr,Ti)O_3$, $SrTiO_3$, $(Ba,Sr)TiO_3$, YBCO, RuO_2, and others.[2,3] YBCO films on silicon, using YSZ as the buffer layer, is being pursued by industry for fabricating high efficiency RF filters for telecommunication systems, magnetic resonance imaging systems, bolometers, high speed computing, and others.[4,5]

There are several advantages to using a precursor nitride film. Cubic nitrides can be epitaxially deposited on reactive metals such as Ni in a straightforward manner. The nitride deposition on Ni is done in a nitrogen-containing ambient, and because Ni does not form a nitride, the strict process control required for the current technique is not needed. No separate partial pressure of H_2O is required, so the process is easy to control and scale-up. Nitrides have been used in the tool industry for many years, and their manufacturability in various configurations has already been demonstrated. High-rate deposition (up to 1 μm/min) of nitride coatings is being developed and used in many industrial applications.

Most metal nitrides exhibit a cubic symmetry with a range of lattice parameters suitable for the coated conductor application. Solid solutions of transition metal nitrides and rare earth nitrides can be used to "tailor" lattice parameters, because of extensive compositional range, with solid solubility among most of them. The compositional flexibility and chemical compatibility among these nitrides can be used favorably to tailor several properties required for nitride deposition and resulting oxide buffer layers.

GROWTH OF YBCO / CeO_2 / ECONO™-YSZ / Ni-RABiTS STACKS

Nitride films are grown in a D.C. reactive magnetron sputtering system with an all-metal-sealed chamber and a sample-insertion load-lock system. Base pressures of $\sim10^{-8}$ Torr are typical. The magnetron sputtering sources (US GUN I) utilize 5 cm-diameter, 3.2 mm-thick sputtering targets. Background pressures of 15-20 mTorr of >99.999 % pure Ar-N_2 were maintained during growth. Sputtering targets consisted of 99.95 % V, 99.95 % Ti, Zr, Y, and 99.95 % $Y_{0.2}Zr_{0.8}$ alloy.

YZN buffer layers were deposited on nickel RABiTS substrates, and were converted to YSZ by an *ex-situ* oxidation step. Standard pulsed laser deposition (PLD) was used to deposit ~20 nm thick ceria layers on selected ECONO™-deposited YSZ / RABiTS substrates to achieve good lattice match with X-ray diffraction (XRD) φ-scans of the layers of a representative stack, shown in Fig. 2, reveal the in-plane texture of the layers. Figure 2(a) shows a (111) φ-scan of a typical textured Ni-RABiTS, with a full-width at half-maximum (FWHM) of 9.2°. The (111) φ-scan of a YSZ layer on this substrate is shown in Fig. 2(b), with a FWHM of 10.0°. Local cube-on-cube epitaxy has resulted in only a slight decrease in preferred orientation of 0.8°. The φ-scan of the PLD-deposited CeO_2 layer, Fig. 2(c), shows an *improvement* in texture, with a FWHM of 9.5°. The (001) YBCO layer maintains this texture, as shown in the φ-scan in Fig. 2(d), with a 9.5° FWHM. The overall decrease in texture from Ni-RABiTS to YBCO is 0.3°. These data demonstrate the high quality of YBCO that can be grown on an ECONO™-grown buffer layer.

High Temperature Superconductors

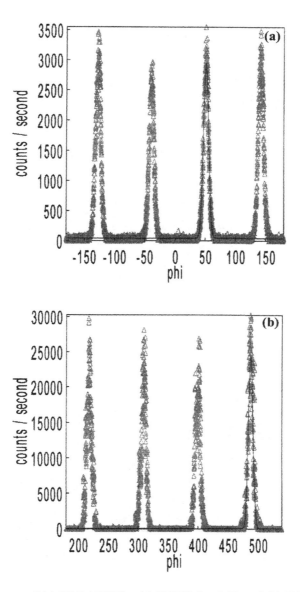

Figure 2. φ-scans of (a) Ni RABiTS with FWHM = 9.2° and (b) ECONO™-deposited textured YSZ layer with FWHM = 10.0°, showing the excellent retention of texture of the buffer layer.

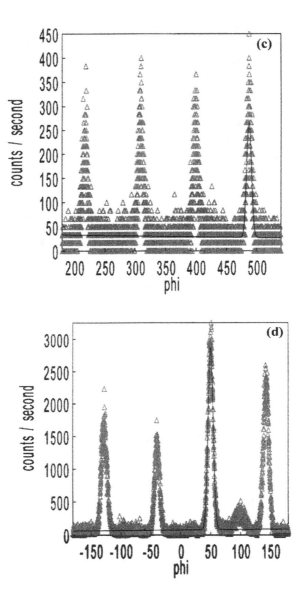

Figure 2 (continued). φ-scans of (c) the CeO2 nickel-diffusion barrier, with FWHM = 9.5°, and *improvement* in texture of 0.5°, and (d) the (001) YBCO film, with FWHM = 9.5°. The decrease in texture through the entire stack is only 0.3°.

High Temperature Superconductors

The results of electrical transport measurements on a typical film stack of YBCO / CeO$_2$ / ECONO™-YSZ / Ni-RABiTS, with a 200 nm-thick YBCO layer thickness, are shown in the I-V plot in Fig. 2. Film width was 3 mm. $I_c = 6$ A at 1 µV/cm criterion. J_c of the film is calculated to be 1 MA/cm^2. A second, identically-processed stack showed the same I-V behavior and J_c values.

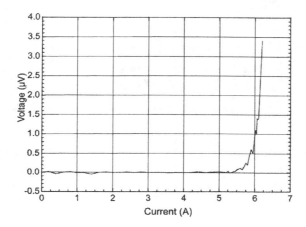

Figure 3. I-V plot measured o a 200 nm-thick YBCO film, grown on a CeO$_2$ / ECONO™-YSZ / Ni-RABiTS. $J_c = 10^6$ A/cm^2 at 77 K, measured by whole body transport current measurement in self field using a 1 μV/cm criterion.

GROWTH OF YSZ ON SILICON BY ECONO™ PROCESS

Many potential applications of the ECONO™ process to provide an epitaxial oxide template on silicon exist. Thus, the ECONO™ process has also been used to deposit epitaxial YSZ on silicon. A cross-sectional TEM image of a YZN film grown epitaxially on silicon, half-way through the oxygen-nitrogen exchange process, is shown in Fig. 4(a). The conversion front appears quite planar. This is significant considering that YSZ and YZN do not have the same crystal structure, and that a 50 % difference in molar volume exists between the two phases. (The difference in lattice parameter is 13 %.) The two phases are in Fig. 4(b) and (c), taken from the same area, and also in the high-resolution TEM (HRTEM) image in Fig. 4(d). The quality of the YSZ film is strongly dependent upon the initial rate of the oxidation process. The first atomic layer(s) in the YZN film surface to undergo oxide conversion are critical for maintaining epitaxial orientation of the rest of the film.

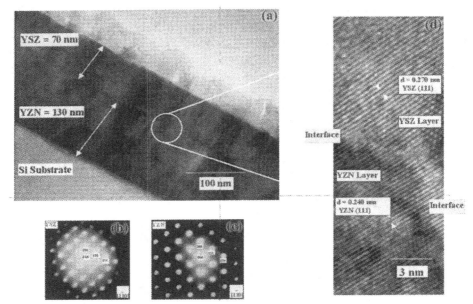

Figure 4. Cross-sectional TEM characterization of a YZN / Si film in the middle of the conversion process to YSZ. (a) The planar growth front is apparent. Epitaxy is maintained despite a 50% difference in molar volume (13 % difference in lattice parameter) between the two phases, as shown by the electron diffraction patterns in (b) and (c), and the HRTEM image of the interface in (d).

The nitrogen distribution within the same film was studied by parallel energy dispersive spectroscopy (PEELS) analysis in the TEM. Figure 5(a) is a cross-sectional TEM image and Fig,. 5(b) is an elemental map of nitrogen over the same area, acquired by setting an electron prism and an aperture to collect only intensity immediately after the nitrogen energy loss edge. A nitrogen-signal line scan is shown at the bottom of Fig. 5(b), clearly revealing a gradient in nitrogen concentration across the YSZ layer. The exact mechanism of nitrogen diffusion through YSZ has not yet been determined.

High Temperature Superconductors

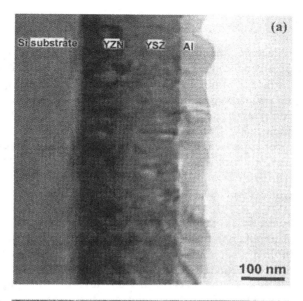

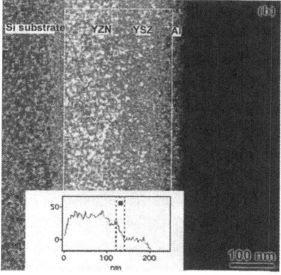

Figure 5. (a) Cross-sectional TEM image and (b) nitrogen elemental map of the same area, collected using PEELS. Brighter regions contain higher concentrations of nitrogen. A nitrogen-concentration line scan is shown at the bottom of (b).

CURRENT RESEARCH FOCUS

The ultimate objective of using the ECONO™ process is to develop a suitable process that produces consistent, reliable, and high texture-quality buffer films with appropriate surface morphology on desirable metal substrate tapes in a continuous manner – with the goal of producing economical superconducting wire. We are currently in the process of upgrading our film deposition system to enable the production of ECONO™ YSZ on lengths of RABiTS. Several buffer layer concepts are being pursued by researchers worldwide with an effort to meet both the cost and performance requirements for tapes in long lengths. The performance criterion is defined by the texture quality (both in and out of plane) and the microstructural characteristics of the film (YSZ film density and surface morphology).

Successful fabrication of biaxially textured superconducting wire based on the coated conductor technology requires optimization of the cost/performance of the HTS conductor. From a superconducting performance standpoint, a long, flexible, single crystal-like wire is required. From a cost and fabrication standpoint, an industrially scalable, low-cost process is required. It is now clear that, although it is fairly straightforward to fabricate long lengths of biaxially textured metals or alloys, it is quite difficult to deposit high-quality oxide buffer layers using low-cost processes. Buffer layers must: (a) provide an effective chemical barrier for diffusion of deleterious elements from the metal to the superconductor, (b) have a high degree of crystallinity via epitaxy with the biaxially textured metal template, and (c) provide a good epitaxial template for superconductor growth.

The above arguments clearly suggest the need for alternate buffer technologies to satisfy the requirements for economic production of superconducting tapes. Our results from Phase I and subsequent efforts show the proposed approach of "epitaxial oxidation of cubic nitrides" is a viable approach for meeting the cost and performance objectives.

The optimum thickness of the nitride layer must be determined. It is desirable to keep the thickness to a minimum for lower production costs, but the YSZ layer must be thick enough to be a good diffusion barrier between the metal and YBCO during YBCO processing. However, this thickness has not been optimized for highly dense YSZ as produced in the ECONO process, so it is reasonable to expect that thickness requirements will be diminished over those of typical RABiTS / buffer layer / YBCO structures. Furthermore, a 50 % increase in thickness of the buffer layer occurs when YZN is converted to YSZ due to the higher molar volume of YSZ compared to YZN. Thus, a 200 nm dense YZN film converts to a 300 nm dense YSZ film.

High Temperature Superconductors

References

[1] A. P. Malozemoff, S. Annavarapu, L. Fritzemeier, Q. Li, V. Prunier, M. Rupich, C. Thieme, W. Zhang, A. Goyal, M. Paranthaman, and D. F. Lee, "Low-Cost YBCO Coated Conductor Technology," presented at the Eucas Conference Sitges, Spain, September 14-17, 1999.

[2] S. Jun, S. K. Young, J. Lee, and W. K. Young, Appl. Phys. Lett. **78**, 2542 (2001).

[3] Q. X. Jia, C. Kwon, and P. Lu, Integr. Ferroelec. **24**, 57 (1999).

[4] Fall 2001 BMDO Update, BMDO-funded Superconductor Research Leads to Revolutionary New Commercial Products (2001).

[5] Proceedings of the SPIE, The International Society for Optoelectronic Engineering: High-Temperature Superconducting Detectors, Bolometric and Nonbolometric, Los Angeles, CA (1994).

LATEST DEVELOPMENTS IN USING COMBUSTION CHEMICAL VAPOR DEPOSITION TO FABRICATE COATED CONDUCTORS

Adam C. King, Shara S. Shoup, Marvis K. White, Steve L. Krebs, Dave S. Mattox, Todd A. Polley, Natalie Darnell
MicroCoating Technologies
5315 Peachtree Industrial Blvd
Atlanta, GA 30341

Ken R. Marken, Seung Hong, Bolek Czabaj
Oxford Superconducting Technology
600 Milik St.
PO Box 429
Carteret, NJ 07008-0429

ABSTRACT
    The non-vacuum Combustion Chemical Vapor Deposition (CCVD) technique has been utilized to deposit both buffer layers and high temperature superconductors (HTS) on metal substrates that were prepared using the Rolling Assisted Biaxially Textured Substrate (RABiTS^{TM}) process [1]. Previously CeO_2 capped $SrTiO_3$ deposited by CCVD on RABiTS^{TM} substrates have enabled pulsed laser deposition (PLD) yttrium barium copper oxide (YBCO) with current carrying capability of $>1MA/cm^2$ [2]. MCT has partnered with Oxford Superconducting Technology who supplies the substrates and is in the process of scaling the textured metal to 100 plus meter lengths. Other buffer layers, such as, gadolinium oxide and lanthanum zirconate have also been investigated using CCVD and are reported here. Results indicate high quality epitaxial films of Gd_2O_3 can be grown on nickel and nickel 3% tungsten. Reproducibility, uniformity and scalability of the Gd_2O_3 buffer layer on this nickel alloy have also been proven. Previously it has been reported that CCVD YBCO deposited on single crystal was capable of achieving the benchmark of $>1MA/cm^2$ [2]. Currently, epitaxial YBCO superconductors were deposited onto a ceria capped $SrTiO_3$ buffer layer on Ni and on a ceria capped Gd_2O_3 buffer layer on Ni-3%W. Experimentation to pass high currents through the HTS films is in progress. Further research to eventually develop long length HTS tapes created solely using

the non-vacuum CCVD method will continue at MicroCoating Technologies (MCT).

INTRODUCTION

The use of high temperature superconductors has proliferated over the past few years in several different applications such as transmission cables, motors, etc [3]. Typically, these applications have used bismuth strontium calcium copper oxide (BSCCO) superconductors, but the cost of these superconductors still limits their overall use in the industry. Research on producing lower cost and high performance HTS systems, such as the YBCO superconductor, have continued with some success. Depositing the YBCO material onto a flexible metal tape has proven to be challenging, as has scaling any system to produce high-performance long lengths of this tape. Ideally, once the YBCO system has been commercially scaled, the target cost of the superconducting tape will be $10-50/kA-m [4]. Many of the costly vacuum-based deposition systems will not be able to achieve this low cost for the production of YBCO, but the non-vacuum CCVD process has this potential.

CCVD can potentially *reduce costs at least tenfold* in comparison to these other systems. As CCVD does not rely on a vacuum to regulate the atmosphere within which the buffer layers and the following superconductor will be deposited, the capital expense is dramatically reduced. Due to the fact that the CCVD process is not a batch process, there exists a significant advantage over most vacuum-based methodologies. The scaling of a typical CCVD system is quite easy and inexpensive. Long lengths of continuous coated tape can be easily achieved for both the buffer and superconducting layers. Adding additional nozzles at a minimal overall expense can also easily increase the deposition rate of said materials. CCVD is a viable processing option for the production of long length HTS tapes.

EXPERIMENTAL DETAILS

In the traditional CCVD process (Fig. 1) [5], precursors are dissolved in a solvent that typically acts as the combustible fuel. This solution is atomized to form submicron droplets by means of the Nanomiser™ technology, proprietary to MCT. These droplets are then convected by an oxygen stream to the flame where they are combusted. A substrate is coated by simply drawing it through the flame plasma. The heat from the flame provides the energy required to evaporate the droplets and for the precursors to react and to vapor deposit on the substrates. Substrate temperature is an independent process parameter that can be varied to actively control the film's microstructure, epitaxy, and oxidation of the nickel-based substrate. In the case of the buffer layer depositions, a localized reducing atmosphere is also used to prevent nickel oxidation, and a solvent system with a

High Temperature Superconductors

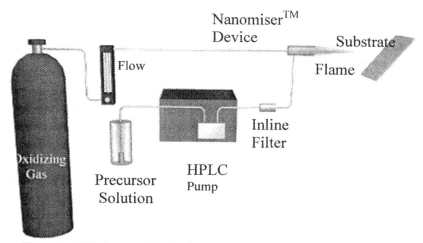

Figure 1. CCVD system showing inexpensive equipment and ease of setup.

lower carbon content is used to minimize carbon deposition in such an atmosphere. Dynamic seals allow for continuous substrate passage between the open air and the localized reducing atmosphere.

One of the largest challenges in tailoring the CCVD system for deposition of YBCO was the removal of all carbon sources from the process. Once formed, barium carbonate is difficult to remove and detrimental to the performance of the superconductor coating. The process was modified to allow for use of aqueous solutions in the deposition of YBCO. A new Nanomiser™ device for the atomization of aqueous solutions and the use of a hydrox flame has resulted in successful YBCO. The locally reducing atmosphere, similar to what is used to deposit buffer layers, is also used to aid in the deposition of YBCO on the metal substrates. The challenge presented now is growing the complex oxide in a reducing atmosphere.

DISCUSSION
Buffer Scaling and Optimization

Experiments have turned from nickel to a new substrate, a nickel alloy that contains 3% tungsten. The switch in substrate emphasis from pure Ni to Ni-3%W was due to its superior strength over Ni and Oxford's success at producing smooth, cube-textured Ni-3%W with lower omega and phi full width half maximum (FWHM) values than for Ni. Several buffer layers have been tested on this substrate including $SrTiO_3$, CeO_2, Gd_2O_3, $LaMnO_3$ and $LaZrO_3$. At the present time, the material that appears to grow with the best orientation is Gd_2O_3. It is also possible to deposit an epitaxial ceria cap on the Gd_2O_3 layer. The X-ray diffraction (XRD) results shown below in Figure 2 and Figure 3 indicate the out-of-plane and in-plane orientation of the sample is quite good. The CeO_2 (111) Phi scan indicates FWHM of 8.36°.

Once it was determined that short samples of the CeO_2/Gd_2O_3 architecture could be deposited onto the nickel alloy, it was then necessary to determine if

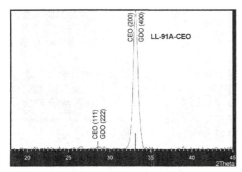

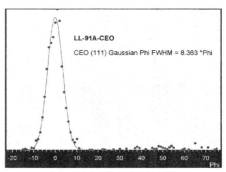

Figure 2. Theta-2theta scan of a $CeO_2/Gd_2O_3/Ni$-3%W architecture

Figure 3. Phi scan of a $CeO_2/Gd_2O_3/Ni$-3%W architecture

long lengths of this material could be produced with a relatively uniform coating over the length of the sample. For these trials, meter lengths of the CeO_2/Gd_2O_3 architecture were produced on Ni-3%W and then tested at ORNL's user facility with their reel-to-reel x-ray system. The results averaged along the meter are shown in Table I indicating reproducibility was achieved with the CCVD process. Another depiction of this reel-to-reel data is shown in Figure 4.

Table I. Reproducibility of 4 separate meters of CCVD $CeO_2/Gd_2O_3/Ni$-3%W

Sample	Out-of-Plane % Misorient.	Omega FWHM	In-Plane % Misorient.	Phi FWHM
1	1.42	6.84	0.00	6.84
2	1.33	6.90	0.00	6.82
3	1.53	6.80	0.00	6.62
4	1.54	6.80	0.00	6.58

The microstructure of these buffer layers was also investigated. For this purpose, a scanning electron microscope was utilized to determine the homogeneity of the films and if any major surface defects existed. A typical

High Temperature Superconductors

micrograph of a CeO_2/Gd_2O_3 deposition is shown in Figure 5. The ceria layer forms a dense, homogeneous film over the gadolinium oxide layer. The surface particles are CeO_2/Gd_2O_3 as indicated by energy dispersive spectroscopy (EDS). This surface is a suitable template for the growth of epitaxial as described below. At this time, minimal depositions of YBCO by both PLD and CCVD have been deposited onto these new buffer layers, but preliminary results indicate that

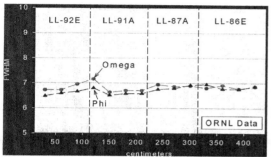

Figure 4. X-ray data over lengths of $CeO_2/Gd_2O_3/$ Ni-3%W architecture buffered tape

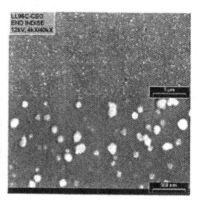

Figure 5. SEM Micrograph of CEO/GDO/Ni-3%W

this architecture is capable of suppressing nickel oxide growth and nickel/tungsten diffusion. The evidentiary support for this is shown in Table II. This table contains analyses from X-ray Photoelectron Spectroscopy (XPS) indicating penetration depth of nickel (presumed to be nickel oxide) and tungsten. The last four samples indicated in Table II are from meter runs mentioned earlier. Although the thinner buffer layers did provide an effective barrier to nickel and tungsten diffusion, the overall thickness of these films will most likely to be increased to provide a more effective buffer layer for later YBCO depositions (LL-77A-YBCO is the only sample that has a YBCO layer deposited on it).

Further development of other buffer layers deposited by CCVD is showing promise for the future. One such layer deposited onto the Ni-3%W substrate is $LaZrO_3$. This buffer layer has been deposited only very recently but has shown epitaxial growth on the Ni-3%W substrate. Initial results show an out-of-plane orientation of ~8.06% misorientation which can be seen in Figure 6. Brief analysis indicates that the composition is slightly off in regards to the ratio of La to Zr, which can easily be adjusted. Based upon past experiments, once the correct composition is achieved, the misorientation should be lowered. Experiments involving $LaZrO_3$, other buffer layers, and various architectures of

High Temperature Superconductors

those buffer layers, will continue. Testing of YBCO films on these new CCVD substrates is currently underway.

Table II. Barrier effectiveness results by XPS of the CeO₂/Gd₂O₃/Ni-3%W architecture

Sample	CeO₂ Thickness (nm)	Gd₂O₃ Thickness (nm)	% Gd₂O₃ Penetration by Ni (nm)	%Gd₂O₃ Penetration by W (nm)
LL-86G	None	243	9	Unmeas.
LL-76G-1	None	180	14	Unmeas.
LL-76G-2	None	315	11	14
LL-77A-YBCO (ORNL-PLD)	32	280	45	8
LL-86E	37	75	39	69
LL-87A	43	84	56	80
LL-92E	38	80	35	73
LL-95C	35	90	28	66

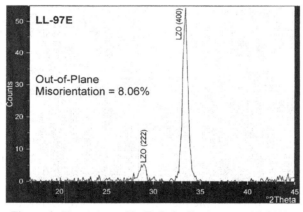

Figure 6. X-ray data of LaZrO₃ buffer layer on Ni-3%W

YBCO depositions

Over the period of the past several months, significant progress has been made in depositing CCVD YBCO on various CCVD buffer layers deposited on RABiTS™ textured substrates. It has been reported earlier that YBCO deposited by CCVD had been able to reach the benchmark of >1MA/cm² on single crystal substrates. More recently, extensive work has been performed on depositing YBCO on several different CCVD buffered metal tape substrates. CCVD YBCO has been epitaxially grown on CCVD CeO_2 capped $SrTiO_3$ buffer layers on Ni and also on CCVD CeO_2 capped Gd_2O_3 buffer layers on Ni-3%W. Both depositions were performed in a localized reducing atmosphere to limit the growth of nickel oxide into the buffer layer during the YBCO deposition. Lower temperatures also proved to be necessary to decrease the amount of nickel oxide and also to facilitate epitaxial, c-axis oriented growth. The temperature range where the best films were grown in was 700-750°C. Initial experiments show that with elevated temperatures, a-axis growth was present. The results of the CCVD YBCO layers deposited in the 700-750°C temperature range are shown below. YBCO grown on the $SrTiO_3$ buffer layer on Ni shows good orientation with clear evidence of (003) and (005) peaks as seen in Figure 7. The second XRD plot depicted along with the CCVD sample, is a YBCO PLD sample used as a for comparative purposes. The PLD YBCO and CCVD YBCO samples show approximately the same level of NiO growth while the CCVD YBCO layer appears significantly thicker. In Figure 8, the results of the YBCO depositions on CeO_2 capped Gd_2O_3 on Ni-3%W, also show good orientation. A (103) pole figure was performed on the YBCO grown on Gd_2O_3, which is shown in Figure 9. The phi FWHM of this film was 9.68°. A typical micrograph of a CCVD YBCO layer can be seen in Figure 10. The micrograph shows the crystalline grains and the dense microstructure of the CCVD grown YBCO layer.

Though superconducting, these YBCO layers have not yet demonstrated $1MA/cm^2$ capability. Some of the concerns with the efficacy of the YBCO layers arise due to their thickness, their possible oxygen deficiency, possible stresses within the film due to lattice strain and obtaining the optimal composition. The thickness issue can be addressed by simply depositing longer which does not tend to have a deleterious effect on the epitaxy of the films up to relatively thick films. Thickness of the current CCVD YBCO films are ~200nm. The oxygen deficiency of the superconducting layers is still in question but simply because the depositions are done in a locally reducing atmosphere. Initial results show that it may be necessary to use an oxygenation step on these films, but more research is needed. It does appear however, that the largest steps in progress can be made in the region of composition and relieving of stresses within the film.

For some time, EDS data obtained from scanning electron microscopy (SEM) was used to determine whether or not the correct 1:2:3 ratio was being achieved. Some problems with this methodology arose when the YBCO was deposited onto STO. The peak intensities of yttrium overlap those of strontium, and the peak intensities of barium somewhat overlap those of titanium. For these reasons, a derivation of the curves did not yield a relevant compositional analysis. Therefore, in order to obtain more accurate and precise results of compositional changes, an inductively couple plasma mass spectrometer (ICP-MS) was utilized. Initial findings, as compared to PLD deposited films exhibiting critical current densities >1MA/cm² and used as standards, did in fact show that CCVD composition was incorrect. Further research will be performed to optimize the composition.

The other most successful route thus far to increasing the current carrying capability of the films was through a high temperature anneal in a low partial pressure of oxygen to alleviate lattice strain. The results from these initial tests prove to be quite promising. Post-anneal electrical testing demonstrated that the YBCO film was able to carry a relatively low current at 77K (critical current density = ~40,000 A/cm²). In previous experiments, obtaining the initial current was the most important step, as further parameter refinement of the deposition can quickly lead to higher current densities.

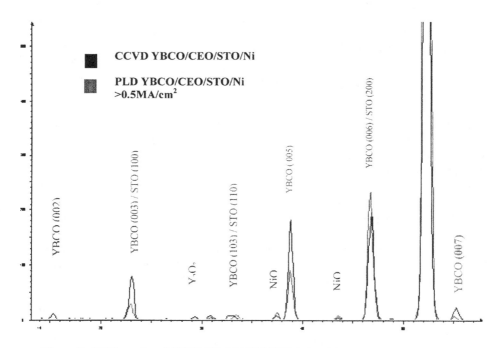

Figure 7. XRD results of CCVD YBCO/CEO/STO/Ni architecture compared to a PLD YBCO standard showing a thicker CCVD YBCO layer with comparable NiO.

High Temperature Superconductors

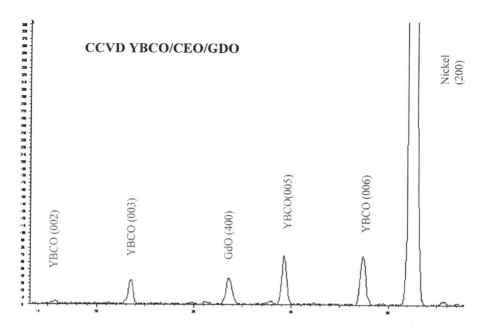

Figure 8. XRD results of a typical CCVD YBCO/CEO/GDO architecture.

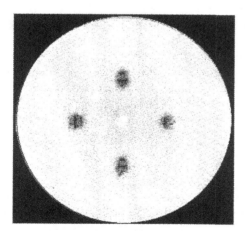

Figure 9. Pole figure of (103) YBCO/CEO/GDO/Ni-W

Figure10. SEM Micrograph of a CCVD YBCO layer on CEO/GDO/Ni-W

CONCLUSION

Significant progress has been made in the use of CCVD as an effective process for depositing buffer layers on textured metal tapes and for depositing YBCO on those buffer layers. CCVD buffer layers deposited on textured nickel tape are high quality templates for >1MA/cm^2 PLD films. CCVD YBCO has reached this benchmark on single crystals and will be pursued on CCVD buffered metal substrates. New developments in the textured substrate have led to research on buffer layers that will allow high current YBCO growth on these substrates. Gd_2O_3 depositions have proven to be the most promising, showing excellent growth and uniformity over long lengths of tape. In the near future, YBCO depositions will evaluate how well the gadolinium oxide layer can act as a buffer layer. CCVD depositions of YBCO have also continued to progress as depositions switched from single crystal substrates to nickel and nickel alloys. Initial results of the YBCO material exhibit great epitaxy and with some improvements in density, composition, film stress and perhaps oxygen content, these films will also be able to reach the industry benchmark.

ACKNOWLEDGEMENTS

MicroCoating Technologies would like to acknowledge the Department of Energy on their continued support of our technology and research and ORNL for their help in depositing PLD films on our CCVD and the use of their user XRD facility.

REFERENCES

1. A. Goyal, D.P. Norton, J. Budai, M. Paranthaman, E.D. Specht, D.M. Kroeger, D.K. Christen, Q. He, B. Saffian, F.A. List, D.F. Lee, P.M. Martin, C.E. Klabunde, E. Hatfield, and V.K. Sikka, Appl. Phys. Lett. **69**, 1795 (1996).

2. 2001 Annual Peer Review, F.A. List and D.F. Lee, Oak Ridge National Laboratory report, Washington D.C., August 2001.

3. L.R. Lawrence Jr., C. Cox, D. Broman, High Temperature Superconductivity: The Products and their benefits.

4. Department of Energy Coated Conductor Development Roadmapping Workshop, St. Petersburg, FL, January 2001.

5. A. T. Hunt, W. B. Carter, and J. K. Cochran, Jr., Appl. Phys. Lett. **63** (2), 266 (1993).

GROWTH OF YBCO THICK FILMS ON ND$_2$CUO$_4$ BUFFERED SUBSTRATES

Xiaoding Qi, Zainovia Lockman, Masood Soorie, Yura Bugoslavsky, David Caplin and Judith L. MacManus-Driscoll

Department of Materials and Centre for High Temperature Superconductivity, Imperial College, London SW7 2BP, UK

ABSTRACT

A new buffer, Nd$_2$CuO$_4$, has been developed for the liquid phase epitaxy (LPE) growth of YBCO thick films. Nd$_2$CuO$_4$ is an ideal buffer for the LPE growth of YBCO thick films, since it has a good lattice and thermal-expansion match to YBCO, minimum reaction with the high-temperature CuO:BaO solution, and is not poisoning to superconductivity. Nd$_2$CuO$_4$ buffers were also grown from the liquid phase using a number of novel techniques. Nd$_2$CuO$_4$ films grown on the surface oxidized Ni foils showed an excellent biaxial texture and a good surface smoothness suitable for subsequent YBCO growth. LPE growth of YBCO on such Nd$_2$CuO$_4$/NiO/Ni substrates showed that dense nucleation of YBCO on Nd$_2$CuO$_4$ was possible without any REBCO seed layer and there was little reaction between the Nd$_2$CuO$_4$ buffer and the high-temperature CuO:BaO solution. T$_c$'s of 90 K and transport J$_c$'s over 10^5 A/cm^2 (77K) have been achieved in such YBCO/Nd$_2$CuO$_4$/NiO/Ni coated conductors with the superconducting layer YBCO about 1-2 μm thick Hence, there is a great potential to fabricate YBCO coated conductors via a fast all-LPE route.

INTRODUCTION

YBa$_2$Cu$_3$O$_{7-x}$ (YBCO) has a better intrinsic current carrying property in magnetic fields than the Bi-based materials, but this better performance can only be realized when all the crystallites in a film are very well aligned along a same direction, so that the superconducting current can be transferred across the crystallites. In order to achieve a good texture in the YBCO film, a highly textured metallic substrate and its buffers are required. There are currently two main methods to achieve the required texture on the metallic substrates or buffers. The first is called IBAD (ion beam assisted deposition) [1], which gives a highly textured buffer layer on a polycrystalline, randomly oriented Ni-based substrate, and the second is called RABITs (rolling assisted biaxial texturing of substrates) [2], which produces a highly textured Ni-based substrate by mechanical working. At present, the buffers and top YBCO layer are dominantly grown by vapor deposition methods such as pulsed laser deposition (PLD). These vapor deposition methods have a relatively slow growth rate, typically several hundred nanometers per hour, and require a high vacuum system. Although they produced very good samples in short length with Jc's as high as 3 MA/cm^2, none of them has been justified to be viable for making long length coated conductors at reasonable cost.

On the other hand, liquid phase epitaxy (LPE), which has widely been used to grow semiconductor and oxide garnet films during 1970s and 1980s [3], has a much faster growth rate, typically in the order of 1 micron per minute. LPE growth of YBCO on single crystal substrates has demonstrated that Jc's over 1 MA/cm^2 can be maintained in the films of thickness up to about 10 microns [4]. This is particularly useful for making thick films for coated conductor applications. However, there is a problem yet to be solved before LPE can be used to grow YBCO on the metallic substrates. That is the relatively high growth temperature, at which most of the metallic substrates and buffers react strongly with the BaO:CuO high temperature solution. A lot of efforts have been made to reduce the growth temperature [5]. However, by reducing the growth temperature, it also reduces the solubility of YBCO in the liquid, which will make the growth more difficult. Therefore, a better method is to search for a special buffer, which is particularly resistant to the attack of the high temperature liquid. Here, we report the development of a new buffer, Nd_2CuO_4, and the initial results of LPE growth of YBCO on Nd_2CuO_4 buffered, surface oxidized Ni substrates.

Nd_2CuO_4 is a tetragonal crystal with a=b≈0.394 nm. It has a lattice mismatch about 2.3% to YBCO, which can be reduced further by doping some smaller rare-earth (RE) ions like Yb^{3+} or Ce^{4+}. By doping Ce^{4+}, this compound becomes conductive. A conductive buffer layer is desired for the coated conductor application because in the event of any breakdown in the superconducting layer, a conductive buffer can reduce the impulse significantly. Nd_2CuO_4 forms in the pseudo binary system of Nd_2O_3-CuO and the phase diagram is shown in Fig. 1, which was determined by differential thermal analysis (DTA), high temperature crystallization and X-ray diffraction (XRD). Nd_2CuO_4 has a much larger solubility in the self flux CuO compared to YBCO in the 2BaO:5CuO flux. Therefore, it is relatively easy to grow from the liquid phase. Since its composition falls under the RE-Ba-Cu-O system, Nd_2CuO_4 neither poisons superconductivity nor reacts strongly with the barium cuprate solution at high temperature. It is a suitable buffer for LPE of YBCO.

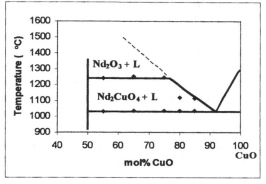

Fig.1 Pseudo-binary phase diagram of Nd_2CuO_4-CuO

EXPERIMENTS

The substrates used in this study were RABITs Ni with a biaxial texture about 10°. The first buffer was NiO, which was formed in flowing O_2 at 1200 °C for 12 hours. NiO was found to grow pseudomorphically on the (002) Ni foils with a biaxial texture around 10° and a thickness around 40 μm.[6]. On top of NiO was Nd_2CuO_4, which was prepared by a number of liquid phase processing methods [7]. The first method was the conventional top dipping LPE [3]. A high temperature solution of 20%Nd_2O_3:80%CuO was prepared at 1200 °C and then quickly cooled down to about 1150 °C, at which the substrates were dipped into the liquid and rotated at 60 rpm. After growing in liquid for about 5 minutes, the substrates were pulled clear and brought to room temperature in about half an hour.

In the second method, a precursor with composition around 30%Nd_2O_3:70%CuO was screen-printed on the substrates using an organic binder and then heated up to 1200 °C to reach a single liquid state. This temperature was kept for about 20 minutes and then cooled down to room temperature at 300 °C/hr. Because of the large surface area to volume ratio and preferential evaporation of CuO, the liquid became supersaturated rapidly, leading to the nucleation and growth of Nd_2CuO_4 on the substrates..

In a third method, a growth system similar to the zone-refining furnace was used. A hydraulically pressed and sintered source rod of 40%Nd_2O_3:60%CuO was fed into the focus of an infrared beam to produce a molten zone at one end. When the increasing size of the molten zone exceeded the limit of the surface tension, the liquid dropped down onto the substrate below, which was maintained at a suitable temperature (~1100 °C) for the epitaxial growth induced from the highly textured substrate. Because of the extremely good wetting property of the high-temperature cuprate solution, the liquid drops quickly spread out on the substrates. Actually, the wetting was so good that not only the top side of the substrate was completely covered, but also the back side of the substrate.

LPE growth of YBCO on the Nd_2CuO_4/NiO/Ni substrates was carried out in 1% pO_2 using the top dipping method. The experimental details about LPE of YBCO can be found elsewhere [8]. The high-temperature solution used in this experiment was prepared at 1010 °C from the composition Y:Ba:Cu= 1:12:30 and the LPE growth took place at 950 °C. Typical growth time in the liquid was around 60 seconds. After the growth, the films were annealed in flowing O_2 for 120 hours.

RESULTS AND DISCUSSIONS

Although surface oxidized NiO had a good biaxial texture, its surface was quite rough. The typical roughness was around 200nm. Considering such a roughness, liquid phase processing is probably the best way to grow Nd_2CuO_4 on top of NiO. Especially, the high-temperature cuprate solution had an extremely good wetting property, which could easily creep into all the groves between the NiO grains. Shown in Fig.2 is the scanning electron microscope (SEM) image of the cross section of the Nd_2CuO_4/NiO/Ni structure. The interface between

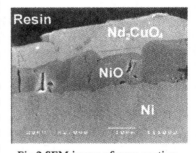

Fig.2 SEM image of cross-section

the Nd_2CuO_4 and NiO was quite smooth and Nd_2CuO_4 could grow into all the gaps.

X-ray diffraction (XRD) showed that the Nd_2CuO_4 films grown on NiO/Ni were c-oriented with high crystallinity and a very good biaxial texture [7]. XRD pole figure showed that the full width at half maximum (FWHM) was only about 5°, which was better than the original texture in NiO. Nd_2CuO_4 can grow relatively easily on NiO because there is a good lattice match between these two compounds, as illustrated in Fig.3. The

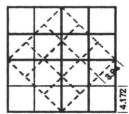

Fig.3 Epitaxial relation between Nd_2CuO_4 and NiO

High Temperature Superconductors

a/b-axis is about 0.394nm for Nd_2CuO_4 and 0.417 nm for NiO, therefore, 3 unit cells of Nd_2CuO_4 can grow along two units of the [110] axis of NiO with lattice mismatch as close as 0.3%. Such an epitaxial relation has been confirmed by the XRD pole figure as well as the observation of surface morphology.

Although three liquid phase processing methods have been tried to grow Nd_2CuO_4, the second method was so far studied in most detail. Most of the results presented above were measured from the samples made by this method. The first method had the simplicity and the ease of control, but it was found that the available supersaturation was not high enough to achieve a dense nucleation on NiO, therefore it was difficult to achieve 100% coverage, although the texture obtained was excellent. The great advantage of the third method was that before the growth, the high temperature solution was prepared at a different place from the substrate, therefore the contact time between the substrate and liquid was much shorter and hence, it could minimize any potential reaction at the interface. In addition, in this method a higher temperature could be used to prepare liquid and at the same time, the temperature of the substrate could be kept below 1200 °C, which was the requirement not to destroy the texture in NiO. By using a higher temperature to make liquid, a smaller amount of flux was required, so the growth could be easier and quicker. Recently, we found that it might not need to oxidize Ni before the growth, a textured NiO was possible to form during the process of making Nd_2CuO_4 by this method. Therefore, there is a potential to make Nd_2CuO_4/NiO/Ni in one process. More detailed study about this method is currently being carried out.

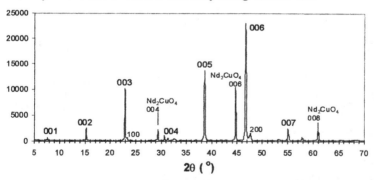

Fig.4 XRD θ-2θ scan of the sample YBCO/Nd_2CuO_4/NiO/Ni

Considering the much lower Y solubility in the liquid, the top YBCO layer on Nd_2CuO_4/NiO/Ni was grown by the conventional top dipping LPE. The growth of

YBCO on the Nd_2CuO_4 was found to be relatively straightforward. It was found that dense nucleation of YBCO on Nd_2CuO_4 was possible without any REBCO seed layer and there was little reaction between the Nd_2CuO_4 buffer and the high-temperature CuO:BaO solution. A XRD θ-2θ scan of the sample $YBCO/Nd_2CuO_4/NiO/Ni$ is shown in Fig.4, which indicates that the films was dominantly c-oriented except small amount of a/b-oriented grains, which was also observed on the surface by optical microscopy. High X-ray counts in the Fig.4 (scanning rate=0.5 sec/step) also show a good crystallinity of the YBCO film, which is one of the benefits of films grown from the liquid phase. The optical micrograph of the surface is shown in Fig.5. The magnified inset at bottom-left corner shows a good alignment of the grains. XRD pole figure conformed the excellent biaxial texture in the YBCO thick films with FWHM=2.5°. The 45° tilted grains were consistent with the epitaxial relation discussed above. T_c's of 90 K and transport J_c's over 10^5 A/cm^2 (77K) have been achieved in such $YBCO/Nd_2CuO_4/NiO/Ni$ coated conductors with the superconducting layer YBCO about 1.5-2 μm thick. A R-T plot is presented in Fig.6, which shows a sharp superconducting transition with "zero" resistance at 86 K, suggesting that there was little contamination of Ni.

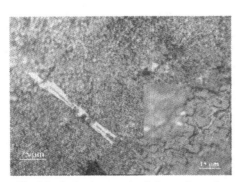

Fig.5 Optical micrograph of the surface of $YBCO/Nd_2CuO_4/NiO/Ni$

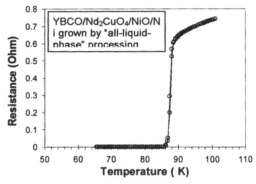

Fig.6 R(T) curve of $YBCO/Nd_2CuO_4/NiO/Ni$

CONCLUDING REMARKS

A new buffer, Nd_2CuO_4, has been developed for the LPE growth of YBCO coated conductor. There are a few potential liquid phase processing methods that can be used for fast production of Nd_2CuO_4 on surface oxidized RABiTs Ni. LPE growth of YBCO showed that dense nucleation of YBCO on Nd_2CuO_4 was possible without

High Temperature Superconductors

any REBCO seed layer and there was little reaction between the Nd_2CuO_4 buffer and the high-temperature CuO:BaO solution. T_c's of 90 K and transport J_c's over 10^5 A/cm^2 (77K) have been achieved in such $YBCO/Nd_2CuO_4/NiO/Ni$ coated conductors with the superconducting layer YBCO about 1-2 μm thick Therefore, there is a great potential to fabricate YBCO coated conductors via a fast all-LPE route.

ACKNOWLEDGEMENTS

We are indebted to EPSRC of the UK for funding this work. Dr. W. Goldacker and Dr. R. Nast of Karlsruhe Forschungszentrum are acknowledged for provision of the (200) Ni foils. Prof. Don McK Paul and Dr. Geetha Balakrishana at Warwick University are thanked for providing crystal growth facilities for the development of a novel horizontal LPE process.

REFERENCES

1. Y. Iijima, N. Tanabe, O. Kohno, Advances in Superconductivity IV, Proc.4th Int. Symp. Supercond. (ISS), pp.679-82, 1992, Tokyo, Springer Verlag
2. J.D. Budai, R.T. Young, B.S. Chao, Appl. Phys. Lett. **62**(15),1836-38(1993)
3. R.L. Moon; Liquid Phase Epitaxy, In: B.R. Pamplin, editor, Crystal Growth, 2nded., Oxford, Pergamon Press, 1980, pp.421-61.
4. S. Miura, K. Hashimoto, F. Wang, Y. Enomoto and T. Morishita; Physica C **278**, 201(1997).
5. X. Qi and J.L. MacManus-Driscoll, *Current Opinion in Solid State & Materials Science*, **5**, 291-230(2001)
6. Z. Lockman, X. Qi, A. Berenov, R. Nast, W. Goldacker and J.MacManus-Driscoll, Physica C: Superconductivity, **351**, 34-37(2001)
7. X. Qi, M. Soorie, Z. Lockman and J L MacManus-Driscoll, Journal of Materials Research, **17**(1), 1-4(2002)
8. X. Qi and J.L. MacManus-Driscoll; J. Crystal Growth **213**, 312(2000).

MICROSTRUCTURAL CHARACTERISATION OF HIGH J_C, YBCO THICK FILMS GROWN AT VERY HIGH RATES AND HIGH TEMPERATURES BY PLD

A. Berenov, N. Malde, Y. Bugoslavsky, L.F. Cohen and J. L. MacManus-Driscoll
Center for High Temperature Superconductivity, Imperial College of Science Technology and Medicine, Prince Consort Road, London SW7 2BP, U.K.

S. J. Foltyn
Superconductivity Technology Center, Los Alamos National Laboratory, Los Alamos, New Mexico 87545, U.S.A.

ABSTRACT

In order to achieve scaleability in processing of IBAD conductors for commercial applications, rapid YBCO film growth rates are required. In this work scanning Raman spectroscopy, x-ray diffraction, scanning electron microscopy and vibrating sample magnetometry have been used to study high rate grown YBCO films. The films were grown by PLD on (100) $SrTiO_3$ using a high power industrial laser at growth temperatures from 750°C to 870°C and growth rates up to 4 micron/minute. The films showed high degree of in-plane and out-of-plane alignment. Two YBCO layers with extended "c" lattice parameters observed in the films were caused by a cation disorder. High values of T_C and J_C (90 K and 10^6 A/cm^2, respectively) were observed despite presence of structural disorder in the films.

INTRODUCTION

Over the last year significant efforts have been made to demonstrate the feasibility of long length coated HTS conductors. Different preparation techniques (e.g. IBAD, LPE, CVD) are being evaluated for possible scalability with the aim to achieve fast growth rate and attractive values of critical current densities (>1 MA/cm^2). However fast growth rates can lead to the crystalline disorder (especially Y/Ba cation disorder [1]), poor crystallinity (both in-plane and out-of-plane) and possible formation of metastable phases during deposition, with reduction in both Tc and Jc.

Recently, the Los Alamos group has reported preparation of YBCO thick films using an industrial laser [2]. Independently, Jo et al. [3] and Suh et al. [4] reported preparation of YBCO thin films at deposition rates around 0.7 μm/min by electron beam co-evaporation and pulsed laser deposition (PLD), respectively. Jo et al. [3] observed that YBCO phase was composed from two distinctive layers. Whereas Suh at al. [4] reported that fast deposition rate at relatively low substrate temperature of 700°C led to cation disorder in Y/Ba sub lattices and the formation of poorly conductive cubic YBCO phase. Ohmatsu et al. [5] reported fabrication of long length YBCO/YSZ/Ni tapes by fast growth PLD in a reel-to-reel system. In order to form epitaxial YSZ bluffer layer on non-textured Ni substrate, the inclined substrate deposition was used. A linear dependence between the YBCO deposition rate and the power of the laser was observed with highest deposition rate of 3.8 μm/min achieved at 200W. In addition an improvement of in-plane texture of YSZ was observed at high laser power. Satisfactory values of J_c 2×10^5 A/cm^2 within 11m tape length were reported.

In this work YBCO thick films were grown at fast deposition rates. In order to decrease the amount of cation disorder in the films, high growth temperatures were employed. The microstructure and superconducting properties of the films were studied.

EXPERIMENTAL

1-μm thick YBCO films were grown by PLD on (100) single crystal SrTiO$_3$ substrates with the high growth rate of around 4 micron/min [6]. The growth temperatures were measured by optical pyrometer and are presented in Table 1. Several pieces of the films were etched in 1% bromine solution in methanol.

The films were characterised by XRD, SEM and Raman spectrometry. Magnetisation measurements were performed using a vibrating sample magnetometer and critical current densities were calculated using Bean model [7].

Table 1. Preparation conditions and properties of studied YBCO thick films.

Sample	Growth Temperature °C	'c' parameter, Å		$\frac{(005)}{(007)}$ ratio	T_C (onset), K	J_C at 77K (0 T), A/cm^2
PD574B	751	11.695	11.679	6.95	92.1	1.1×10^6
PD571	802	11.708	11.691	6.40	91.5	1.8×10^6
PD572	861	11.739	11.714	4.02	91.5	8.7×10^5
PD576	872	11.743	11.721	3.94	90.2	1.8×10^6

High Temperature Superconductors

RESULTS AND DISCUSSION

XRD analysis of the films showed intensive (*00l*) and small (*103*) YBCO reflections. In addition, epitaxialy grown $BaCuO_2$ was present in the films grown at the two highest temperatures, namely 861°C and 872°C. The YBCO peaks were asymmetric and presumably consisted of two overlapping peaks as shown in Figure 1. The degree of asymmetry and thus the relative intensity of the overlapping peaks seem to be temperature dependent. Bromine etching decreased the intensity of the peak located at higher angle (see inset in Figure 1). Consequently two YBCO layers were present in the films: a top layer with shorter "c" axis and a bottom layer with larger "c" axis. Similarly Jo *et al.* [8] observed a faulted surface YBCO layer (c=11.68Å) and a well crystalline YBCO layer close to the film/substrate interphace (c=11.74 Å) in the films grown by electron beam co-evaporation. Those distinctive layers caused significant asymmetry of YBCO x-ray reflections.

Two sets of (00l) YBCO reflections were obtained by deconvolution of XRD data and two values of 'c' lattice parameters were calculated using Nelson-Riley extrapolation function (Table 1). "c" lattice parameter for both YBCO phases increased with the growth temperature (Figure 2). This increase can be due to changes in oxygen stoichiometry and/or cation disorder in the films. The method

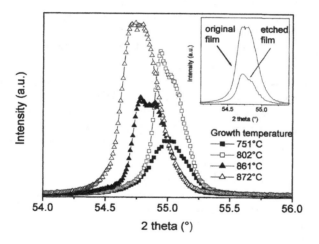

Figure 1. XRD plots of (*007*) YBCO peak of the films. Inset shows (007) YBCO peak of the film grown at 872 °C before and after bromine etching.

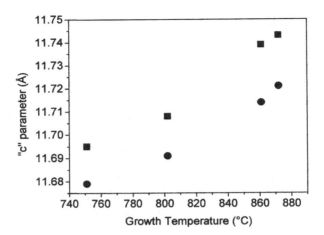

Figure 2. "c" lattice parameters of YBCO thick films as a function of growth temperature.

to assess the amount of disorder in YBCO by measuring integrated intensity of (*005*) and (*006*) peaks [9] could not be used as the (*006*) peak overlaps with (*002*) $SrTiO_3$ peak. However, the ratio of integrated intensity of (*005*) and (*007*) peaks is also valid for estimating cation disorder. The (*005*)/(*007*) ratio decreased with the increase of '*c*'-lattice parameter suggesting that a significant amount of cation disorder was present in the films (Table 1). Hence, the enhanced '*c*' value is not due to deoxygenation of the films. The Raman measurements and magnetisation measurements did not reveal any significant changes in oxygen contents as discussed later. From the rocking curves, the FWHM of the (*007*) peak was 0.6° for all the films. The Φ scans of (*012*) peaks showed sharp a-b alignment with FWHM of 0.6-1.0°. Additionally splitting of the peaks during the Φ scans was observed in the films grown at the temperatures higher than 800°C. This splitting of the peaks was attributed to the YBCO twinning caused by orthorhombic-tetragonal phase transformation [10].

The SEM image of the films revealed dense uniform surface covered with spherical outgrowths. The number and average diameter of the outgrowths increased with the growth temperature. Some a-b aligned grains of YBCO were also present.

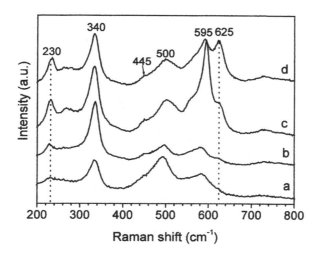

Figure 3. Room temperature Raman spectra of the YBCO films. Growth temperatures were 751°C (a), 802°C (b), 861°C (c), 871°C (d).

The Raman spectra of the films are shown in Figure 3. The modes observed at 147 (not shown), 340, 445 and 500 cm^{-1} correspond to the vibrations along the c-axis of Cu, planar oxygen in CuO_2 (out of phase), planar oxygen in CuO_2 (in-phase) and apical oxygen in orthorhombic YBCO, respectively. The origin of 595 cm^{-1} mode is caused by a loss of transitional symmetry upon defect formation in YBCO [1]. The intensity of the two peaks at 230 and 625cm^{-1} increased with growth temperature (Figure 3). These peaks were assigned to a $BaCuO_2$ [1,11]. A significant variation of the peak intensities across the sample surface was observed in the films grown at 861 and 872°C. This lack of reproducibility was attributed to the presence of partial melting leading to inhomogeneity. The position of the apical oxygen vibration was in the range 494 – 496 cm^{-1} and did not significantly change across the film surface.

The values of T_C and J_C at 77K and 0 T are given in Table 1. The onset of T_C only slightly decreased when the growth temperature was increased (Figure 4). The magnetic transition became broader presumably due to the increase of inhomogeneity in the films caused by cation disorder. At the same time no clear trend of J_C with the growth temperature was observed, although J_C values remained ~ 1 MA/cm^2. Field dependences of critical current densities of the films are shown in Figure 5. Slight enhancement of J_C in magnetic field of up to 4 T was observed in the film grown at 872°C.

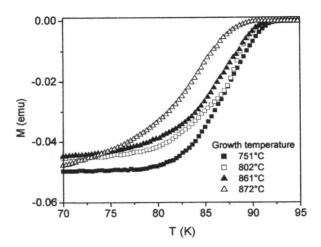

Figure 4. Temperature dependence of the magnetisation moment. Films were cooled in zero field and warmed in 1 mT.

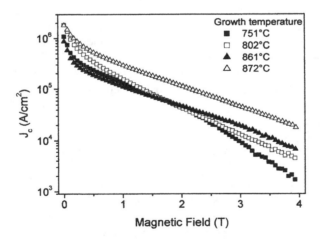

Figure 5. Magnetic field dependences of critical current density of YBCO films at 77K. Field was applied perpendicular to the film surface.

High Temperature Superconductors

CONCLUSIONS

YBCO films were prepared by high temperature (751 - 872°C) PLD at growth rate of upto 4 micron/min. The films showed high degrees of in-plane and out-of-plane texture. The $BaCuO_2$ (quenched liquid) phase was observed in amounts increasing with temperature. The relationship between the ratio of integrated intensity of (005) and (007) x-ray peaks and 'c'-lattice parameters indicated the presence of increasing cation disorder with temperature. Observed high values of T_C (around 90 K) and J_C (~ 1 MA/cm^2) were independent of growth temperature and/or the amount of cation disorder.

ACKNOWLEDGMENTS

Financial support from EPSRC is gratefully acknowledged.

REFERENCES

[1] G. Gibson "A Probe of Structural Defects in YBCO Thin Films using XRD and Raman Microscopy", PhD. Thesis, University of London, 1999.
[2] S.R. Foltyn, E.J. Peterson, J.Y. Coulter, P.N. Arendt, Q.X. Jia, P.C. Dowden, M.P. Maley, X.D. Wu and D.E. Peterson, "Influence of deposition rate on the properties of thick $YBaCu_3O_{7-\delta}$ films", *Journal of Materials Research* **12** [11] 2941-2946 (1997)
[3] W. Jo, L.S-J. Peng, W. Wang, T. Ohnishi, A.F. Marshall, R.H. Hammond, M.R. Beasley and E.J. Peterson, "Thermodynamic stability and kinetics of Y-Ba-Cu-O film growth at high rates in atomic and molecular oxygen", *Journal of Crystal Growth* **225** 183-189 (2001).
[4] J.D. Suh, G.Y. Sung, K.Y. Kang, H.S. Kim, J.Y. Lee, D.K. Kim, C.H. Kim, "Cubic Y-Ba-Cu-O thin films by high speed pulsed laser deposition", *Physica C* **308** 251-256 (1998).
[5] K. Ohmatsu, K. Muranaka, S. Hahakura, T. Taneda, K. Fujino, H. Takei, Y. Sato, K. Matsuo, Y. Takahashi, "Development of in-plane aligned YBCO tapes fabricated by inclined substrate deposition", *Physica C* **351-360** 946 –951 (2001).
[6] S. Foltyn, Q. Jia, T. Holesinger, P. Arendt, H. Kung, Y. Coulter, J. F. Smith, in DOE Annual Peer Review Proceedings, Washington (2000).
[7] Bean C.P., *Physical Review Letters* **8** 250 (1962).
[8] W. Jo, T. Ohnishi, J. Huh, A.F. Marshall, R.H. Hammond, M.R. Beasley, "Growth of high critical current $YBa_2Cu_3O_{7-\delta}$ films with liquid-fluxed microstructures by the in-situ electron beam method"; pp. 77-79 in Extended Abstracts of 2001 International Workshop on Superconductivity co-sponsored by ISTEC and MRS, Hawaii, 2001.

[9] J.L. MacManus-Driscoll, J.A. Alonso, P.C. Wang, T.H. Geballe, J.C. Bravman, *Physica C* **232** 288 (1994).

[10] D. Schweitzer, T. Bollmeier, B. Stritzker and B. Rauschenbach, "Twinning of $YBa_2Cu_3O_7$ thin films on different substrates", *Thin Solid Films* **280** 147-151 (1996).

[11] P. Gomez, J. Jimenez, P. Martin, J. Pigueras and F. Dominguez-Adame, "Cathodoluminescence and microRaman analysis of oxygen loss in electron irradiated $YBa_2Cu_3O_{7-x}$" *Journal of Applied Physics*, **74** [10] 6289-6292 (1993).

DEVELOPMENT OF COATED CONDUCTORS ON BIAXIALLY TEXTURED SUBSTRATES: THE INFLUENCE OF SUBSTRATE PARAMETERS

R.I. Tomov, A. Kursumovic, M. Majoros, R.Hühne, B.A. Glowacki, and J.E. Evetts, Department of Materials Science and IRC in Superconductivity, University of Cambridge, Pembroke Street, Cambridge, CB2 3QZ UK

A. Tuissi, and E. Villa, Istituto per la Tecnologia dei Materiali e dei Processi Energetici del Consiglio Nazionale delle Ricerche - *CNR TEMPE*, Cso P.Sposi 29, Lecco, Italy

S. Tönies, Y. Sun, A.Vostner and H.W. Weber, Atomic Institute of the Austrian Universities, Stadioalle 2, A-1020 Wien, Austria

ABSTRACT

Standard $YBa_2Cu_3O_{7-y}$(YBCO)/oxide buffer architecture was deposited on textured Ni-based substrates by Pulsed Laser Deposition (PLD). The influence of substrate properties on texture development and transport characteristics was investigated. Issues concerning optimization of deposition parameters, substrate oxidation and possible current limiting mechanisms are discussed. XRD characterization shows biaxial alignment of the YBCO layers with out-of-plane and in-plane best FWHM values of 4° and 6.5° respectively. The film morphology was studied using AFM and SEM. DC SQUID magnetometry and direct transport measurements were together used to evaluate the critical current density J_c.

NTRODUCTION

The development of coated conductors requires deposition of biaxially aligned high-temperature superconductor films onto flexible metal substrates. The development of Rolling-Assisted-Biaxially Textured substrates (RABiTS) by Oak Ridge National Laboratory[1] initiated worldwide efforts of producing high irreversibility line superconductor based (YBCO) coated conductors on textured metal tapes. The main requirement in RABiTS technology is the deposition of a high quality ceramic oxide buffers (CeO_2, YSZ, RE_2O_3) transferring the substrate texture to the HTS films as well as preventing Ni contamination of the superconductor. Critical currents larger than 1 MA/cm^2 at 77K and zero magnetic fields have been already reported for different buffer structure combinations [2-4].

Nickel is well suited, as a substrate material for YBCO coatings due to easy formation of a sharp cube texture and compatibility to commonly used buffers layers. However the low tensile strength of Ni, its poor oxidation resistance, its ferromagnetism and low electric resistivity makes Ni difficult to process and unfavorable for use in magnetic fields or for AC applications. The alloying of Ni causes in most cases a decrease of Curie temperature and improvement of the mechanical properties[5]. Good results have been published recently on utilization of robust and non-magnetic NiV and NiCr[6,7] and ferromagnetic commercialy available in km length NiFe tapes[8]. Although good substrate texture can be achieved by cold rolling and subsequent heat treatment the situation with Ni-based alloy tapes appears to be more complicated due to two major problems - lower oxidation resistance leading to the formation of a number of different non-reducible oxides (V_2O_5, Cr_2O_3, $NiFeO_4$) and substrate grooving. Grooves can propagate through the buffer layers and into the superconductor disturbing the texture as well as serving as a fast channels for oxygen diffusion. As shown by T.A. Gladstone et al.[9] grain boundary grooves in the Ni-alloy tapes tend to be deeper than in pure Ni due to the higher re-crystallization temperature of the alloys.

In this paper we report deposition of a standard coated conductor architectures (YBCO/Y_2O_3(YSZ)/CeO_2) onto two mechanically robust Ni-alloy substrates possessing very different characteristics – (i) newly developed ternary alloy NiCrW which is non-magnetic and highly resistant to oxidation[10] and (ii) commercial NiFe tape, which is ferromagnetic and highly susceptible to oxidation. The PLD deposition of the first buffer layer was modified by using segmented Pd:CeO_2 target. Issues concerning oxidation resistance and the influence of substrate grooving on the final conductor quality are discussed. To emphasize the particular features of the substrates a comparison was made with the results achieved on pure Ni tapes with pure CeO_2 target.

EXPERIMENT
Substrate preparation and characterization

Biaxially textured (90 μm) NCrW and Ni tapes were obtained by recrystallization after heavy cold rolling in accordance to the RABiTS method. Recrystallization anneal has been carried out under high vacuum (p<10^{-5} mbar) at temperatures of 1000°C and 700°C respectively. Commercially available NiFe tape of thickness 25μm was cold-rolled and annealed in a factory environment by Carpenter Technologies (UK)[8]. The pole figure analysis confirmed sharp macrotexture (NiCrW -$\Delta\varphi$= 9° and $\Delta\psi$= 9°; NiFe -$\Delta\varphi$= 9° and $\Delta\psi$= 9°; Ni -$\Delta\varphi$= 7° and $\Delta\psi$= 8°) for all tapes. AFM analyses revealed a smooth surface of pure Ni tape with shallow grooves (see Fig1a.). As expected relatively rough morphology of NiCrW and NiFe is accompanied by deep grooving (typically 1-1.5 μm wide, and approximately 30 nm deep for NiFe and 80 nm deep for NiCrW) (see Fig.1b and c). A large number of trenches are clearly present on NiFe tape running in the rolling direction. These trenches have a typical trough to peak modulation of 100 nm. Such deep grooves and trenches can prevent proper cube-on-cube CeO_2 growth and also serve as channels for oxygen diffusion. Scanning electron microscopy (SEM) characterizations show the existence of pinholes and twins on the surface of NiCrW tape with an average grain size of 20-30 μm.

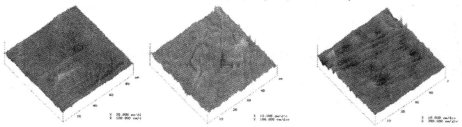

(a) Ni-$\Delta\phi$= 7°; $\Delta\psi$= 8°   (b) NiCrW-$\Delta\phi$= 9°; $\Delta\psi$= 9°   (c) NiFe-$\Delta\phi$= 9°; $\Delta\psi$= 10°
RMS = 7 nm   RMS = 15 nm   RMS = 25 nm
Fig1. AFM characterization of Ni, NiCrW and NiFe tapes.

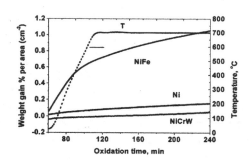

Fig. 2. TG signals in 5 mbar of O_2 pressure for textured metallic substrates.

Thermogravimetric (TG) analysis for oxidation resistance study was carried out for a short oxidation time of about 3h representing the typical processing duration. Many parameters that influence this kind of measure allow us only a qualitative discussion of the results. Moreover the surface texture of the samples is another aspect that could be important at the initial stage of the oxidation process. In Fig. 2, the mass gain % versus time curves of NiFe, NiCrW and pure Ni tapes are plotted on secondary axis and the temperature curve as a function of time is also reported. After an initial transient the steady state protective oxidation takes place. However, at 700°C in oxygen at a pressure of 5 mbar, NiCrW shows a very low oxidation rate, better than pure Ni, and as expected NiFe shows much higher total mass gain.

EXPERIMENTAL.

All layers were deposited *in-situ* on textured Ni, NiCrW and NiFe tapes by PLD without breaking the vacuum conditions. PLD was performed using a KrF excimer laser (Lambda Physik Compex 200) with laser radiation at 248 nm in 20 ns pulse[8]. The substrates were pre-annealed in forming gas (4%H_2/Ar) for 30 min at 600°C in order to reduce the native NiO formed during storage and handling. The initial critical nucleation of CeO_2 films (~ 10 - 20 nm) on Ni tapes was performed in 5 Pa 4%H_2/Ar with the rest of the buffer (60 nm) deposited in vacuum. The deposition of Pd:CeO_2 on NiCrW and NiFe was done in vacuum. Biaxially aligned YSZ and Y_2O_3 buffer layers were deposited at temperatures between 700°C and 800°C respectively. The YBCO thin films were deposited at in 10 - 20 Pa oxygen at 730 - 760°C.

Surface morphology and roughness were characterized by SEM and AFM. The texture was evaluated by X-ray diffraction - θ - 2θ scan, ω – scan and φ – scan measurements. The superconducting properties were measured by direct transport and DC SQUID magnetometer.

RESULTS AND DISCUSSIONS

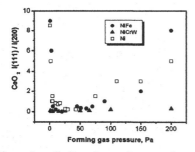

Fig. 3. CeO_2 intensity peak ratio I_{111}/I_{200} vs. pre-annealing 4%H_2/Ar pressure for NiFe, NiCrW and Ni tapes.

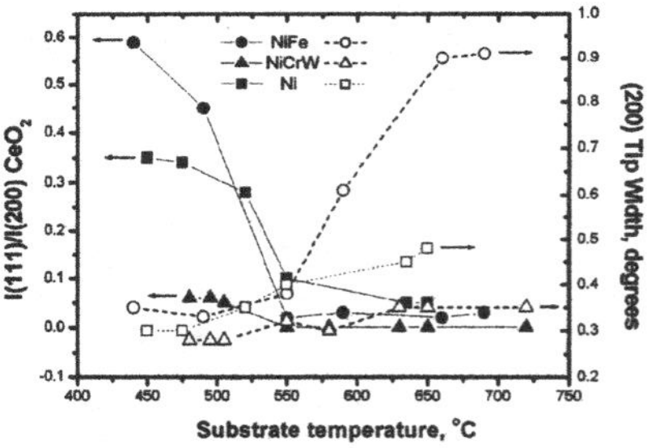

Fig. 4. Optimization of CeO_2 deposition substrate temperature for NiFe, NiCrW and Ni tapes (Intensity peak ratio I_{111}/I_{200} – left; (200) Tip width – right).

The quality of the first buffer appeared to be crucial for the rest of the conductor architecture and strongly dependent on pre-annealing conditions (forming gas pressure). The influence of the forming gas pressure is demonstrated on Fig.3. The quality of CeO_2 epitaxy (minimum of intensity ratio $I_{(111)}/I_{(200)}$) is reduced at low and high forming gas pressure for Ni and NiFe tapes while NiCrW tape appeared to be unaffected. At low pressure the hydrogen content is insufficient to reduce effectively native NiO. We believe that the worsening of the buffer quality at high pressures is a result of possible gas impurities that may suppress NiO reduction. In agreements with the tapes' oxidation characteristic a narrow optimum pressure window of 10-15 Pa was found for NiFe tape while for pure Ni the window was significantly wider (20-45Pa). Although predominantly (200) oriented CeO_2 films deposited on Ni systematically consisted of a mixture of (111) and (200) oriented phase for wide range of forming gas pressure and temperatures with low repeatability. Such a growth behavior could be determined by the initial nucleation of oxygen-deficient phase of $(001)Ce_2O_3$ whose oxygen sublattice is well matched to $(111)CeO_2$ as proposed by Boikov *et al* [11]. The deposition in vacuum (base pressure of ~10^{-7} Torr and $P_{(H2)}/P_{(H2O)}$~ 10^2) by implementing $Pd:CeO_2$ segmented target allowed us to perform controllable deposition of CeO_2 buffer layer with proper in-plane and out-of-plane orientation on NiCrW and NiFe tapes. The optimization study has demonstrated that preferentially oriented $(200)CeO_2$ films can be grown over relatively wide region of temperatures (480°C-650°C) with the CeO_2 $I(111)/I_{(200)}$ ratio reaching below 10^{-2} for a deposition temperature of 550°C (see Fig. 4). As the deposition temperature increases, the intensity of the (111) CeO_2 peak decreases. However, at the same time, a shoulder develops on the low angle side of the (002) CeO_2 peak. This might be the result of the nucleation of oxygen deficient material approaching the stability limit of CeO_2 at these temperatures. The effect is more obvious for NiFe tape probably due to higher affinity of Fe to oxygen at increased temperatures.

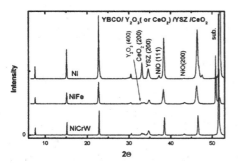

Fig. 5. XRD spectra of $YBCO/Y_2O_3(CeO_2)/YSZ/CeO_2$ architecture where the first buffer is deposited with initial nucleation in forming gas + vacuum on Ni and in vacuum on NiFe, NiCrW.

The spectra of whole architecture (see Fig. 5) reveals weak (200) NiO peak for films deposited on NiCrW and NiFe tape and a mixture of (111) and (200) NiO for films deposited on pure Ni. The as-deposited CeO_2 and YSZ films appeared to be crack-free presenting features originating from the substrate grain boundaries and propagating through the buffer layers[8]. Therefore we consider the diffusion through the buffers and along the grain boundaries as the main routes for oxygen diffusion during YBCO deposition. AFM characterization of the whole architecture deposited on NiCrW (see Fig. 6a) reveals step-like growth of YBCO films with roughness RMS value below 30 nm over a 10 μm area. A SEM examination of YBCO shows a dense microstructure typical for high quality YBCO films (see Fig. 6b). Note also the appearance of NiO outgrowths propagating along the substrate grain boundaries in the regions of deep grooves as well as through the pinholes observed in the substrate. They can be considered as a source for local nickel contamination of the YBCO film as well as a current limiting factor "sealing" the boundary and providing strains in the whole architecture.

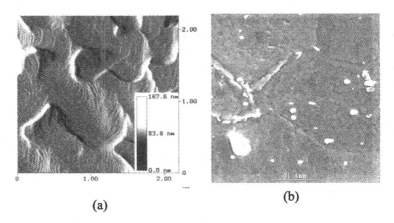

(a)   (b)

Fig. 6 AFM (a) and SEM (b) images of YBCO deposited on
Y$_2$O$_3$/YSZ/CeO$_2$/ NiCrW architecture.

A significant narrowing in the out-of-plane texture is observed for YBCO (see Fig. 7) with the rocking curve through (005) YBCO peak yielding an full with of half maximum (FWHM) value of 4°. The in-plane alignment of the epitaxial YBCO/Y$_2$O$_3$/YSZ/CeO$_2$/NiCrW structure was determined by XRD φ-scans through the YBCO (103), Y$_2$O$_3$(222), YSZ(111), CeO$_2$(111) and NiCrW(111) peaks. The respective in-plane FWHM values of 6.5°, 9.4°, 9.3°, 8.6°, 9.4° (see Fig. 7) indicate good epitaxy of the oxide layers with the biaxially textured metal tape. The same deposition procedure performed on NiFe tape leads to oxidation of the tape during YBCO and formation of cube-on-cube NiO(200) at the interface region preserving the texture of the overlaying buffer stack. The slight decrease of in-plane texture quality is observed mainly due to non-perfect sub-growth of the NiO template, however YBCO FWHM values of $\Delta\omega(005) = 4.3°$ and $\Delta\varphi (103) = 8.8°$ still reveal a good epitaxy.

High Temperature Superconductors

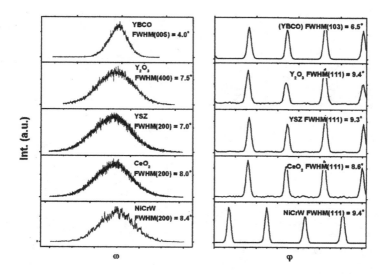

Fig. 7 Out – of - plane (ω - scans) and in - plane texture (φ - scans) of YBCO/Y$_2$O$_3$/YSZ/CeO$_2$/NiCrW architecture.

The utilization of segmented target is beneficial for both NiCrW and NiFe tapes but the effect is more pronounced on the tape with lower oxidation resistance. We believe that due to the Pd gettering around various substrate defects (grooves, pinholes, surface irregularities) the high rate local (111)NiO oxidation is suppressed. More detailed investigation on the role of Pd doping is published elsewhere [8]. The comparison of two identical architectures deposited from pure CeO$_2$ and segmented Pd:CeO$_2$ targets on NiFe tape is made on Fig. 8. Cross-section Focused Ion Beam (FIB) and SEM images show that the use of Pd leads to a significant reduction of the NiO outgrowths, sharper interface features and denser microstructure. Although effective in suppression of local oxidation Pd is not able to prevent NiO propagation taking place along macro-defects like the deep grooves originating from the substrates. The formation of a NiO along the grain boundary interface induces stresses related to the volume change and leads to cracking and delamination.

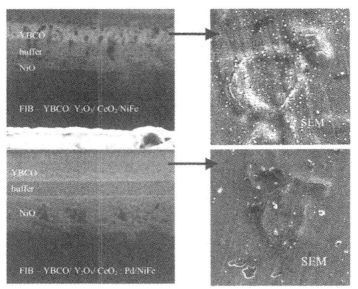

Fig. 8 FIB cross-section of $YBCO/Y_2O_3/CeO_2$ and $YBCO/Y_2O_3/CeO_2$:Pd architectures deposited on NiFe tape (left) and SEM images of YBCO layers (right).

The resistance vs. temperature measurements show a $T_{c(onset)}$ =90K and ΔT_c=5K for NiCrW and NiFe tapes and T_c(onset)=90K and ΔT_c=2K for pure Ni tape indicating the importance in NiO propagation and consequent the existence of partial replacement of Cu atoms with Ni, which automatically leads to a degradation of T_c. Fig. 9a shows the field dependence of transport J_c (77K) measured for the triple buffer architecture on NiCrW and NiFe tapes. J_c values of 0.2 MA/cm^2 and 0.06 MA/cm^2 have been measured for NiCrW and NiFe respectively. Such a difference in critical current density for comparable YBCO macro-texture quality reflects the influence of the substrate properties on the film quality. Magnetic measurements have been performed at 5K on both triple and double buffer architectures ($YBCO/Y_2O_3/CeO_2$) deposited on NiCrW and NiFe see Fig. 14b). The performance of the both architectures deposited on NiCrW is similar reaching best values of critical current density of 0.55-0,6 MA/cm^2 at 0.2 T. In the same time measurements performed on NiFe tape revealed that J_c of triple buffer architecture (0.09 MA/cm^2; 3T) is systematically higher than that of double buffer architecture (0.06 MA/cm^2; 3T) over the used range of magnetic field (3 to 5 T) (see Fig. 9b).

High Temperature Superconductors

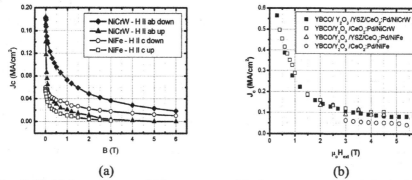

Fig. 9. Field dependence of (a) transport critical current measured on triple buffer architecture and (b) SQUID current measurements of triple and double buffer architectures deposited on NiCrW and NiFe substrates.

CONCLUSIONS

In summary, biaxially aligned superconducting YBCO films were deposited *in situ* on textured Ni, NiFe and NiCrW substrates by pulsed laser deposition. Sharp in-plane and out-of-plane texture of all the buffers as well as YBCO was developed. The crystalline quality and the alignment of the CeO_2 buffer layer determine the quality of the overall heterostructure. PLD deposition in vacuum instead in forming gas implementing segmented $Pd:CeO_2$ target allowed repeatable growth of the first buffer with good out-of-plane and in-pane texture. Apart from the large-angle grain boundaries other problematic features such as NiO formation along the defects (boundaries, pinholes) and cracking along substrates grain boundaries were found as possible current limiting factors. Despite a lower oxygen resistance and ferromagnetic properties of NiFe tape the promising results demonstrated in this study plus its good mechanical strength and commercial availability in km lengths are a strong incentive for continuing research. The superior oxygenation resistance and excellent magnetic and mechanical properties along with the reported results make NiCrW tape a strong candidate for coated conductor applications.

ACKNOWLEDGMENTS

This research was carried out in the frameworks of the Brite Euram programmes CONTEXT (Contract No. BRPR-CT97-0607) and MUST (Contract No. BRPR-CT97-0331). The support from ABB-Research, Zurich is also gratefully acknowledged. M. Majoros acknowledges the AFRL/PRPS Wright-Patterson Air Force Base, Ohio for the financial support.

REFERENCES:

[1]A. Goyal, D.P. Norton, J.D. Budai, M. Paranthaman, E.D. Specht, D.M. Kroeger, D.K. Christen, Q. He, B. Saffian, F.A. List, D.F. Lee, P.M. Martin, C.E. Klabunde, E. Hartfiels and V.K. Sikka, "High critical current density superconducting tapes by epitaxial deposition of YBCO thick films on biaxially textured metals", *Appl.Phys.Lett.*, **69** 1795-1798 (1996).

[2]A. Goyal, D.P. Norton, D.M. Kroeger, D.K. Christen, M. Paranthaman, E.D. Specht, J.D. Budai, Q. He, B. Saffian, F.A. List, D.F. Lee, E. Hartfiels, P.M. Martin, C.E. Klabunde, J. Mathis and C. Park, "Conductors with controlled grain boundaries: An approach to the next generation, high temperature superconducting wire", *J.Mater.Res.*, **12** 2924-2940 (1997).

[3]F.A.List, A. Goyal, M. Paranthaman, D.P. Norton, E.D. Specht, D.F. Lee, D.M. Kroeger , "High Jc YBCO films on biaxially textured Ni with oxide buffer layers deposited using electron beam evaporation and sputtering", *Physica C* **302** 87-92 (1998).

[4]Y. Iijima and K. Matsumoto, "High-temperature-superconductors: technical progress in Japan", *Supercond. Sci. Technol.* **13** 68-81 (2000).

[5]C.L.H. Thieme, S. Annavarapu, W. Zhang, V. Prunier, L. Fritzmeier, Q. Li, U. Schoop, M.W. Rupich, M. Gopal, S.R. Foltyn and T. Holesinger, "Non-magnetic substrates for low cost YBCO coated conductors", *IEEE Trans. Appl. Supercond.* **11** [1] 3329-3332 (2001).

[6]F. Fabri, C. Annino, V. Boffa, G. Celentano, L. Ciontea, U. Gambardella, G. Crimaldi, A. Mancini, T. Petrisor, "Properties of biaxially oriented Y_2O_3 based buffer layers deposited on cube textured non-magnetic Ni-V substrates for YBCO coated conductors", *Phisica C* **341-348** 2503-2504 (2000).

[7]B. Holzapfel, L. Fernandez, M.A. Arranz, N. Reger, B. de Boer, J. Eickemeyer and L. Schultz "Heteroepitaxial growth of oxide buffer layers by pulsed laser deposition on biaxially oriented Ni and Ni-alloy tapes", *Inst.Phys.Conf.Ser.* **167** 419-422 (1999).

[8]R.I. Tomov, A. Kursumovic, M. Majoros, D-J. Kang, B.A. Glowacki and J.E. Evetts, "Pulsed laser deposition of epitaxial $YBa_2Cu_3Cu_{7-y}$ / oxide multilayers onto textured NiFe substrates for coated conductor applications", *Supercond. Sci.Technol.* **15** 598-605 (2002).

[9]T.A. Gladstone, J.C. Moore, A.J. Wilkinson, and C.R.M. Grovenor, "Grain boundary misorientation and thermal grooving in cube-textured Ni and Ni-Cr tape", *IEEE Trans. Appl. Supercond.* **11** [1] 2923-2926(2001).

[10]CNR TeMPE – Italian Patent Application N°: MI2000 A000975 05/05/2000.

[11]Yu.A. Boikov and T. Claeson, "High tunability of the permittivity of $YBa_2Cu_3O_{7-\delta}/SrTiO_3$ heterostructures on sapphire substrates" *J. Appl. Phys.* **80** 3232-3236 (1997).

YBa$_2$Cu$_3$O$_{7-\delta}$ FILMS THROUGH A FLUORINE FREE TMAP MOD APPROACH

Yongli Xu and D. Shi
Department of Materials Science,
University of Cincinnati, Cincinnati,
OH 45221, USA

A. Goyal, M. Paranthaman, N.A. Rutter, P.M. Martin, and D. M. Kroeger
Oak Ridge National Laboratory, Oak Ridge, TN 37831, USA

ABSTRACT

YBa$_2$Cu$_3$O$_{7-\delta}$ (YBCO) films were fabricated via a fluorine free metal trimethylacetate (TMA) proponic (P) acid and Amine based sol-gel route (TMAP) and spin-coat deposited on single crystal (001) LaAlO$_3$ (LAO) and (001) SrTiO$_3$ (STO) with a focus on optimizing the processing parameters in the non-fluorine chemical solution deposition (CSD). Trimethylacetate salts of copper and yttrium and barium-hydroxide were used as the precursors, which were dissolved in proponic acid and Amine based solvents. A 180 ppm oxygen partial pressure and water vapor atmosphere were employed for the pyrolysis at 745°C for one hour. A critical transition temperature (T$_{c0}$) of 90K and a critical transport current density (J$_c$) of 0.5 MA/cm^2 (77 K and self-field) were demonstrated for the YBCO film on (001) oriented LAO substrates with a thickness of 300 nm. A FWHM of 0.6° for the (005) omega scan shows good out-of-plane texture. In (113) phi scans, the peaks are 90° apart with a FWHM of 1.4° showing good in-plane textures. A good in-plane texture is also shown in (113) and (115) pole figures. The TMAP approach is promising for high current density and high film quality.

INTRODUCTION

Chemical solution deposition (CSD) is one of the deposition techniques used to fabricate YBCO films. CSD has some advantages over vapor deposition of YBCO such as precise control of composition, high speed, and low-cost. Metal Organic Deposition (MOD), one of the popular CSD methods, involves the coating of an organic precursor solution on a substrate followed by thermal decomposition to form the final desired compound. Epitaxial nucleation and growth can occur when the process is carried out on lattice matched single crystal substrates. Trifluoroacetate (TFA) MOD is well established as a promising method for the fabrication of high J_c (over 1 MA/cm^2) YBCO films[1,2] and has been applied on biaxially textured metal substrates like RABiTS[3,4]. The interest in fluorine-containing precursors for YBCO arises because it is believed that non-fluorine precursors might result in the formation of stable $BaCO_3$ at the grain boundaries[5]. The use of TFA salts appears to avoid the formation of $BaCO_3$ because the stability of barium fluoride is greater than that of barium carbonate and fluorine can be removed during the high temperature anneal (>700°C) in a humid, low oxygen partial pressure environment[2]. Nonetheless several factors maintain interest in a fluorine-free precursor MOD approach. The most important being that removal of fluorine at high temperatures is a non-trivial process. There appear to be many issues related to fluid-flow and complicated reactor designs may be required for scale-up. In this work, we report on the fabrication of high quality YBCO films using a fluorine-free precursor. The transport critical current density of these films is over 0.5 MA/cm^2 at 77 K and self-field.

EXPERIMENTAL

Films were prepared by the following procedure[6,7]: yttrium trimethylacetate, barium hydroxide, and copper trimethylacetate were dissolved into proponic acid and amine solvents in a controlled stoichiometry (Y:Ba:Cu=1:2:3) to form a dark green solution with a total ionic concentration of 0.5-0.8 M. After filtration, this solution was applied onto the substrates in a spin-coater at speed of 2000-5000RPM. A 200-250°C baking step was used between multi-coatings on a hot plate to make thicker films. Burn out was carried out in the quartz tube furnace at 400°C for 10 hours in a humid pure oxygen atmosphere. These intermediate films were then subject to a high temperature anneal in which they were heated to 700 - 860°C at 10-40°C/min and held in a low oxygen partial pressure for 1-2hrs. Following the hold, the furnace atmosphere was switched to dry, and later to dynamic dry oxygen. All films were slow-cooled in oxygen to 500°C where they were held for 0.5-3hrs and furnace-cooled to room temperature in flowing oxygen. Humid gas was introduced into the furnace in the burned-out stage and prior turning on pure oxygen in the annealing stage. J_c was measured in the usual four-point transport method at 77 K and zero field. In addition, for selected samples, texture was analyzed by x-ray diffraction, surface morphology was characterized by scanning electron microscopy (SEM) and TGA was used for the thermal decomposition analysis.

RESULT AND DISCUSSION:

Fig. 1. shows Thermal Gravimetric Analysis (TGA) results for the copper trimethylacetate and yttrium trimethylacetate precursors. For the copper trimethylacetate precursor, the weight reduces quickly when the temperature increases beyond 200°C, and then become constant after 250°C, which

High Temperature Superconductors

demonstrates the full decomposition of copper trimethylacetate at that temperature. For yttrium trimethylacetate precursor, the decomposition occurs over a wider range and is finished at about 520°C. Weight losses are structure-related thermal processes and reflect the binding energies. Different atmospheres, for example, air and pure oxygen, have minor effects on the shapes and temperatures of TGA curves both for copper trimethylacetate and yttrium trimethylacetate precursor.

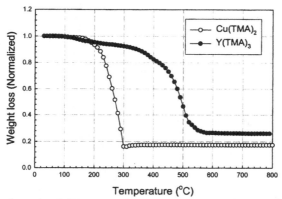

Fig. 1. Weight loss in TGA analysis for copper trimethylacetate and yttrium trimethylacetate precursors, atmosphere: N_2 with P_{O_2}=100-200ppm, ramp: 10°C/min.

CuO is easily evaporated at elevated temperature in dry atmosphere. The decomposition temperature for copper trimethylacetate (250°C) is lower than those of yttrium trimethylacetate (520°C) and barium hydroxide[8] (490°C). In considering CuO formation during the decomposition of copper trimethylacetate, keeping the baking temperature below 250°C is reasonable to prevent the loss of CuO between multi-coatings. But when the baking temperature is less than 150°C, later coatings dissolve away former ones easily. So normally 200-250°C for 3-5 minutes was used for the baking process.

In the burn out stage, several temperatures (310-600°C) were selected to control reactions of the components in precursor decomposition. 400°C was used as an optimized temperature for the burn out stage because rough surface morphologies of YBCO film were observed routinely on samples burned out below 400°C (for example, 310°C) and above 400°C (for example, 500°C and 600°C). For this research, four temperatures 725°C, 745°C, 780°C, and 820°C were used for high temperature annealing. Fig. 2 shows the θ-2θ scans of YBCO films annealed at 745°C and 820°C respectively.

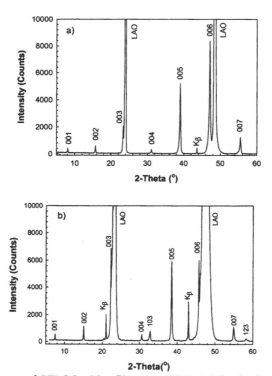

Fig. 2. θ-2θ scans of YBCO thin film on (001) LAO single crystal substrates burned out at 400°C in wet O_2 and high temperatures annealed in low oxygen partial pressure (Po_2=180ppm) for 50min wet and 10min dry at different temperatures: (a) 745°C, (b) 820°C

Only substrate peaks can be detected for the films burned out at temperatures below 400°C, indicating that the film is amorphous. YBCO (00*l*) peaks can be identified for samples annealed at 700°C and are stronger after annealing at 725°C for one hour. The (103) peak is seen when the temperature is beyond 780°C, especially for the sample annealed at 820°C as shown in Fig. 2 b). All (00*l*) YBCO peaks are shown in Fig. 2 a). The absence of other YBCO peaks confirms that the YBCO has good out-of-plane texture. The values of FWHM (Full Width at Half Maximum) of the (005) rocking curves are small and nearly constant (0.6°) for the samples annealed at 725°C, 745°C, and 780°C, demonstrating high quality *c*-oriented texture of these YBCO films. In the (113)

High Temperature Superconductors

phi-scan, Fig. 3, peaks are 90° apart with a FWHM of 1.4° showing good in-plane texture for a sample annealed at 745°C for one hour. This sample also had the highest YBCO intensities both for ω-scans and ϕ-scans.

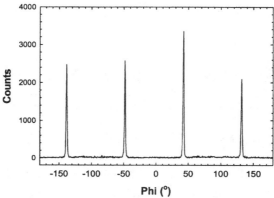

Fig. 3 The 113 ϕ-scan of YBCO films on LAO (001) substrate burned out at 400°C and annealed at 745°C for one hour in humid Ar/O$_2$ atmosphere with Po$_2$=180ppm

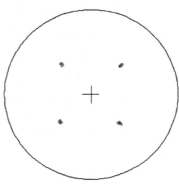

Fig. 4 The YBCO (115) pole figure of 400°C burned out film and annealed at 745°C/1hr in humid Ar/O$_2$ atmosphere with Po$_2$=180ppm

Pole figures were obtained to get a more complete examination of the degree of epitaxy. For 745°C annealed sample, only four small poles located on the correct positions on the pole figure plot, Fig. 4, confirms that the YBCO phase grows epitaxially on the LAO substrate. The T$_c$ measurement shows a sharp transition with zero resistance at 90 K for the sample annealed at 745°C for one hour, but broader transitions were detected for samples annealed at 725°C and 780°C. A semiconductor transition behavior was shown for the 820°C annealed sample. The transport critical current density (J_c) of 0.5 MA/cm^2 was obtained at 77 K in self-field for the 745°C annealed sample, whilst very limited currents can go through the YBCO films annealed at 725°C and 780°C. No current can go through the film annealed at 820°C for one hour.

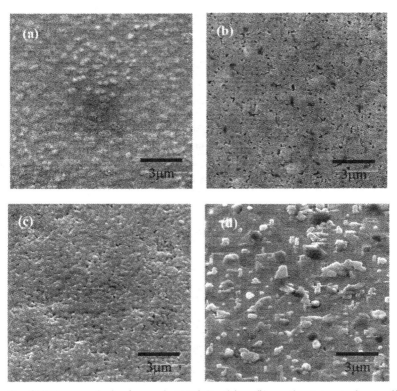

Fig. 5. SEM morphologies of YBCO thin films heat-treated at different temperatures. (a) 600°C burn out in wet oxygen; (b), (c), and (d): 400°C burn out in wet oxygen and annealed at 725°C/1hr, 745°C/1hr, and 820°C/1hr respectively in P_{O2}=180ppm wet atmosphere.

Though no crystallization is detected on the sample heat treated below 600°C, the SEM top view morphology, Fig. 5 (a), shows the nuclei like dots started at 600°C or below. No composition difference can be identified by EDS (Energy Dissipation Spectra) on dots and other areas. Associated with the decomposition

High Temperature Superconductors

of the copper trimethylacetate and yttrium trimethylacetate precursors in the TGA analysis (Fig. 1), it is believed that these dots are the decomposition products formed initially in the green films. As the decomposition of the copper trimethylacetate occurs at relative low temperature, and the resistivity (calculated from resistance of quenched sample during 600°C burn out) can go as low as 0.1-0.01ohm-cm, so correlating these dots with CuO is reasonable. The distribution density of these dots is comparable with that of grains; thus they might be the places for the further reactions and/or nucleations of YBCO phase. When increasing treatment temperature, these dots increase in size and finally connect to each other forming the grain like areas shown in Fig. 5 (b).

The surface morphology of the 725°C and 745°C annealed samples is relatively smooth, although some pores decorate the surface. From the SEM micrograph Fig. 5 (b), it is clear to identify that the microstructure of the 725°C annealed sample is not fully developed. This is also confirmed both by XRD data and the results of electrical property measurements. The underdeveloped YBCO phase is the reason for the low J_c and relatively broad transition temperature in the T_c measurement. Irregularly shaped surface segregates are observed commonly on samples annealed higher than 820°C as shown in Fig. 5 (d). The rough surface is obviously related to the over reaction among the components and with the environment such as humid gas, or to the decomposition of YBCO phase as reported in the literature[9]. With the rough surface coupled with texture changes shown in XRD data, no good electrical properties can be expected on samples annealed at temperature beyond 820°C.

The morphology of the 745°C annealed sample is similar to that reported by R. Feenstra et al[9]. in e-beam derived YBCO films with post annealing. Enhanced growth of plate-like grains, smooth surface, absence of large pores and second phase particles, and the absence of a-axis oriented grains indicate the potential for high critical current density.

SUMMARY

A fluorine-free Metal Organic Deposition (MOD) method was employed for the fabrication of $YBa_2Cu_3O_{7-\delta}$ films. A T_c of 90 K and a transport critical current density (J_c) of 0.5 MA/cm^2 (77 K and self-field) were demonstrated for the YBCO film on (001) oriented LAO substrate annealed at 745°C for one hour in humid furnace gas with oxygen partial pressure of 180ppm. There is no evidence that $BaCO_3$ is formed as an intermediate compound during decomposition of the precursors in well-prepared samples. SEM morphology reveals that feature development on top of the film occurs much earlier than crystallization takes place (as revealed by XRD data). Annealing temperature is critical for the YBCO phase development in low oxygen partial pressure and humid furnace gas. Undeveloped phases and poor texture at relatively low annealing temperature are the reasons for the low current densities. The YBCO phase seems to decompose at relatively high annealing temperatures (820°C) resulting in a rough surface and a semiconductor behavior.

ACKNOWLEDGEMENT

This work was sponsored by the U.S. Department of Energy, Office of Energy Efficiency and Renewable Energy - Superconductivity Program for Electric Systems. The research was performed at the Oak Ridge National Laboratory, managed by U.T.-Battelle, LLC for the USDOE under contract No. DE-AC05-00OR22725.

REFERENCE

[1] A. Gupta, R. Jagannathan, E.I. Cooper, E.A. Geiss, J.I. Landman, and B.W. Hussey, Appl. Phys. Lett. **52**(24), 2077 (1988)

[2] Paul C. McIntyre, Michael J. Cima, John A. Smith, Jr., and Robert B. Hallock, J. Appl. Phys. **71**(4), 1868-77, (1992)

[3] A. Goyal, D.P. Norton, J.D. Budai, M. Paranthaman, etc., Appl. Phys. Lett. **69**(12) (1996) 1795. A. Goyal et al., US Patents: 5, 739, 086; 5, 741, 377; 5, 846, 912, 5, 898, 020.

[4] A.P. Malozemoff, S. Annavarapu, L. Fritzemeier, Q. Li, V. Prunier, M. Rupich, C. Thieme, W. Zhang, A. Goyal, M. Paranthaman, and D.F. Lee, Superconductor Sci. & Technol. **13**, 473-476 (2000).

[5] F.Parmigiani, G. Chiarello, and N.Ripamonti, Physical Review B, **36**(13) (1987) 7148.

[6] Yongli Xu, Donglu Shi, Shaun McClellen, Relva Buchanan, Shixin Wang and L.M. Wang, IEEE Trans. Appl. Supercond. **11**(1), 2865-2868, (2001)

[7] Donglu Shi, Yongli Xu, Shaun McClellen, and Relva Buchanan, Physica C **354** (2001) 71-76

[8] Peir-Yung Chu and Relva C. Buchanan, J. Mater. Res., **9**(4), 844-51, (1994)

[9] R. Feenstra, T.B. Lindemer, J.D, Budai, and M.D. Galloway, J. Appl. Phys. **69** (9), 1 May (1991) 6569-85

INCLINED SUBSTRATE PULSED LASER DEPOSITION OF YBCO THIN FILMS ON POLYCRYSTALLINE AG SUBSTRATES

M. Li, B. Ma, R. E. Koritala, B. L. Fisher, S. E. Dorris, K. Venkataraman,*
and U. Balachandran
Energy Technology Division
*Chemical Technology Division
Argonne National Laboratory, Argonne, IL 60439, USA

ABSTRACT

Films of $YBa_2Cu_3O_{7-x}$ (YBCO) with c-axis orientation were directly deposited on nontextured silver substrates by inclined substrate pulsed laser ablation. The structure of the YBCO films was characterized by X-ray diffraction 2θ-scans, Ω-scans, and pole-figure analysis. A good alignment of the c-axis of the YBCO films was confirmed by the Ω-scans, in which the full width at half maximum of the YBCO(005) was 3.8°. A sharp interface between the YBCO film and Ag substrate was observed by transmission electron microscopy. The surface morphology of the film, examined by scanning electron microscopy, reflected the recrystallization of the Ag substrate. Raman spectroscopy was used to evaluate the quality of the YBCO films. The superconducting transition temperature (T_c) and the critical current density (J_c) of the films were determined by inductive and transport measurements, respectively. T_c = 91 K with a sharp transition and J_c = 2.7 $\times$ $10^5 A/cm^2$ at 77 K in zero external field were achieved on a film with 0.14-μm thickness.

INTRODUCTION

A promising material for future electric power applications is $YBa_2Cu_3O_{7-x}$ (YBCO)-coated conductors [1-4]. Recently, YBCO thin films grown on silver tapes have attracted much interest because of their high superconducting transition temperature (T_c) and high critical current density (J_c). Silver is an ideal substrate candidate due to its compatibility with YBCO and its inertness to oxidation [5-6]. The electrical properties have been improved in the YBCO films by silver doping up to 20 at.% [7-8]. Unlike other substrate materials such as Ni-based alloys, in which buffer layers are necessary to prevent Ni diffusion into the YBCO film, YBCO can be directly deposited on silver substrate without a buffer layer. To achieve high J_c, textured silver sheets are often chosen as substrates. YBCO films grown on these textured substrates have achieved J_c of $\approx 5 \times 10^5$ A/cm^2 [9-11]. However, a relatively complex process is usually involved in preparing well-textured silver substrate for YBCO deposition.

We have explored direct deposition of YBCO on nontextured silver substrate for simplifying the fabrication process. In our studies, nontextured polycrystalline silver sheet was used as the substrate, and YBCO films were directly deposited on these substrates by inclined substrate pulsed laser deposition (IS-PLD). Inclined substrate deposition is a promising technique for growing biaxially textured films on nontextured substrates [3-4], and improvement in film quality is expected from this technique. This work shows the feasibility for fabricating high-J_c YBCO films on nontextured polycrystalline silver substrate.

EXPERIMENTAL

Inclined substrate PLD was used to grow YBCO films. A schematic diagram of the deposition system is shown in Figure 1. Substrates with silver paste were mounted on a tiltable heater. Different from the conventional arrangement in which the surfaces of the target and the heater were parallel, the heater can be tilted at an angle with respect to the target surface. The substrate inclination angle, α,

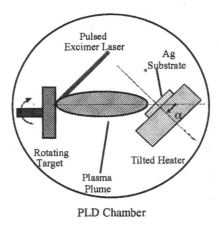

Figure 1. Schematic diagram of inclined substrate pulsed laser deposition system

High Temperature Superconductors

substrate normal with respect to the evaporation plume direction, was typically set at 55°, which is believed to be a favorable angle to produce (001) oriented textured film. To compare the dependence of α, 35° and 72° were also tested. Nontextured polycrystalline silver coupons, 5×10 mm in dimensions and 0.1 mm in thickness, were used as substrates after mechanical polishing with diamond paste up to the 1-μm level. The substrates were heated programmatically to a desired deposition temperature. Substrate temperatures were measured by a thermocouple mounted on the surface of the heater.

An excimer laser (Lambda Physik, Compex 201) with a wavelength of 248 nm and pulse width of 25 ns was used for the deposition. Conditions included a pulse repeat rate of 8 Hz and oxygen pressure of 200 mtorr. The laser was focused at a rotating target with an energy density of 2~3 J/cm^2. A commercial YBCO target (Superconductive Components Inc., purity of 99.999%), 2.5 cm in diameter and 0.64 cm in thickness, was used. The distance between the target and the substrate was 7 cm. The typical thickness of the YBCO films was 0.14 μm.

The structures of these films and the substrates, including the out-of-plane and in-plane textures, were examined by X-ray diffraction (XRD) 2θ-scans, Ω-scans, and pole-figure analysis. Transmission electron microscopy (TEM) and scanning electron microscopy (SEM) were used to observe the interface structure and the surface morphology of the samples, respectively. The quality of the films was also characterized by Raman spectral analysis. An inductive method was used to measure T_c. The J_c was determined by the standard four-point method on samples with typical size of 5 × 10 mm using a criterion of 1 μV/cm.

RESULTS AND DISCUSSION

The crystalline structure of the YBCO films grown on polycrystalline silver substrates was determined from the XRD 2θ-scan pattern, shown in Fig. 2(a). This XRD pattern consists of a set of sharp and strong peaks, which have been identified as YBCO(00l) peaks. No a-axis orientation peaks were observed in this pattern, indicating good c-axis alignment in the YBCO films. A full- width- at-half- maximum (FWHM) of ≈0.2° was measured for the YBCO(005) peak, suggesting a high-quality crystalline structure. The c-axis alignment was investigated by Ω-scans, shown in Fig. 2(b). The Ω-scans of the YBCO(005) peak gave a FWHM value of 3.8°, indicating a small orientation dispersion of the c-axis. To characterize the in-plane texture, X-ray pole-figure analysis of the YBCO(103) plane was performed, as shown in Fig 2(c). It reveals that the intensity of the diffraction of the selected plane is fairly uniformly distributed over the whole

360° φ-angle range, indicating no in-plane texture. This finding is expected from the polycrystalline texture of the substrate.

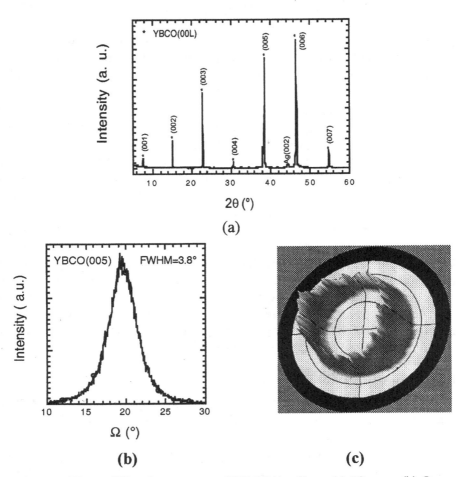

Fiure 2. X-ray diffraction patterns of YBCO/Ag films: (a) 2θ-scan, (b) Ω-scan of YBCO(005), and (c) pole-figure of YBCO(103)

Raman spectra were also used to obtain information about the phase integrity, cation disorder, and second-phase formation such as CuO and $BaCuO_2$. Figure 3 shows the Raman spectrum of the as-grown YBCO films, which exhibit a strong Raman band at 340 cm^{-1}, along with a weak band centered at about 500 cm^{-1}. The strong Raman band at ≈ 340 cm^{-1} indicates that a c-axis-oriented YBCO film was formed. The existence of the weak band at around 500 cm^{-1} suggests minimal c-axis misalignment of the YBCO film [12-13], in this case, in-plane texture disorder, to some extent. Lack of other bands suggests that the YBCO film has a stoichiometric composition.

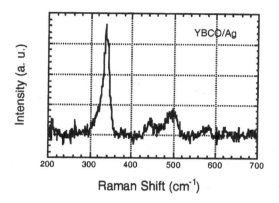

Figure 3. Raman spectrum of YBCO films grown on the Ag substrates

The T_c of the YBCO films grown on silver substrates was measured by an inductive method. Figure 4 shows a typical superconducting transition temperature pattern of the YBCO films. T_c values of 91 K with narrow transition widths of 3.0 K were achieved for the YBCO film deposited on silver substrate at

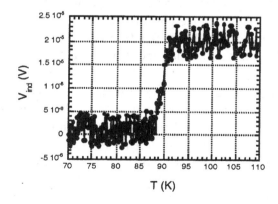

Figure 4. T_c pattern of YBCO/Ag films measured by inductive method.

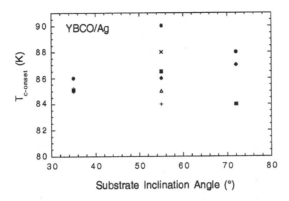

Figure 5. T$_c$ of YBCO films as function of substrate inclination angle

755 °C with a substrate inclination angle of 55°. The onset T$_c$ values and the transition widths of the YBCO films deposited at inclined substrate angles of 35°, 55°, and 72° covered a range, but the best values were obtained on samples deposited with an inclined substrate angle of 55°. Figure 5 shows the T$_c$ distribution over different inclined substrate angles. The J$_c$ of the YBCO films was measured by the four-point transport method over an entire 5×10 mm sample. Before the J$_c$ measurement, the YBCO/Ag samples were coated with a silver layer of about 2-μm thickness by e-beam evaporation and then annealed in flowing ultra-high purity oxygen at 400 °C for 0.5 h. A J$_c$ of ≈2.7 × 10^5 A/cm^2 was achieved at 77 K with zero external field for a 0.14-μm-thick YBCO film deposited at 755 °C with an inclination angle of 55°. The high J$_c$ values of the samples may be partly attributed to the possibly improved microstructure of the films and the enhanced surface mobility of the deposited atoms due to the IS-IS-PLD process. Further studies concerning the growth of YBCO films on Ag substrates are under way.

The structure of the Ag substrates was also examined by XRD. Figure 6(a) shows the 2θ-scan pattern of the silver substrates. More than three peaks were found in the region of 2θ = 25-85° in both the as-polished and the post-deposition substrates, suggesting a polycrystalline structure for the substrates. The FWHMs of the peaks from the pattern of the post-deposition substrate were much smaller than those of the as-polished one, indicating the size of the crystalline grains of the as-polished substrates were much smaller than those of the post-deposition substrates. The orientation distribution of the grains was examined with X-ray

High Temperature Superconductors

diffraction Ω-scans. Figure 6(b) shows the Ω-scan of the Ag(200) plane. A broad peak with FWHM=13° was observed for the as-polished substrates, indicating a poor alignment and continual orientation divergence of the Ag(200) plane. For the post-deposition substrate, several peaks with different intensities and FWHM values emerged in the pattern, suggesting a discontinuous orientation of the Ag(200) plane. Similar patterns of discontinuous orientation were also observed for the other lattice planes of the substrates. The transition of the orientation distribution of Ag grains from continuous to discontinuous may have contributed to the recrystallization during the deposition.

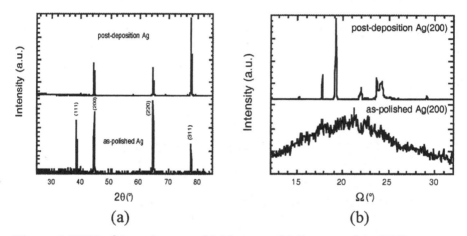

(a)   (b)

Figure 6. XRD of Ag substrates: (a) 2θ-scans; (b) Ω-scans of Ag(200)

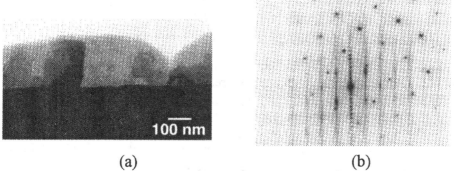

(a)   (b)

Figure 7. TEM investigation of YBCO/Ag interface: (a) TEM image; (b) electron diffraction pattern

TEM observation revealed a clear, sharp interface for the YBCO/Ag sample, as illustrated in Fig 7(a). Figure 7(b) is a selected area electron diffraction pattern from the interface shown in Fig. 7(a) with the Ag oriented along the [001] zone axis. Further studies on the orientation relationship of the YBCO films over the Ag substrate in the interface are under way.

The surface morphology of the YBCO films was examined by SEM. In the low magnification mode, different irregular-mosaic-shaped areas were observed. These areas have clear, sharp borders, but different contrasts and micromorphologies. They can be categorized by contrast to black areas, bright areas, and some intermediate areas, as illustrated in Fig. 8. The black areas have a denser surface and better connection among micrograins, along with a few short-bar-like grains randomly distributed. The bright areas have a less dense surface with some short-bar-like grains lying along a particular direction. However, energy dispersion spectral analysis showed no obvious compositional difference among these areas.

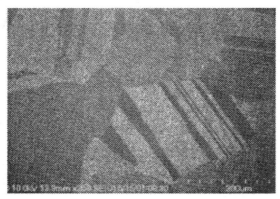

Figure 8. SEM image of the as-grown YBCO films in low magnification mode

This surface morphology may directly relate to the substrate morphology. On a structurally and compositionally uniform substrate, a uniform film at the microscopic scale is expected to grow, since other growth parameters in PLD are all macroscopic quantities. Since silver has a face-centered cubic crystalline structure and the recrystallization temperature is low, the silver substrate is expected to undergo recrystallization, and large grains may form during the film deposition [5-6]. This may influence the morphology of the deposited YBCO films. To find the exact correlation between the morphology of the films and the orientation of the silver grains, as well as the film formation mechanism, more work is needed.

CONCLUSIONS

YBCO films with c-axis orientation were directly grown on polycrystalline silver substrates by inclined substrate pulsed laser deposition. X-ray pole-figure measurement showed that the in-plane texture of the YBCO films was randomly distributed. Raman spectroscopy confirmed the c-axis orientation and the stoichiometric composition of the YBCO film. Recrystallization of the substrates during deposition was confirmed by XRD Ω-scan. TEM revealed a sharp interface between the YBCO film and the Ag substrate. SEM revealed dense and smooth YBCO films with areas having a variety of irregular-mosaic-shaped morphologies. This effect was attributed to the recrystallization of the substrate. An as-grown YBCO film with thickness of 0.14 μm achieved onset T_c of 91 K, transition width of 3.0 K, and J_c of $2.7 \times 10^5 A/cm^2$. These studies have demonstrated a promising approach for fabricating high-T_c and high-J_c YBCO films in a simple and economic way for practical application.

ACKNOWLEDGMENT

The authors would like to thank Dr. D. J. Miller for his XRD pole-figure measurement of the YBCO film samples. This work is supported by the U.S. Department of Energy (DOE), Energy Efficiency and Renewable Energy, as part of a DOE program to develop electric power technology, under Contract W-31-109-Eng-38.

REFERENCES

[1]Norton, D. P., Goyal, A., Budai, J. D., Christen, D. K., Kroeger, D. M., Specht, E. D., He, Q., Saffian, B., Paranthaman, M., Klabunde, C. E., Lee, D. F., Sales, B. C., and List, F. A., *Science*, 274, pp. 755-757 (1996).
[2]Iijima, Y., Hosaka, M., Tanabe, N., Sadakata, N., Saitoh, T., Kohno, O., and Takeda, K., *J. Mater. Res.*, 12(11), pp. 2913-2916 (1997).
[3]Bauer, M., Semerad, R., and Kinder, H., *IEEE Transactions on Applied Superconductivity*, 9(2), pp. 1502-1505 (1999).
[4]Ma, B., Li, M., Jee, A. J., Koritala, R. E., Fisher, B. L., and Balachandran, U., *Physica C*, 366, pp. 270-274 (2002).
[5]Budai, J. D., Young, R. T., and Chao, B. S., *Appl. Phys. Lett.*, 62(15), 1836-38 (1993).
[6]Zhou, M., Guo, H., Liu, D. M., Zuo, T. Y., Zhai, L. H., Zhou, Y. L., Wang, R. P., Pan, S. H., and Wang, H. H., *Physica C*, 337, pp.101-105 (2000).

[7]Wang, R., Zhou, Y., Pan, S., He, M., Chen Z., and Yang G., *Physica C*, 328, pp. 37-43 (1999).

[8]Gladstone, T. A., Moore, J. C., Henry, B. M., Speller, S., Salter, C. J., Wilkinson, A. J. and Grovenor, C. R. M., *Supercond. Sci. Technol.*, 13, pp.1399-1407 (2000).

[9]Liu, D., Zhou, M., Hu, Y., and Zho, T., *Physica C*, 337, pp. 75-78 (2000).

[10]Salamati, H., Babaei-Brojeny, A. A., and Safa, M., *Supercond. Sci. Technol.*, 14, pp. 816-819 (2001).

[11]Mendoza, E., Puig, T., Varesi, E., Carrillo, A. E., Plain, J., and Obradors, X., *Physica C*, 334, pp.7-14 (2000).

[12]Ferraro, J. R., and Maroni, V. A., *Appl. Spectrosc.*, 44, pp. 351-366 (1990).

[13]Gibson, G., Cohen, L. F., Humphreys, R. G., and MacManus-Driscol, J. L., *Physica C*, 333, pp. 139-145 (2000).

High Temperature Superconductors

THE GROWTH MODES AND TRANSPORT PROPERTIES OF YBaCuO PREPARED BY BATCH AND CONTINUOUS LIQUID PHASE EPITAXY

A. Kursumovic, Y. S. Cheng, R.I. Tomov, R. Hühne, B.A. Glowacki, and J.E. Evetts

IRC in Superconductivity and Department of Materials Science & Metallurgy, University of Cambridge, Cambridge CB2 3QZ, UK

Growth modes for YBaCuO on single crystal and metallic (Ni-based) substrates are reported. Seed layers with different microstructures were prepared by PLD to explore the influence of seed structure on LPE film growth. It was found that the YBaCuO seed layer shows a different degree of thermodynamic stability depending on its quality, being stable in the region where its apparent grain size exceeds the size of the critical nucleus in the BaCuO flux. Growth rates were measured and compared with a kinetic model developed previously. Transport measurements show a transition temperature around 90K and critical current density ~0.5 MAcm^{-2}.

INTRODUCTION

Liquid Phase Epitaxy (LPE) offers a fast processing route to fabricate thick films of REBa$_2$Cu$_3$O$_{7-\delta}$ (RE-123, where RE is a rare earth or yttrium). The technique involves the undercooling of a molten flux (e.g., BaO-CuO), supersaturated with a rare earth, below the peritectic temperature, in the presence of a suitable substrate in order to nucleate and grow RE-123 phase. A forced convection is often induced by relative motion of substrate and crucible in order to increase the growth by delivery of supersaturated solution to the growth interface [1].

The process occurs very close to the equilibrium; hence supersaturation is relatively small compared with vapour deposition resulting in slow nucleation. However, the growth rate is fast compared to vapour phase techniques, and is mainly restricted by the surface kinetics [2, 3]. Therefore nucleation of epitaxial film is very much dependent on the film-substrate-matching. Deposition on single crystal oxide substrates involves some dissolution before the growth starts. On the other hand, deposition on metallic substrates is much more complex primarily due to severe interaction between the substrate and the flux. The reaction can often lead to undesirable properties, like due poisoning of RE-123 film with the substrate metal. Therefore, a buffer structure is employed with several functions, but primarily to keep the film chemically unaffected by the substrate and to isolate the substrate from both oxygen (that would degrade the substrate surface) and corrosive flux itself. Furthermore, the buffer structure or an additional RE-123 thin film has a seeding role in the case of large substrate mismatch which is usually the case.

EXPERIMENTAL

A fully instrumented thermal-gradient furnace has been constructed for batch (short samples) and continuous (long samples) LPE experiments (see Fig. 1). Selected substrates can be lowered into a saturated flux. Variable speed rotations of the substrate controls forced convection in the flux so that flux that is depleted of rare earth next to the growing film is continuously replaced by supersaturated flux from the bottom of the crucible. In the case of long samples forced convection of the flux can be achieved by crucible rotation. However, at present only deposition in a stationary case is reported. It is estimated that (slow) natural convection as well as diffusion establishes supersaturation in the melt subjected to thermal gradient [2, 4].

In the case of Y-123 growth, a mixture of BaO and CuO was used in the ratio Ba:Cu=3:5 as a starting solvent material. Sintered Y_2BaCuO_5 (Y-211) pellets were placed at the bottom of an YSZ crucible as a source of yttrium. The mixture is preheated above 1050°C (~50°C above the peritectic temperature T_P~1000°C in air) to ensure rapid formation of a homogeneous $Ba_3Cu_5O_8$ flux. The bottom of the crucible where RE dissolution occurred was kept at 1012°C for the case of Y-123 growth, while the top of the crucible, where growth occurred was kept at 990°C. Temperatures were monitored inside the crucible walls during the run. The temperature inside the melt was calibrated against the temperature inside the crucible walls. Other RE used here (Sm and Nd) had higher peritectic temperature. The RE supersaturation was carried out below the peritectic temperature.

Figure 1 shows a schematic drawing of the crucible-sample arrangement for combined batch and long tape deposition. The flow velocities are of the order of 0.1 to 1 mms^{-1} [4] which is often taken as stationary from the kinetics point of film growth. In the case of small samples (batch) the flow velocities can be considerably increased by sample rotation [2]. However, in the case of long tapes the flow can be increased by various means of crucible rotation [4] as provided with this apparatus.

For batch deposition epi-polished $NdGaO_3$ (110) and MgO (100) single crystals with dimensions $10\times10\times1$ mm^3 and cut metal tapes of Ni-based alloys of

50-100 μm thickness and 10 mm width, were used. For continuous deposition long (~10 cm) textured Ni-based tapes were used. On the other hand single crystal samples were also attached onto moving tape carrier to simulate continuous deposition. The seeding Y-123 films were deposited on MgO and metallic substrates by Pulsed Laser Deposition (PLD) method [5].

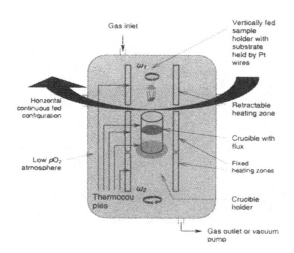

Fig.1. Schematic drawing of the multi-functional LPE apparatus used in this study.

The thickness of films was measured by an optical microscope, and verified by Scanning Electron Microscopy (SEM). X Ray Diffraction (XRD) and Atomic Force Microscopy (AFM) were used for checking texture and surface morphology, respectively.

RESULTS

In depth analysis of Re-123 growth on closely matched substrates is extensively reported previously [2]. Deposition of a uniform RE-123 film on MgO and metallic samples requires a seed that is most naturally a thin RE-123 film or close matching thin buffer/seed film. The seeding films were deposited by PLD deposition [5]. The Y-123 seed films are shown in Figs 2-a and 2-b. Films of proper microstructure (Fig 2-a) as thin as 20-40 nm were adequate for subsequent LPE growth (shown later in Fig 3-a). However, even much thicker but "porous" Re-123 films (Fig. 2-b) dissolved in the flux before LPE growth could start. More details on RE-123 seeding film stability will be published separately. Deposition on a wide range of oxide and metallic substrates was investigated to determine substrate and buffer layer stability, and the effectiveness of different seed layers. Substrates investigated included NdGaO₃, MgO and a range of metallic substrates with different buffer layer systems and Nd-123 and Y-123 seed layers. The range of seed layers investigated had been limited by the requirement to supersaturate the flux with the RE element in the seed. For example a thin Nd-123 seed layer would dissolve in a flux that was not saturated with Nd before the LPE layer would be able to able to nucleate.

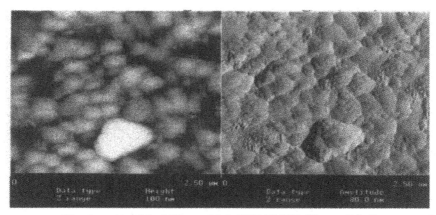

Fig. 2-a. A stable 100 nm thick PLD Y-123 film on MgO

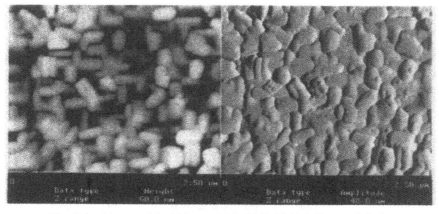

Fig. 2-b. An unstable 100 nm thick PLD Y-123 film on MgO.

Figure 3–a shows a typical spiral surface morphology, from spiral mediated growth for MgO sample rotating at 100 rpm. Basically the same picture is obtained for "continuous" deposition, but smaller grains were observed, due to thinner film deposited in the latter case [6]. A continuous Y-123 film deposited on SOE Fe-Ni alloy [7] is shown in Fig 3-b. LPE film morphology is strictly dependent on the substrate-buffer architecture. Well-oriented but relatively rough films are obtained at present as shown in Fig 3-c on well textured substrate. However, inadequate texture results in an RE-123 film of corresponding texture.

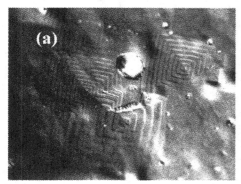

Fig. 3-a. Optical image of LPE Y-123 film on aY-123 on MgO (100) substrate

Fig. 3-b. Continuously deposited Y-123 on SOE Ni-Fe tapes.

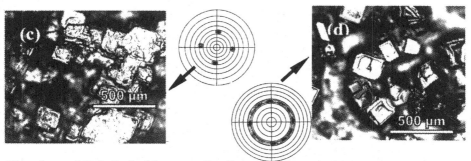

Figs. 3-c and 3-d. Optical images of well textured but rough LPE-grown Y123 on triple-buffer tapes with different seed textures. (c) 4-fold symmetry seed on an Ni-based alloy and (d) 12-fold symmetry seed on pure Ni. Insets are (103) pole figures of the Y123 seeds.

The kinetics of film growth was investigated for the case of a stationary crucible for batch and continuous case. In the case of continuous moving substrates, a velocity of the order of 1mm s^{-1} was employed. Figure 4-a illustrates such a case for two different rare earths. However, for a continuous deposition a 5 µm thick film was made in about 60 s.

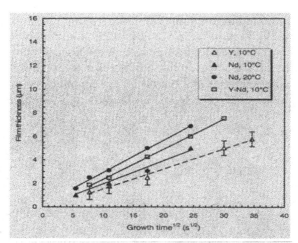

Fig. 4-a. Film thickness (H) of YBCO, NdBCO, and (Y,Nd)BCO as a function of the square root of growth time for c-oriented films grown on NdGaO$_3$ substrates from unstirred solutions (degree Celsius in agenda represents degree of undercooling).

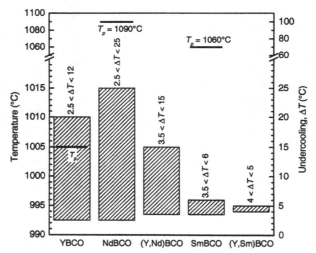

Fig. 4-b. Undercooling working windows for various REBCO systems grown on NdGaO$_3$ substrates at 990°C. T$_p$ denotes the peritectic temperatures for the REBCO systems [8, 9]. ΔT represents degree of undercooling.

High Temperature Superconductors

There is a limit on the growth rate that can be achieved expressed in terms of supersaturation or undercooling. The summary of results on growth with different rare earths, expressed in undercooling terms, is shown in Fig 4-b. An undercooling created by cooling from a temperature within the shaded area to the growth temperature of 990°C would result in c-oriented growth. The undercooling ranges of Sm- and Nd-systems lie within their primary crystallisation fields.

Results on transport properties show high critical temperature values ($T_c\sim92K$) for the case of $NdGaO_3$ substrate. However, on MgO substrates T_c was reduced for thinner films, increasing as films become thicker. Critical currents were approaching 0.5 MAcm^{-2} for Y-123 films on $NdGaO_3$. More detailed transport characteristics are published elsewhere [10].

GROWTH RATE ANALYSIS

A series of growth experiments have been carried out on $NdGaO_3$ substrates, to demonstrate LPE with the desired c-axis orientation and determine the layer growth rate as well as the working window. $NdGaO_3$ substrate was chosen here because of very close matching with Y-123 (~0.2%) that promotes easy Y-123 nucleation. The substrate was kept in the solution ~10°C (Fig. 4-b) below the peritectic temperature. In the case of a stationary substrate the film growth had familiar $\sqrt{t}$ growth kinetics. Growth under substrate rotation was studied elsewhere [2]. For long times, the film thickness in the unstirred solution was found to increase in time as:

$$H(t) \approx \frac{(C_L - C_e)}{C_{xt}}\left(2\sqrt{\frac{Dt}{\pi}} - \frac{D}{k}\right) \qquad (1)$$

where C_L and C_e are solute (RE) concentrations in the bulk of the solvent at the RE-211 dissolution temperature and the equilibrium one at the film growth temperature respectively (D is the diffusion coefficient and k is the surface kinetics coefficient).

This is the function that fits the data for Y-123 in Fig. 4-a as $H(t)\approx(\tan\alpha)\sqrt{t}+H(0)$. The slope $\tan\alpha \approx 0.46\text{-}0.49$ $\mu m\cdot s^{-0.5}$, (for Y-123) is independent of the surface kinetics, and would have the same interpretation if growth were only limited by flux bulk diffusion. The diffusion coefficient can be calculated as: $D=0.25\tan^2\alpha\cdot\pi[C_{xt}/(C_L-C_e)]^2$. With the Y supersaturation of the solution [2] to be $C_L-C_e=0.17$ at% and the Y concentration in the crystal $C_{xt}=8$ at%, the calculation gives $D=3.7\cdot10^{-10}$ m^{2}s^{-1}. The negative intercept in (4-a):

$$H(0) \approx -\frac{D}{k}\cdot\frac{(C_L - C_e)}{C_{xt}} \qquad (2)$$

indicates finite incorporation kinetics at the growing interface found to be $k \approx 4 \cdot 10^{-6}$ ms^{-1}. This intercept would disappear only in the case of $k \rightarrow \infty$, i.e. infinitely fast incorporation kinetics at the interface, which is usually not the case. However, it should be noted that present analysis should take care of faster RE saturation "ageing" in the flux when undercooling from below the peritectic temperature, since there are always RE-123 particles in the flux acting as seeds. Therefore, only Y-123 curve in Fig. 4-a shows clear intercept.

In the stirred solution, after an initial transient stage there is a steady state regime with a stagnant (with respect to the substrate) diffusion boundary layer between the substrate and the liquid. This layer, according to Cochran's approximation in the case of a stirring substrate rotation at an angular velocity ω, has a thickness:

$$\delta = 1.6 D^{1/3} v^{1/6} \omega^{-1/2}, \tag{3}$$

where $v = \eta / \rho$ is the kinematic viscosity η is the (dynamic) viscosity and ρ is the density. In the case of linear relative movement with a velocity, u, between the substrate and the liquid as in the case of the continuous growth, the diffusion boundary layer at a certain distance L from the leading edge has the thickness:

$$\delta = 2.16 D^{1/3} v^{1/6} (L/u)^{1/2}, \tag{4}$$

Hence, the diffusion boundary layer for any substrate rotation rate (ω), or for linear liquid flow over the substrate, can be estimated from the kinematic viscosity and the diffusion coefficient. Additional crucible rotation would induce further steering and increase the steady state growth rate [4, 11].

DISCUSSION

Reproducible LPE growth of REBaCuO films without a seed layer is so far only successful on very closely matched substrates such as NdGaO$_3$. Substrates with larger lattice mismatch need a seed layer to grow an LPE film of desired microstructure. It was found that the YBaCuO seed layer, deposited by PLD, shows a different degree of thermodynamic stability depending on its quality. The PLD films that appear "dense" are stable (Fig. 2-a) and act as good seeds less than 50 nm thickness. Our investigations so far suggest that individual grain development from a "uniform" film, due to stress relaxation, at higher temperatures result in seemingly "porous" film, (Fig. 2-b). These individually acting grains, being of smaller size than critical nuclei at the growth temperature, dissolve in the flux. The morphology of LPE films grown on stable PLD seeds reveals a similar growth mode in the steady state regime, i.e. dislocation mediated spiral growth as shown in Fig 3-a. On metallic substrates, the LPE films show initially grain size determined by the grains of the metallic substrate.

High Temperature Superconductors

Unfortunately, these relatively rough films tend to keep a lot of flux physically attached to the surface (Fig. 3-c). It was found that the initial microstructure of LPE film follows the texture of the metallic tape-buffer-seed architecture.

The results show in the case of rotating substrate that c–axis growth is achieved only in a certain temperature window. Lower temperature side of the working window is substrate dependent. It is influenced in two ways. Firstly, nucleation starts slowly at low undercooling, hence could be hindered as non-nucleating. Secondly, melting (etch-back) of the substrate occurs in the case that the flux is not saturated by the substrate main components such as Ga in the case of $NdGaO_3$. Although thick LPE films with "perfect" crystallinity and epitaxy are grown on single crystal substrates [6], this melt back usually occurs before nucleation starts under low super saturation resulting eventually in a-b plane oriented growth.

However, the higher limit is kinetically favouring growth in the a-b plane. That could be the reason why mixture of rare earths has narrower working window from the undercooling point of view, although their supersaturation is not reduced. However, at present we have no explanation for such a narrow working window in the case of Sm-123 growth. Hence too high supersaturation, brought about by undercooling or mixing rare earths, leads to kinetically preferred a-b plane growth. In the case of continuous deposition, regions for both c-axis and a-b plane oriented films were found coinciding with the results obtained by rotating substrate experiments.

Transport measurements are reported showing a transition temperature around 90K and critical current density ~0.5 $MAcm^{-2}$.

ACKNOWLEDGEMENTS

This research was funded by the EPSRC. YSC acknowledges the Schiff Foundation and ORS Award for support.

LITERATURE

[1]   C. Klemenz, I. Utke, H.J. Scheel, "Film orientation, growth parameters and growth modes in epitaxy of $YBa_2Cu_3O_x$", *Journal of Crystal Growth*, **204**, 62-68 (1999).

[2]   A. Kursumovic, Y.S. Cheng, B.A. Glowacki, J. Madsen[+] and J.E. Evetts, "Study of the rate limiting processes in liquid phase epitaxy of thick YBaCuO films", *Journal of Crystal Growth* **218**, 45-56 (2000).

[3]   T. Izumi, N. Hobara, T. Izumi, M. Kai, K. Hasegawa, C. Krauns, Y. Nakamura, T. Watanabe and Y. Shiohara, " Fabrication of YBCO film on metal tape by LPE processing", *Proc. 5th ISTEC-MRS Intl. Workshop in*

Superconductivity (Honolulu HI, USA, June 2001) 237-240.

[4]   H. J. Scheel, "Accelerated crucible rotation: A novel stirring technique in high-temperature solution growth", *Journal of Crystal Growth*, **13/14**, 560-565 (1972).

[5]   R. I. Tomov, A. Kursumovic, M. Majorosh, D-J. Kang, B. A. Glowacki, and J. E. Evetts, "Pulsed laser deposition of epitaxial $YBa_2Cu_3O_{7-y}$ / oxide multilayers onto textured NiFe substrates for coated conductor applications", *Superconductor Sci. Technology*, **15**, 598-605 (2002).

[6]   A. Kursumovic, Y.S. Cheng, A.P. Bramley, B.A. Glowacki and J.E. Evetts, "Microstructure Development in Thick YBCO Films Grown by Liquid Phase Epitaxy" *Inst. Phys. Conf. Ser. No. 167*, **1** 147-150, 2000 IOP Publ. Ltd.

[7]   N. A. Rutter, B. A. Glowacki, J. H. Durell, J. E. Evetts, H. te Lintelo, R. De Gryse and J. Denul, " Formation of native cube textured oxide on a flexible NiFe tape substrate for coated conductor applications", *Inst. Phys. Conf. Ser.* No. 167, Vol. 1, 407-410, 2000 IOP Publ. Ltd

[8]   M. Muralidhar, H. S. Chauhan, T. Saitoh, K. Kamada, K. Segawa, and M. Mukarami, "Effect of mixing three rare-earth elements on the superconducting properties of RE Ba2 Cu3Oy, *Superconductor Sci. Technology*, **10**, 663-670 (1997).

[9]   M. Tagami, C. Krauns, M. Samida, N. Nakamura, Y. Yamada, T. Umeda and Y. Shiohara, " Concentration of RE (Y, Pr, Sm, Gd, Dy, Yb) elements in liquid of Ba-Cu-O liquid/REBa2CuO7-d or RE2BaCuO5 two phase region", *J. Inst. Metals,* **60**, 353-359 (1996).

[10]   A. Vostner, Y. Sun, S. Tönies, H.W. Webber, **Y.S. Cheng**, A. Kuršumović and J.E. Evetts, "Irreversibility Properties of Coated Conductors Deposited by LPE on Single Crystalline Substrates", *Proc. 10th Intl. Workshop on Critical Currents* (Göttingen, Germany, June 2001) 291-293.

[11]   E. O. Schuls-DuBois, "Accelerated crucible rotation: Hydrodynamics and stirring effect", *Journal of Crystal Growth*, **12**, 81-87 (1972).

EFFECT OF TRANSVERSE COMPRESSIVE STRESS ON TRANSPORT CRITICAL CURRENT DENSITY OF Y-Ba-Cu-O COATED Ni and Ni-W RABiTS TAPES[*]

Najib Cheggour, Jack W. Ekin, and Cameron C. Clickner
National Institute of Standards and Technology, Boulder CO 80305, USA

Roeland Feenstra, Amit Goyal, Mariappan Paranthaman, and Noel Rutter
Oak Ridge National Laboratory, Oak Ridge TN 37831, USA

ABSTRACT

Transport properties of yttrium-barium-copper-oxide (YBCO) coatings on both pure-nickel and nickel-3at% tungsten rolling-assisted-biaxially-textured substrates (RABiTS) were tested under transverse compressive stress up to 180 MPa. Transport critical-current densities (J_c) were measured on samples having an initial J_c in the range of 1 MA/cm^2 at 76 K and self magnetic field. Transverse compressive stress can cause significant degradation of J_c in YBCO deposited on *pure*-nickel RABiTS unless appropriate work-hardening treatment or a sufficient frictional support are provided to the sample. Preliminary results obtained for YBCO on nickel-3at% tungsten *alloy* RABiTS suggest that the electromechanical properties of these conductors may be improved if such material is used for the substrate. Stress-strain characteristics were measured on nickel and nickel-3at% tungsten RABiTS substrate materials at 76 K. The tensile yield strength, Young's modulus, and proportional limit of elasticity of both materials were determined. Scanning electron microscopy was used to also investigate the microstructure of the coated-conductor samples after electromechanical testing. Mechanical cracks with multi-patterned orientations were observed in the ceramic layers. The data indicate that the formation of cracks in the ceramic layers initiated by the *yielding of the substrate material*, is the primary cause for the degradation of J_c in the samples investigated.

---

[*]Contribution of the National Institute of Standards and Technology, not subject to copyright.

INTRODUCTION

Highly textured yttrium-barium-copper-oxide (YBCO) films deposited on buffered flexible metallic substrates exhibit critical-current densities (J_c) that are in excess of 1 MA/cm^2 at liquid-nitrogen temperature and self magnetic field[1-3]. These coated conductors may therefore potentially be used in the construction of electrical devices such as underground power-transmission lines, transformers, motors, generators, and magnetic separators. Strong grain-alignment is essential for achieving such high J_c values[4-9]. This requirement can be met by using, for example, the rolling-assisted-biaxially-textured substrates (RABiTS) or the ion-beam-assisted-deposition (IBAD) technologies. Since the ceramic coatings are inherently brittle, the tolerance of the coated conductors to mechanical stresses can affect the ultimate performance of devices made of these tapes. Hence the study of their electromechanical properties is crucial to both guide conductor development and provide critical data for the design of specific applications[1,10-15]. When a power-transmission cable is bent, each tape-element is subjected to bending strain and also to transverse compressive stress as the tapes in the different layers of the cable are pushed against each other and against the wall of the conduit. In this work we present experimental results on the effect of transverse compressive stress on J_c, obtained in thin (0.3 μm)YBCO films deposited on pure-nickel and nickel-3at% tungsten RABiTS. The microstructure of the samples after electromechanical testing is shown and reveals cracking of the ceramic films due to stress application. We also report stress-strain characteristics of pure-Ni and Ni-3at%W substrate materials and compare their tensile yield strength, Young's modulus and proportional limit of elasticity. These results will provide insight into the role played by the yield strength of the substrate in determining the electromechanical properties of these conductors, as well as provide a preliminary comparison between YBCO coatings on pure-Ni and Ni-3at%W substrate materials.

EXPERIMENTAL PROCEDURE

Nickel and Ni-3at% tungsten rods were thermomechanically processed to obtain biaxially textured substrates[16]. The buffer layers were grown on the substrates by reactive evaporation and RF magnetron sputtering. Thereafter the YBCO layer was deposited on the buffered substrates using the *ex situ* BaF$_2$ method[17]. Samples with pure-Ni (sample 1 to 3) had a structure, from outside in, as YBCO | CeO$_2$ | yttria-stabilized zirconia (YSZ) | CeO$_2$ | Ni. Sample 4 incorporated a Ni-3at%W substrate and had an architecture as YBCO | CeO$_2$ | YSZ | Y$_2$O$_3$ | Ni-W. The tapes were then coated with a layer

High Temperature Superconductors

of silver (1 μm thick) and annealed in oxygen. An additional silver layer (10 μm thick) was deposited by thermal evaporation to insure low contact resistivity between the sample the current leads. For samples 1 to 3, the thickness of the pure-Ni subtrate was 50 μm, the buffer layers less than 1 μm, and the superconductor layer about 0.3 μm. Sample 4 had a Ni-3at%W substrate 50 μm thick, while the thickness of the buffer and YBCO layers were kept unchanged. The samples had a width of about 3 mm to 3.5 mm and a length of about 2.5 cm.

The apparatus for conducting J_c vs. transverse stress (σ_T) measurements has been described elsewhere[18]. Two copper-leads and a pair of voltage-taps were soldered to the sample. The sample was mounted onto a flat stainless steel block (bottom anvil), located at the bottom end of the apparatus. The current-leads were designed and soldered to the apparatus in such a way that one copper-lead was stationary and the other was free to flex[19]. The flexible copper-lead insured that the sample was stress-free during the cooling of the apparatus from room temperature to the operating temperature. The top anvil, also flat and made of stainless steel, was attached to the probe via a biaxially gimbaled pressure-foot so that this anvil conformed to the sample and bottom anvil surfaces. Measurements were carried out in liquid nitrogen at 76 K. Voltage-current curves were taken as a function of σ_T up to 180 MPa in self-field, and values of J_c determined at an electrical-field criterion of 1 μV/cm. The values of J_c were measured to within ± 1 % and transverse compressive stress to within ± 2 %. The samples investigated had initial values of J_c in the range of 1 MA/cm^2 at 76 K and self-field.

The apparatus for measuring stress-strain characteristics was constructed to investigate very soft substrates[20]. The procedure for mounting samples on the probe was adapted in such a way to avoid work-hardening of the substrates prior to stress-strain measurements. Samples were 50 μm thick, 1 cm wide and 30 cm long. The apparatus was connected to a servo-hydraulic actuator, equipped with a 1300 N load-cell and a linear variable differential transformer (LVDT) to measure stress and strain respectively. The apparatus was inserted, under load-control, into a liquid-nitrogen cryostat to insure stress-free cooling of the sample. To check the accuracy of measurements, the Young's modulus was determined for stainless-steel tapes about 80 μm thick and 3 mm wide and was found to be 176 GPa at 293 K, 200 GPa at 76 K and 210 GPa at 4 K. These values are consistent with those for type 304 stainless-steel sheets. The errors in the determination of the Young's modulus and yield strength are mainly due to the estimation of the sample dimensions, particularly its thickness. This uncertainty can be estimated to about 10%, depending on the thickness variation along the substrate's length.

TRANSVERSE COMPRESSIVE STRESS RESULTS

Two modes of application of transverse compressive stress to the sample were used. In the *monotonic-loading* mode, stress was applied to the sample and gradually increased without releasing the load between measurement steps. In contrast the *load-unload* mode consisted of applying stress of a certain value to the sample, then releasing it (with both anvils keeping physical contact with the sample) before reapplying it at a higher value. Sample 1 and 2 were measured in the *monotonic-loading* mode and *load-unload* mode respectively, and presented very different behavior (Fig. 1). Sample 1 showed no J_c degradation up to about 120 MPa, while Sample 2 presented a significant J_c degradation amounting to about 28 % at 100 MPa and 39 % at 180 MPa. In the *monotonic-loading* mode, the sample has a strong frictional support from the pressing anvils. This support prevents the sample from expanding laterally. In contrast, the frictional support is significantly reduced when operating in the *load-unload* mode. In-plane expansion becomes possible, which may lead to yielding of the substrate material and cracking of the buffer and YBCO layers.

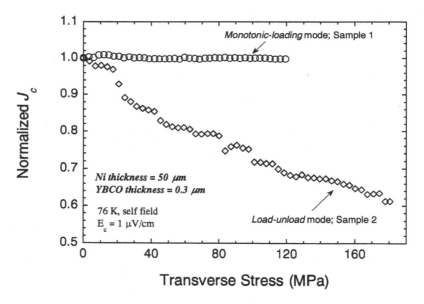

FIGURE 1. Effect of transverse stress on J_c in YBCO films on *pure*-nickel RABiTS. The results obtained in two modes of measurements illustrate the role played by friction between the sample and pressing anvils.

High Temperature Superconductors

In Fig. 2 we show the benefit of work-hardening the sample substrate and present a preliminary comparison between the response to transverse stress of the YBCO coatings on pure-Ni and Ni-3at%W substrate materials. Sample 3 was pressed first monotonically to 160 MPa, then unloaded and measured in the *load-unload* mode up to 180 MPa. In the monotonic pressing, J_c did not degrade (result not shown in Fig. 2). During this operation the pure-Ni substrate work-hardened, which improved substantially the tolerance of this sample to transverse stress as compared to Sample 2 which did not receive the work-hardening treatment. Sample 3 only showed 6 % degradation of J_c at 100 MPa, and 12 % at 180 MPa. Sample 4, which had a Ni-3at%W substrate, exhibited the best electromechanical performance in the *load-unload* mode, as the degradation of J_c amounted only to about 2% at 100 MPa, and 7.5 % at 180 MPa without a work-hardening treatment.

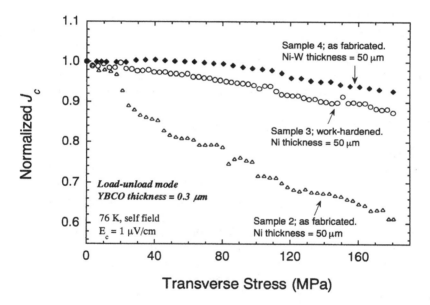

FIGURE 2. Effect of transverse stress on J_c in YBCO films on pure-Ni and Ni-3at%W *alloy* RABiTS. The results illustrate the benefits of work-hardening pure-Ni substrates and provide a preliminary comparison between YBCO coatings on pure-Ni and Ni-3at%W substrate materials.

These results provide evidence that the mechanical properties of the substrate material play a predominant role in determining the response of these samples to transverse compressive stress. This appears to be a valid conclusion at least for the samples measured, which had in common the initial values of J_c that were in the range of 1 MA/cm^2 at 76 K and self-field, and therefore probably a similar grain-alignment structure. More comprehensive data are still required to draw definitive conclusions.

As will be presented below, the tensile yield strength, Young's modulus and proportional limit of elasticity of Ni-3at%W substrates are substantially higher than those of pure-Ni substrates. The mechanical properties of work-hardened pure-Ni substrates are also expected to be superior, as compared to pure-Ni substrates' properties. For the samples measured, the higher these properties are, the more the transverse load needs to be to initiate cracking of the ceramic layers. These results do not rule out the usefulness of soft substrate materials. If strong frictional support can be provided to the tape, even YBCO films on soft substrates may suffer little or no damage from transverse stress up to at least 160 MPa. Nonetheless, it is probably not convenient to provide sufficient frictional support to the tape in many applications. Therefore it is safer to use robust substrate materials in the coated-conductors' architecture. A comparison of the response to axial and bending strains between YBCO coatings on different substrates is also required to guide coated-conductor development.

FATIGUE RESULTS

Subsequent to static transverse-stress measurements, the samples were subjected to *cyclic* transverse-stress up to 180 MPa and 2000 cycles. Load was cycled between a given value (σ_T, generally equal to the maximum load used in the static-load testing) and near zero at a frequency of 0.33 Hz. The results are presented in Fig. 3. The additional degradation of J_c from fatigue tends to increase with the degree of initial damage produced by the previous static transverse-stress testing. As will be presented below, the formation of cracks in the ceramic layers, caused by the application of transverse stress, is the primary cause for the degradation of J_c in the samples investigated. It is therefore possible that the more the sample is pre-cracked, the more it becomes sensitive to fatigue cycles. Most likely, fatigue tends to propagate the existing cracks which results in a further degradation of J_c. When a sample is free from cracks (Sample 1), fatigue does not seem to induce a noticeable damage to the YBCO layer.

High Temperature Superconductors

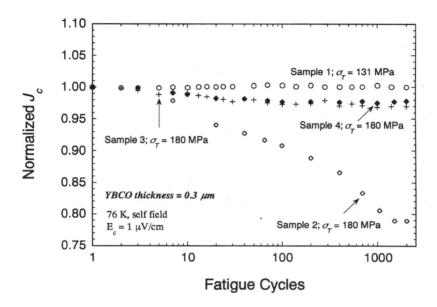

FIGURE 3. Effect of fatigue on J_c in YBCO films on pure-Ni and Ni-3at%W RABiTS. Fatigue tests were carried out on the samples after the static transverse-stress measurements. Fatigue has more effect on J_c in samples containing cracks previous to testing.

MICROSTUCTURE STUDY

After electromechanical testing, silver was etched away to expose the YBCO layer. A solution of $25\%H_2O_2 + 25\%NH_4OH + 50\%H_2O$ was used for silver etching[14]. The microstructure of this layer was examined using scanning electron microscopy (SEM), and regions with cracks were found in it (Figs. 4-6). For samples with pure-Ni substrates, the cracked regions were distributed in patches of a few micrometers to about 600 μm wide (Fig. 4). In the case of Sample 4 that has a Ni-3at%W substrate, cracks were mostly located near the sample edges, with a group of diagonal cracks extending about 400 μm towards the middle of the sample (Fig. 6). Cracks had multi-patterned orientations with respect to the direction of the electrical-current flow. The percentage of the affected areas correlates well with the total J_c degradation in each sample.

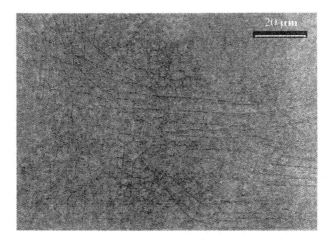

FIGURE 4. Scanning electron micrograph of YBCO film on buffered pure-Ni RABiTS substrate after electromechanical testing, showing a mixed pattern of mechanical cracks throughout the sample, both longitudinal and transverse to the electrical-current flow direction. The vertical axis of the image coincides with the direction of the electrical current applied to the sample.

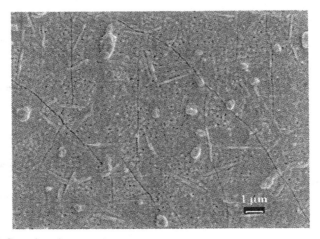

FIGURE 5. Scanning electron micrograph of YBCO film on buffered Ni-3at%W RABiTS substrate after electromechanical testing, showing very fine cracks located mostly near the sample edges. The vertical axis of the image coincides with the direction of the electrical current applied to the sample.

High Temperature Superconductors

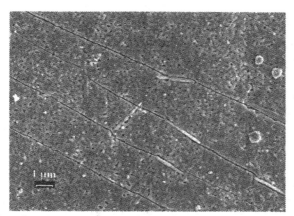

FIGURE 6. Scanning electron micrograph of YBCO film on buffered Ni-3at%W RABiTS substrate after electromechanical testing, showing diagonal cracks with respect to the electrical-current flow direction. The vertical axis of the image coincides with the direction of the electrical current applied to the sample.

MECHANICAL PROPERTIES OF PURE-Ni AND Ni-3at%W SUBSTRATES AT 76 K

Stress-strain characteristics of pure-Ni and Ni-3at%W substrates at 76 K are presented in Figs. 7a and 7b. The samples investigated received the same heat-treatment as the ones used for buffer and YBCO depositions. Three samples of each material were measured. Average values over the three samples of the tensile yield strength, Young's modulus and proportional limit of elasticity are summarized in Table I. The tensile yield strength of Ni-3at%W is more than a factor of three higher than that of pure-Ni, while the Young's modulus and proportional limit of elasticity are nearly a factor of two higher. When the sample was plastically deformed, the tensile load was released and reapplied to evaluate the degree of sample's work-hardening (Figs 7a, 7b). This operation was repeated several times up to a strain of about 1 %. For the samples presented in Figs. 7a and 7b, the Young's modulus of the second to the sixth slope was estimated and an average value determined (Table I). The change of the Young's modulus before and after plastic deformation for pure-Ni substrates is substantial (~64 %) as compared to that for Ni-3at%W substrates (~3 %). Also note that in the plastic regime, the stress increases with strain at a higher rate for pure-Ni than for Ni-3at%W. Therefore the work-hardening of pure-Ni is greater than that of Ni-3at%W. The difference in the mechanical properties between the two materials at 76 K is also depicted in Fig.8 for direct comparison.

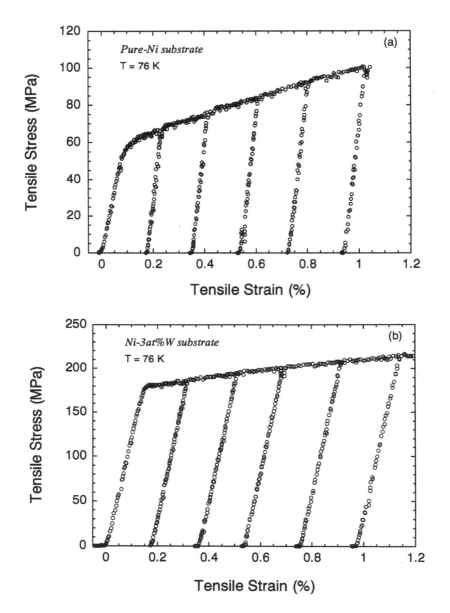

FIGURE 7. Stress-strain characteristics at 76 K of annealed pure-Ni (a) and Ni-3at%W (b) substrates.

High Temperature Superconductors

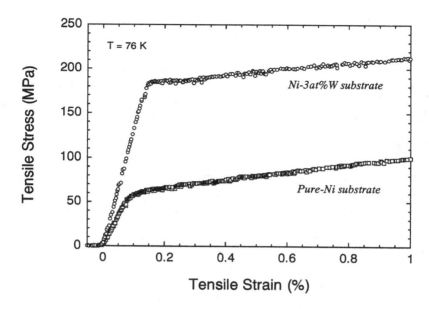

FIGURE 8. Comparison of stress-strain curves at 76 K of annealed pure-Ni and Ni-3at%W substrates.

Table I. Mechanical properties at 76 K of annealed pure-Ni and Ni-3at%W substrates.

	Pure-Ni	Ni-3at%W
Proportional limit of elasticity, ε_p (%)	0.08	0.15
Tensile Yield Strength (MPa)	55	183
Young's Modulus, E (GPa); 1st stress-strain slope	72	128
Young's Modulus, E (GPa); Average of second to sixth stress-strain slopes (Fig. 7)	118	132

CONCLUSION

The critical-current degradation from transverse compressive stress in YBCO coated conductors on buffered pure-nickel and nickel-3at% tungsten RABiTS has been investigated on samples that had initial J_c values in the range of 1 MA/cm^2 at 76 K and self-field. Transverse stress can cause significant degradation of J_c in YBCO deposited on pure-nickel RABiTS unless a sufficient frictional support is provided to the sample. Appropriate work-hardening of pure-Ni substrates or the use of Ni-3at%W substrates improve the electromechanical performance of these conductors.

The microstructure of the samples after electromechanical testing has been studied and has shown cracking of the YBCO layer. For samples on pure-Ni substrates, these defects were randomly distributed. In contrast for the sample on Ni-3at%W, the cracks are more confined near the sample edges. The mechanical properties of pure-Ni and Ni-3at%W substrates have also been presented. These results are given to guide the development of YBCO coated conductors.

ACKNOWLEDGMENTS

This research was supported in part by the U.S. Department of Energy, Office of Energy Efficiency and Renewable Energy, Office of Power Technologies—Superconductivity Program, and the U.S. Department of Energy, High Energy Physics Program.

REFERENCES

[1]X.D. Wu, S.R. Foltyn, P.N. Arendt, W.R. Blumenthal, I.H. Campbell, J.D. Cotton, J.Y. Coulter, W.L. Hults, M.P. Maley, H.F. Safar, and J.L. Smith, "Properties of YBa$_2$Cu$_3$O$_{7-\delta}$ thick films on flexible buffered metallic substrates," *Appl. Phys. Lett* **67**, pp. 2397-2399 (1995).

[2]A. Goyal, D.P. Norton, J.D. Budai, M. Paranthaman, E.D. Specht, D.M. Kroeger, D.K. Christen, Q. He, B. Saffian, F.A. List, D.F. Lee, P.M. Martin, C.E. Klabunde, E. Hartfield, and V.K. Sikka, "High critical current density superconducting tapes by epitaxial deposition of YBa$_2$Cu$_3$O$_x$ thick films on biaxially textured metals" *Appl. Phys. Lett* **69**, pp. 1795-1797 (1996).

[3]D.P. Norton, A. Goyal, J.D. Budai, D.K. Christen, D.M. Kroeger, E.D. Specht, Q. He, B. Saffian, M. Paranthaman, C.E. Klabunde, D.F. Lee, B.C. Sales, and F.A. List, "Epitaxial YBa$_2$Cu$_3$O$_7$ on biaxially textured nickel (001): An approach to superconducting tapes with high critical current density" *Science* **274**, pp. 755-757 (1996).

[4]D. Dimos, P. Chaudhari, J. Mannhart, and F.K. LeGoues "Orientation dependence of grain-boundary critical currents in $YBa_2Cu_3O_{7-\delta}$ bicrystals," *Phys. Rev. Lett.* **61**, pp. 219-222 (1988).

[5]D. Dimos, P. Chaudhari, and J. Mannhart, "Superconducting transport properties of grain boundaries in $YBa_2Cu_3O_7$ bicrystals," *Phys. Rev. B* **41**, pp. 4038-4049 (1990).

[6]Y. Iijima, N. Tanabe, O. Kohno, and Y. Ikeno, "In-plane aligned $YBa_2Cu_3O_{7-x}$ thin films deposited on polycrystalline metallic substrates," *Appl. Phys. Lett* **60**, pp. 769-771 (1992).

[7]A. Goyal, D.P. Norton, D.M. Kroeger, D.K. Christen, M. Paranthaman, E.D. Specht, J.D. Budai, Q. He, B. Saffian, F.A. List, D.F. Lee, E. Hatfield, P.M. Martin, C.E. Klabunde, J. Mathis, and C. Park, "Conductors with controlled grain boundaries: An approach to the next generation, high temperature wire," *J. Mater. Res.* **12**, pp. 2924-2940 (1997).

[8]D.T. Verebelyi, D.K. Christen, R. Feenstra, C. Cantoni, A. Goyal, D.F. Lee, M. Paranthaman, P.N. Arendt, R.F. DePaula, J.R. Groves, and C. Prouteau, "Low angle grain boundary transport in $YBa_2Cu_3O_{7-\delta}$ coated conductors," *Appl. Phys. Lett* **76**, pp. 1755-1757 (2000).

[9]D.T. Verebelyi, C. Cantoni, J.D. Budai, D.K. Christen, H.J. Kim, and J.R. Thompson, "Critical current density of $YBa_2Cu_3O_{7-\delta}$ low-angle grain boundaries in self-field," *Appl. Phys. Lett* **78**, pp. 2031-2033 (2001).

[10]H.C. Freyhardt, J. Hoffmann, J. Wiesmann, J. Dzick, K. Heinemann, A. Isaev, F. Garcia-Moreno, S. Sievers, and A. Usoskin, "YBaCuO thick films on planar and curved technical substrates," *IEEE Trans. Appl. Supercond.* **7**, pp. 1426-1431 (1997).

[11]C. Park, D.P. Norton, J.D. Budai, D.K. Christen, D. Verebelyi, R. Feenstra, D.F. Lee, A. Goyal, D.M. Kroeger, and M. Paranthaman, "Bend strain tolerance of critical currents for $YBa_2Cu_3O_7$ films deposited on rolled-textured (001) Ni," *Appl. Phys. Lett* **13**, pp. 1904-1906 (1998).

[12]C.L.H. Thieme, S. Fleshler, D.M. Buczek, M. Jowett, L.G. Fritzemeier, P.N. Arendt, S.R. Foltyn, J.Y. Coulter, and J.O. Willis, "Axial strain dependence at 77 K of the critical current of thick YBaCuO films on Ni-alloy substrates with IBAD buffer layers," *IEEE Trans. Appl. Supercond.* **9**, pp. 1494-1497 (1999).

[13]J.Yoo, and D. Youm, "Tensile stress effects on the critical current densities of coated conductors," *Supercond. Sci. Technol.* **14**, pp. 109-112 (2001).

[14]J.W. Ekin, S.L. Bray, N. Cheggour, C.C. Clickner, S.R. Foltyn, P.N. Arendt, A.A. Polyanskii, D.C. Larbalestier, and C.N. McCowan, "Transverse

stress and fatigue effects in Y-Ba-Cu-O coated IBAD tapes," *IEEE Trans. Appl. Supercond.* **11**, pp. 3389-3392 (2001).

[15]N. Cheggour, J.W. Ekin, C.C. Clickner, R. Feenstra, A. Goyal, M. Paranthaman, D.F. Lee, D.M. Kroeger, and D.K. Christen, "Transverse compressive stress, fatigue, and magnetic substrate effects on the critical current density of Y-Ba-Cu-O coated RABiTS tapes," *Adv. Cryog. Eng.*, in press (2002).

[16]M. Paranthaman, C. Park, X. Cui, A. Goyal, D.F. Lee, P.M. Martin, T.G. Chirayil, D.T. Verebelyi, D.P. Norton, D.K. Christen, and D.M. Kroeger, "$YBa_2Cu_3O_{7-y}$—coated conductors with high engineering current density," *J. Mater. Res.* **15**, pp. 2647-2652 (2000).

[17]R. Feenstra, T.B. Lindemer, J.D. Budai, and M.D. Galloway, "Effect of oxygen pressure on the synthesis of $YBa_2Cu_3O_{7-x}$ thin films by post-deposition annealing," *J. Appl. Phys.* **69**, pp. 6569-6585 (1991).

[18]J.W. Ekin, "Effect of transverse compressive stress on the critical current and upper critical field of Nb_3Sn," *J. Appl. Phys.* **62**, pp. 4829-4834 (1987).

[19]P.E.Kirkpatrick, J.W. Ekin, and S.L. Bray, "A flexible high-current lead for use in high-magnetic-field cryogenic environments," *Rev. Sci. Instrum.* **70**, pp. 3338-3340 (1999).

[20] C.C. Clickner, J.W. Ekin, and N. Cheggour, to be published.

High Temperature Superconductors

PHASE AND MICROSTRUCTURE CHANGE OF HIGH CRITICAL CURRENT DENSITY TFA-MOD YBCO COATED CONDUCTOR

Yutaka Yamada, Takeshi Araki,
Haruhiko Kurosaki, Toyotaka
Yuasa, Yuh Shiohara and
Izumi Hirabayashi
ISTEC-SRL
Mutsuno 2-chome 4-1, Atsuta-ku,
Nagoya 456-8587, Japan

Yasuhiro Iijima and Takashi Saito
Fujikura Ltd.
1-5-1, Kiba, Koto-ku, Tokyo
135-8512, Japan

Junko Shibata and Yuichi Ikuhara
Tokyo University
2-11-16 Yayoi, Bunkyo-ku, Tokyo
113-8656, Japan

Takeharu Katoh and Tsukasa
Hirayama
Japan Fine Ceramic Center
Mutsuno 2-chome 4-1, Atsuta-ku,
Nagoya 456-8587, Japan

ABSTRACTS

For TFA-MOD process in which we obtained high J_c values up to $12.6 MA/cm^2$ on $LaAlO_3$ substrates and $2.9 MA/cm^2$ on metallic IBAD substrates, phase formation and microstructure change during heat-treatment were investigated using resistivity measurements, AFM and TEM. Precursor consisted of fine CuO and amorphous Y-Ba-Cu-O-F matrix was further oxidized up to 600°C and then transformed into Y123 above 670°C where the intermediate phases of $(Y,Ba)(O,F)_2$ and (Ba,Cu) poor phase such as $Y_2Cu_2O_5$ phase were formed at the same time. A fraction of these phases remained un-reacted on the surface of the film and also resulted in rough surface morphology. Although high J_c have been attained, we have much room for further enhancement in J_c. The magnetic field dependence of J_c was also measured and exhibited superior property up to 9.5T at 77K to that of the conventional NbTi conductor at 4.2K.

1. INTRODUCTION

Among many processes for an YBCO coated conductor, TFA-MOD (metal organic deposition using trifluoroacetate) is one of the most promising processes because of its cost efficiency [1] and high critical current density, J_c. TFA-MOD method is principally consisted of only coating and heat-treatment process without vacuum chamber. It is performed in a non-vacuum atmosphere while other processes such as PLD (Pulsed Laser Deposition) method require a complicated vacuum system. Furthermore TFA-MOD process leads to a high critical current density, J_c, over 1 MA/cm^2 at 77K [1-3]. Many groups obtained promising results using textured Ni substrates [1] and IBAD substrates [3-5]. The J_c reached 1 to 2 MA/cm^2 at 77K and 0T. Recently the effort for a long YBCO conductor is also being carried out by improving the heat-treatment furnace for the continuous production system.

However, understanding of the reaction process in TFA-MOD YBCO still remains ambiguous especially in terms of the Y123 formation and microstructural change. For further development, we need such basic understanding.

In this paper, we report the resistivity measurement results, which give some insight on the Y123 formation mechanism. Together with XRD and TEM data, we describe phase and microstructure change during heat-treatment. Especially we concluded that super-current did not flow in the whole cross section of the YBCO film because much un-reacted phases still remained near the surface. Actually some improvements on microstructure by optimizing heat-treatment process resulted in high J_c values above 12MA/cm^2. The J_c-B characteristic at 77 K was also measured and exhibited a superior property at high magnetic fields. These results imply that TFA-MOD YBCO coated conductor is a strong candidate for a practical coated conductor.

2. EXPERIMENTAL

The coating solutions were made from Y, Ba and Cu acetates [3]. During synthesis, special attention was paid to the reduction of the impurities including H_2O and acetic acid. LaAlO$_3$ (LAO) single crystals of 10 x 10 mm^2 and CeO$_2$/YSZ buffered Hastelloy tapes (IBAD) [6] 5-10 x 10 mm^2 were spin-coated or dip-coated with the coating solutions. The coated films were heat-treated at 200°C to 400°C in wet oxygen gas (calcination process) and fired at 800°C in a wet mixture gas of argon and oxygen (firing process). The oxygen and water concentrations were typically 1000 ppm and 4.2%, respectively. The electric resistivity was measured during the firing process at 800°C. The calcined film on the LAO was patterned and Au wires for voltage taps were attached. Critical currents were measured by the four-probe transport method at 77.3 K and in the magnetic fields up to 9.5 T. The criterion of I_c was 5. V/cm.

3. RESULTS AND DISCUSSION
3.1 Resistivity Measurement and Reaction in TFA Process
In the TFA-MOD process, first we obtained a calcined film heat-treated at 200°C to 400°C in wet oxygen gas. Figure 1 shows the microstructure of the calcined film. This was consisted of YBaCuOF matrix and fine CuO grains about 10nm in diameter. The coating film before calcination was M-CF$_3$COOH (M=Y, Ba, Cu), metal trifluoroacetate. This was transformed to YBaCuOF and CuO. Only Cu was oxidized to CuO at the low temperature.

This calcined film was fired at 800°C for 1hour, which was the best condition to obtain a high J_c in our laboratory. The resistivity of the film during firing was measured as shown in Fig. 2. The heat-treatment was done first in an atmosphere of wet Ar-1000 ppm O$_2$ gas with humidity of 4.2% and then the gas was changed to dry Ar/O$_2$ gas in the terminate stage at 800°C. Finally the sample was oxygenated in a pure O$_2$ gas at 450°C.

The resistivity was first decreased (region A in Fig. 2) due to the evaporation of chemicals including ethylene glycol or sintering of Ag particles of the Ag paste in the voltage taps for the resistivity measurement. Then, the resistivity was increased when the temperature was increased to 600°C (B). This was due to the further oxidation of the matrix of the calcined film. Above 600°C the resistivity first decreased slowly (C) and then decreased rapidly above 670C° (D). The decrease of the resistivity means the formation of the aligned Y123 phase [7] with low resistivity among the phases observed in TFA-MOD process.

From the X-ray measurements [8] and the TEM study [9], the film before complete reaction, heat-treated at 775°C, included CuO, (Y, Ba)(O, F)$_2$ and Y$_2$Cu$_2$O$_5$ phases. As reported in detail by McIntyre et al.[2,8] and Solovyov et al.[10], Y123 was formed through the reaction of H$_2$O and these phases. The microstructure of the resultant film was highly aligned especially in the vicinity of the interface between the film and the substrate. This highly orientation resulted in high J_c as reported by the authors [3-5,7]. Normally the high J_c was obtained at 800°C for 1hour. However, the

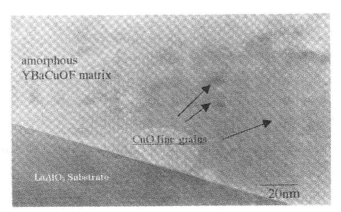

Figure 1. TEM of TFA-MOD YBCO on LaAlO₃ substrate. Matrix is amorphous Y-Ba-Cu-O-F and grains are CuO 10nm in diameter.

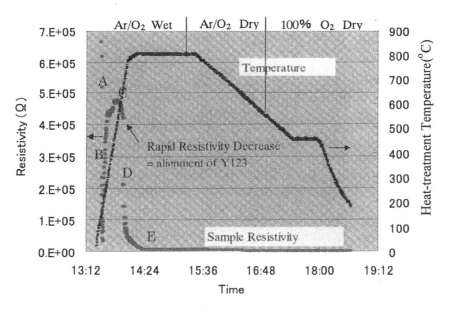

Figure 2. Resistivity change during TFA process.

formation of Y123 and the alignment of Y123 occurred at much lower temperatures around 670 °C from the rapid decrease of resistivity in Fig. 2.

Figure 3 is the XRD data at 700°C for different heat-treatment times. A calcined film showed only CuO or CuO_x peaks in YBaCuOF amorphous matrix as confirmed by TEM in Fig.1. After 10min the matrix YBaCuOF formed $(Y,Ba)(O,F)_2$. In 30min, this peak and the Y123 peak began to develop with the formation of (Ba,Cu) poor phase (compared to Y123) such as $Y_2Cu_2O_5$. This continued up to 60min. In 120min, the $(Y, Ba)(O,F)_2$ peak vanished while the CuO and $Y_2Cu_2O_5$ phases remained. This means the following reaction occurred: YBaCuOF +CuO $\rightarrow$ $(Y,Ba)(O,F)_2$ +CuO + $Y_2Cu_2O_5$ with H_2O and O_2 $\rightarrow$ Y123+ CuO + $Y_2Cu_2O_5$.

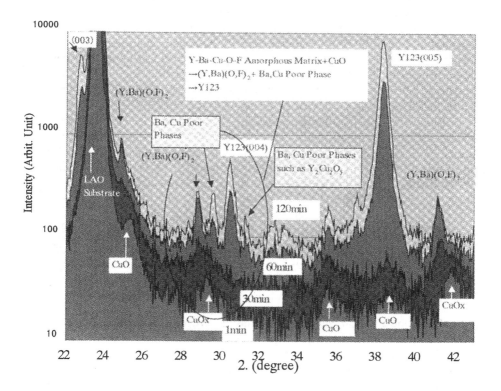

Figure 3. XRD patterns for various heat-treatment times at 700°C.

The un-reacted CuO and (Ba,Cu) poor phase is due that the Y123 formation reaction does not precede because the each grain such as CuO and $Y_2Cu_2O_5$ become too large. Actually large un-reacted $Y_2Cu_2O_5$ and CuO grains were observed by TEM of TFA sample [9] and ex-situ sample [11]. Very recently BaO phase other than these phases was also observed in the vicinity of the surface by TEM [12]. These facts mean that we have still possibility to increase J_c by optimizing the microstructure and thus the heat-treatment process. As described later, we actually obtained improved J_c of $12MA/cm^2$ [13] from the former $7MA/cm^2$ [14]. This was achieved by shortening the calcination time. Appropriate calcination resulted in small CuO grains and subsequent homogeneous Y123 formation reaction in firing process.

At the temperature of 800 °C, the formation of the aligned Y123 grains was completed in a few tens minutes as shown in Fig. 2. The resistivity reached minimum in 10 to 20min (region E). Therefore, the oriented Y123 structure finished forming at this minimum. After this annealing, the gas was changed to dry Ar/O_2 and the temperature was decreased. Then, the resistivity decreased with the temperature decrease and the oxygen content of the gas. The resistance decreased finally to 30 Ω.

Surface morphology was observed by AFM as shown in Figs. 4 (a) and (b) for a calcined film and a heat-treated film at 800°C for 15min. While the surface was very smooth in a calcined film, the surface became rough only in 15min. This means that the reaction to form Y123 was significantly proceeded in 15min, which corresponds to the resistivity data in Fig. 2. The resistivity was much decreased in 15min at 800°C. The rough surface was due to the reaction with H_2O and HF gas phases. For the transport current density, the effective cross sectional area is considered to be small from the above observations of AFM and un-reacted phases in Fig.3. In other words, super-current flows only a small section of TFA-MOD samples. Therefore, we can increase J_c by increasing the fraction of effective cross section for transport current.

From the above results and discussion, we can describe the whole reaction in this TFA-MOD YBCO:
In the calcination from room temperature to 400 °C,
$CF_3COO-M(Y, Ba, Cu) + H_2O+O_2 \rightarrow (Y, Ba,Cu)-(O,F) + CuO$ (precursor).

In subsequent firing process from 400°C to 800°C,

High Temperature Superconductors

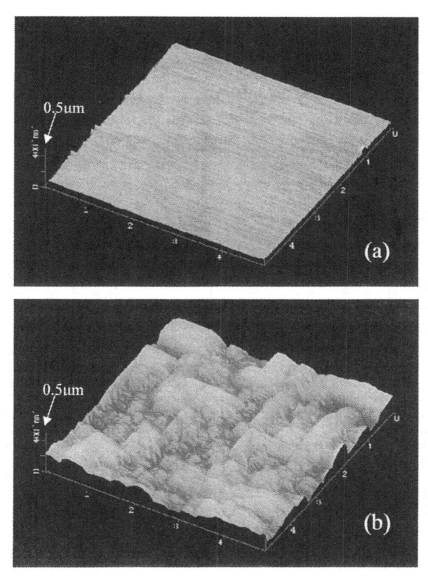

Figure 4. AFM surface morphology of (a) calcined film and (b) heat-treated film at 800°C for 15min.

(Y, Ba, Cu)-(O, F) + CuO (precursor) + H_2O + O_2
→further oxidization to CuO $< 600°C$
→formation of $(Y,Ba)(O,F)_2$ phase and $Y_2Cu_2O_5$ phase
→amorphous YBCO+ HF.
→$YBa_2Cu_3O_{6.1}$ formation and alignment + HF. $> 670°C$

According to the studies by Suenaga et al. [10,11] and Hirayama et al. [9], amorphous YBCO are observed just on the aligned Y123 phase and below the $Y_2Cu_2O_5$, CuO and $(Y,Ba)(O,F)_2$ phases. Thus, amorphous YBCO is considered to be an intermediate phase to form Y123 phase, which may also decrease the resistivity between 600 to 670 °C.

3.2 TFA-MOD YBCO on IBAD

IBAD was well known as one of the effective method to obtain a n aligned buffer layers such as YSZ on the metallic substrate such as Hastelloy [6]. Using dip coating method [15], TFA-MOD YBCO/IBAD coated conductor was fabricated as shown in Fig. 5. The transport J_c of 1.3 MA/cm^2 was obtained at 77K and 0T. Up to now the sample was fabricated in 10cm long [13], which was withdrawn at a speed of 18m/hr during dip coating. The obtained J_c value of the conductor was 1.3 MA/cm^2 at 77K and 0T. On the other hand, spin coating sample exhibited higher value of 2.5MA/cm^2. This ascribes that more homogeneous cross-section may be obtained by spin coating than by dip coating. These high J_c values imply that TFA-MOD YBCO/IBAD is a strong candidate for the practical coated conductor. Figure 6 showed the cross sectional area of TFA-MOD samples on IBAD tape. The thickness of sputtered CeO_2 layer was about 1μm and that of YBCO was 2000Å. The texturing was investigated in the order of the depositing process as shown in Fig. 7. The original IBAD tape exhibited a good texturing, FWHM of Δϕ of 9.1°. After depositing CeO_2 layer, the in-plane alignment was improved to 5.2° of Δϕ. Finally Δϕ was 6.0° after deposition of YBCO. This superior texturing resulted in high Jc of TFA-MOD samples.

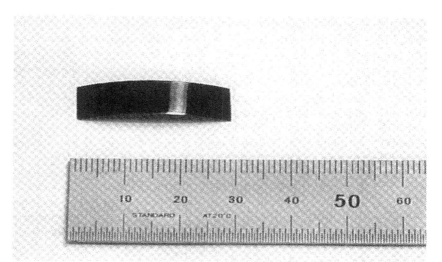

Figure 5. Dip Coated TFA-MOD YBCO on metallic IBAD substrate with J_c of 1.3 MA/cm^2 at 77K and 0T.

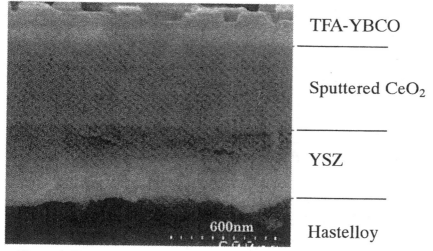

Figure 6. SEM micrograph of the cross sectional area of TFA -MOD YBCO on CeO_2 buffered IBAD substrate.

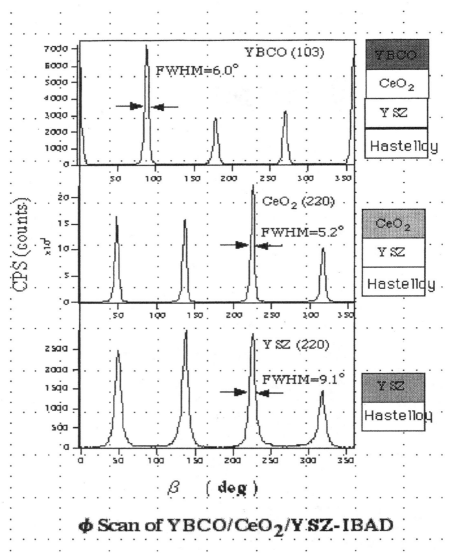

Figure 7. Variation of the degree of texturing for each layer of TFA -MOD YBCO on CeO$_2$ buffered IBAD substrate.

High Temperature Superconductors

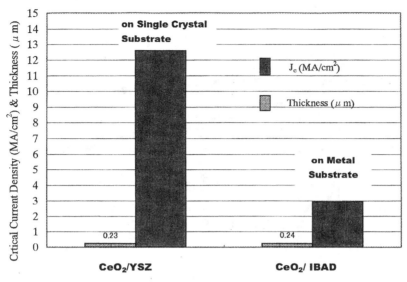

Figure 8. Critical current densities at 77K and 0T achieved in our laboratory for the metallic IBAD and single crystal substrates.

3.3 Critical Current Density

We have reported high J_c values of TFA-MOD YBCO samples up to now. Recently we improved J_c further by optimizing the heat-treatment process, especially calcination stage [13]. Although normal calcination time used by us [3] and McIntyre et al. [2] was more than 20hrs at 200°C to 400°C, the time was shortened to about 9 hours. This resulted in fine CuO grains in a calcined film. This was considered to bring about the homogeneous reaction to form Y123 in the firing stage. Furthermore, through efforts to reduce the amount of impurities such as H_2O and acetic acid in the coating solution, high values up to 12.6 MA/cm^2 were obtained in the case of LAO single crystals substrate as shown in Fig.8. On the other hand, the J_c value on metallic IBAD substrate, 2.9MA/cm^2, was below one fourth of that on the single crystal.

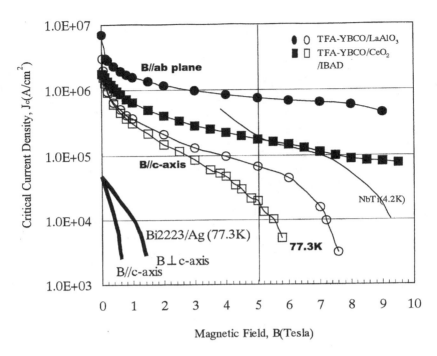

Fig. 9. J_c –B characteristic at 77.3 K for TFA-MOD YBCO on CeO_2 buffered IBAD-YSZ and on single crystal of $LaAlO_3$. The data is compared with the typical values of NbTi wires at 4.2 K and Ag sheathed Bi2223 wires.

Two typical samples on a LAO single crystal and a CeO_2 buffered IBAD-YSZ tape were measured for the magnetic field dependence of J_c at 77 K. The J_c values at 0T were 7 MA/cm^2 and 1.7 MA/cm^2 for the samples on the LAO and the IBAD, respectively. The magnetic field dependence of J_c is shown in Fig. 9, compared with other practical superconducting wires of NbTi and Ag sheathed Bi-2223. The TFA-YBCO/IBAD conductor exhibited high J_c even in a high magnetic field; 20 kA/cm^2 at 5 T parallel to the c-axis and 170 kA/cm^2 at 5 T parallel to the ab-plane. The latter is comparable to J_c of the most popular conductor of NbTi at 4.2 K. The J_c values for the film on LAO was much higher than these values especially

High Temperature Superconductors

for the fields parallel to the ab plane, which was almost $1MA/cm^2$ at 77K and 5T.

4. CONCLUSIONS

In the TFA-MOD process, resistivity measurements, TEM and XRD data suggested that the Y123 phase alignment occurred below 700 °C and was completed at an early stage of the firing at 800 °C from the resistivity minimum. CuO grains well developed first, then, high Jc was obtained after improving the calcination process resulted in a small CuO grains before firing. Then, we obtained high J_c on LAO substrate more than $12MA/cm^2$ at 77K and 0T from the former values of 7 to $8MA/cm^2$. The J_c value on metallic substrate of IBAD exhibited $2.9MA/cm^2$. Using the dip coating favorable for a practical process, 10cm long sample was also obtained. Together with the superior magnetic field dependence of J_c, the TFA-MOD YBCO conductor is considered to be promising as the next generation superconductor.

ACKNOWLEDGEMENTS

This work is supported by the New Energy and Industrial Technology Development Organization (NEDO) as Collaborative Research and Development of Fundamental Technologies for Superconductivity Applications.

REFERECNCES

[1] A.P. Malozemoff, S. Annavarapu, L. Fritzemeier,   Q. Li, V. Prunier, M. Rupich, ,   C. Thieme,   W. Zhang,   A.Goyal,   M. Paranthaman, D.F. Lee, Supercond. Sci. Technol. **13** 473 (2000).

[2] P.C. McIntyre, M.J. Cima and Ng Man Fai, J. Appl. Phys. **68** 4183 (1990).

[3] T. Araki, K. Yamagiwa, S.B. Kim, K. Matsumoto and I. Hirabayashi, Advances in Superconductivity, **XII** 610 (1999).

[4] T. Araki, Y. Takahashi, K. Yamagiwa, T. Yuasa, Y. Iijima, K. Takeda, S.B. Kim, Yutaka Yamada and I. Hirabayashi, IEEE Transactions on Applied Superconductivity, **11** 2869 (2000).

[5] Yutaka Yamada, S.B. Kim, T. Araki, T. Yuasa, Y. Takahashi, H. Kurosaki, Y. Shiohara, I. Hirabayashi, Y. Iijima, and K.Takeda, Physica C, **357-360** (2001) 1007.

[6] Y. Iijima, M. Hosaka, N. Tanabe, N. Sadakata, T. Saitoh, O. Kohno and K. Takeda, Applied Superconductivity **4** 475 (1996).

[7] Yutaka Yamada, T. Araki, H. Kurosaki, S.B. Kim, T. Yuasa, Y. Shiohara, I. Hirabayashi, Y. Iijima, T. Saitoh, J. Shibata, Y. Ikuhara, T. Katoh and T. Hirayama, in Proceedings of CEC/ICMC Conference, Madison, USA, July 16-20 (2001).

[8] P.C. McIntyre, and M.J. Cima, J. Mater. Res. **9** 2219 (1994).

[9] T. Hirayama, J. Shibata, T. Araki, Yutaka Yamada, K. Yamagiwa, I. Hirabayashi, T. Izumi, Y. Shiohara and Y. Ikuhara, in Extended Abstracts of 2001 International Workshop on Superconductivity, 2001, p.251.

[10] V.F. Solovyov, H.J. Wiesmann, Li-jun Wu, M. Suenaga and R. Feenstra, IEEE Trans. on Applied Supercond. **9** 1467 (1999).

[11] Li-jun Wu, Y. Zhu, V.F. Solovyov, H.J. Wiesmann, A.R. Moodenbaught, R.L. Sabatini and M. Suenaga, submitted to J. Mater. Res (2000).

[12] T. Hirayama, private communication.

[13] T. Arakai, to be presented in Applied Superconductivity Conference in Houston, USA, (2002).

[14] Yutaka Yamada, S.B. Kim, T.Araki, T. Yuasa, Y. Takahashi, H. Kurosaki, Y. Shiohara, , I. Hirabayashi, Y. Iijima and K. Takeda, Mat. Res. Soc. Symp. Proc., **659** II4.7.1 (2001).

[15] H. Kurosaki, T. Araki, T. Yuasa, S.B. Kim, Yutaka Yamada, I. Hirabayashi, Y. Iijima and T. Saitoh, in Extended Abstracts of 2001 International Workshop on Superconductivity, p.131 (2001).

GROWTH KINETICS AND TEXTURE OF SOE NiO/Ni AND Ni-BASED ALLOYS RABiTS

Z. Lockman[A], X. Qi[A], W. Goldacker[B], R. Nast[B], B. deBoer[C], A.Kursumovic[D], R. Tomov[D], R. Hühne[D], J.E. Evetts[D] B.A. Glowacki[D] J. L. MacManus-Driscoll[A]

[A] Department of Materials, Imperial College of Science Technology and Medicine, Prince Consort Road, London, SW7 2BP, UK
[B] Forschungszentrum Karlsruhe, Technik und Umwelt, Hermann- von-Helmholtz- Platz 1 76344, Eggenstein-Leopoldshafen, Karlsruhe, Germany
[C] Institut für Festkörper-und Werkstoffforschung, Helmholtzstraße 20, 01069 Dresden, Germany
[D] Department of Materials Science and Metallurgy, University of Cambridge, Pembroke Street, Cambridge CB2 3QZ, UK

Abstract

We report on basic studies of a new coated conductor architectures: $Ni/NiO/Nd_{2-x}Ce_xCuO_4/YBCO$. Various Ni compositions were explored including pure Ni, Ni-Cr, and Ni-Fe, and a new liquid phase processing (LPP) technique to make $Nd_{2-x}Ce_xCuO_4$. Fundamental studies of surface oxidation epitaxy of Ni and Ni-alloys were undertaken. It was shown that a very thin NiO layer can be formed on pure Ni with excellent texture and low surface roughness. Similar, laterally grown surface oxides form for larger thicknesses. For Ni-Cr, a ~5micron thick NiO surface layer with excellent texture and surface smoothness can be produced. For Ni-Fe, a smooth, thin Ni_2FeO_4 was formed underneath a surface (111) oxide which can be removed. The $Nd_{2-x}Ce_xCuO_4$ layer was fabricated by LPP to have a surface roughness of a few 10's of nm. The layer was highly out-of-plane textured and could be completely in-plane textured by adjusting the Ce concentration.

Introduction

Since the discovery of high T_c superconductors in 1986, there has been an enormous effort worldwide to establish a scaleable method for the fabrication of coated conductors on metallic tapes. Although short length samples with J_c over 1 MA/cm^3 have been prepared by a number of vapour deposition methods, e.g. pulsed laser deposition (PLD), they are so far failed to produce long length conductors at a reasonable cost. An alternative route to vapour deposition methods is LPE, which has been widely used for the growth of semiconductor and garnet thin films during 1970s and 1980s. The advantages of LPE compared to vapour methods are the very fast growth rate, typically 1-10 μm/min, and the capability of growing thick films without degrading the structural perfection[1]. However, LPE growth of REBCO thick films on metallic substrate is much more difficult. The high growth temperature largely increases the risk of reaction between the substrate and the film. The solubility of RE's in the BaO:CuO solution is very low, therefore nucleation and growth is more difficult than the traditional LPE of semiconductors and garnets. Compared to PLD and other vapor deposition methods, LPE has a much lower available supersaturation and hence a much smaller driving force for growth. As a consequence, a close lattice matched substrate is required. At present, most of the buffer layers were developed for the use in the vapor deposition methods and therefore are not suitable for LPE growth. Nd_2CuO_4, a tetragonal crystal with a=b≈0.394 nm, is a good potential buffer for LPE. As a part of the RE-Ba-Cu-O system[2] Nd_2CuO_4 neither poisons superconductivity nor reacts strongly with the barium cuprate solution at high temperature. It has a fairly close lattice match and thermal expansion match to REBCO and NiO. In this paper, we report on the growth of $Nd_{2-x}Ce_xCuO_4$ thick films on surface oxidised Ni-alloy substrates by scaleable methods. We also report on the basics of the surface oxidation of different Ni-alloys.

Experimental

100μm thick pure nickel, Ni-10%Cr, Ni-13%Cr and Ni-Fe were prepared by RABiTS method[3]. A summary of all Ni and Ni-alloys used for the SOE is shown in Table I. The tape surface was finished by a standard electro-polishing or chemical etching method (5%HF). The foils were then ultrasonically cleaned in acetone and methanol.

Table I Summary of Ni and Ni-foils used for SOE

Foils	Used by	Thickness (μm)	Supplier
Pure Nickel	Imperial College	80-100	Forschungszentrum, Karlsruhe [4]

Pure Nickel	Cambridge	100	CONTEXT [5]
Ni-10%Cr	Imperial College	100	Forschungszentrum, Karlsruhe [4]
Ni-13%Cr	Imperial College	100	Institut für Festkörper - und Werkstoffforschung [6]
Ni-Fe	Cambridge	50	CONTEXT [5]

Oxidation for pure Ni was done at high temperature (1000-1300°C) in two different oxidation atmosphere a) In air (Cambridge), b) In oxygen (Imperial). Oxidation was carried out in an evacuable high temperature tubular furnace, enabling oxidation in a controlled atmosphere. The progress of oxygenation of the NiO scale was monitored by mass and thickness increase. Mass was measured by a precise microbalance after each partial oxidation. Thickness was measured both by a micrometer and using secondary electron microscope (SEM) or optical microscope. The texture analysis was carried out by x-Ray diffraction techniques: θ-2θ scans, rocking curves and pole figures. Optical microscopy was used to monitor the development of microstructure during film growth.

Oxidation of Ni-Cr alloys was done isothermally in air at 1050°C. Details of the oxidation procedures could be found in our previous study[7].

In order to systematically study high temperature oxidation process of Ni-Fe, a number of Ni-Fe tapes were oxidised using O_2 and Ar/O_2 mixtures in a UHV chamber, tapes being heated resistively using a direct current. There are two methods of carrying out the oxidation which turn out to show important differences. One way is to preheat the tape in vacuum, then introduce the Ar/O_2 gas mixture, or alternatively to introduce the required pressure of gas into the chamber, then ramp up the current. Oxidation experiments were carried out at temperatures between 500°C and 1100°C, in each case preheating the tape before introducing the oxidising atmosphere. Note that the temperature quoted is the initial temperature of the bare Ni-Fe tape, and will be reduced when the surface is oxidised. Below around 950°C a continuous oxide is grown, which is dark grey in appearance at sufficient thicknesses. At higher temperatures however, macroscopic oxide grains become visible on the surface due to secondary recrystallisation and abnormal grain growth.

Results and Discussion

1. SOE of RABiTS Pure Nickel: At low temperatures, NiO film growth is affected by the grain boundaries in the "parent" Ni-phase which has an average grain size of about 30 μm. The growth is faster around grain boundaries creating humps on the NiO surface. The grain size of the oxide film is in the order of 1 μm. At higher temperatures, the grain size of the NiO film increases being >20 μm at 1250°C on textured Ni. The NiO growth has a specific flat surface that is a consequence of predominantly volume diffusion of growing species. At very high temperatures, >1300°C, highly concave growth around grain boundaries makes the film very rough[8].

1a) Air Oxidation Studies in Cambridge: A working window for SOE NiO/Ni was established from the ratio of XRD intensities I(200)/I(111)NiO. For samples oxidised for approximately two hours in air at different temperatures the results are shown in Fig. 1-a. As can be seen, there is a pronounced peak around 1260°C, at somewhat higher temperature than found in oxygen oxidation (refer next section on oxygen oxidation) presumably due to using air rather than 100% oxygen[9]. As expected sharp in-plane alignment is achieved at the "peak" of the working window (Fig. 1-b). In Fig. 1-c the pole figure shows that an acceptable in-plane orientation can be achieved in the range of <20°C. It should be noted, however, that at temperatures >1270°C, increased (220) peak intensities are present that add-up to overall deterioration of the NiO/Ni epitaxy.

The sample mass per unit area, Δm/A, for NiO/Ni films grown for two hours, both in pure oxygen and in air, is shown in Fig. 2. The sample mass increases due to formation of NiO film in the same manner as oxide thickness. These data are corresponding to data on epitaxial development shown in Fig. 1. The oxide film thickness was calculated here, but corresponds closely to experimentally measured values that were performed on some of the samples. As can be seen, an almost perfect epitaxy that was achieved at around 1260°C corresponds to ~25 μm thick NiO layer. The kinetics of NiO film growth in air at different temperatures versus time was investigated in more detail in Ref[10]. This kinetics was found shown to obey expected parabolic ($\sqrt{t}$) law. The activation energy calculated from the kinetic data was 220 kJ/mol for temperatures above 1100°C, corresponding to that for volume diffusion of Ni in NiO in excellent agreement with literature data. On the other hand a grain growth led to the epitaxial development. This grain growth was attributed to the grain boundary diffusion[10]. There are three distinct temperature regimes affecting the growth. At lower temperatures (<1000°C), close to grain boundary dominated diffusion in NiO, a predominant (111) orientation was found in this regime. At higher temperatures (1150-1270°C) a flat oxide surface was obtained on both textured and as-rolled Ni tapes, which indicate uniform volume diffusion during the growth. However, both (111) and (100) oriented NiO grains are nucleated initially and continue to grow competing with each other. This lateral growth occurs by competitive growth of the (001) and (111) NiO grains. Matsumoto et al.[11] obtained cube on cube (NiO on Ni) epitaxial relationship with FWHM for (111) Ni and (111) NiO typically 6-8 and 12-14 degrees respectively.

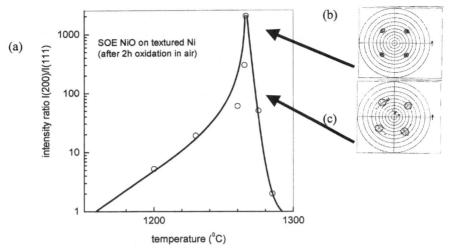

FIG. 1. (a) Effective working window for SOE NiO/Ni in air; (b) pole figure for a narrow temperature range of excellent in-plane texture, I(200)/I(111)>1000, and (c) and satisfactory in-plane texture , with I(200)/I(111)>50.

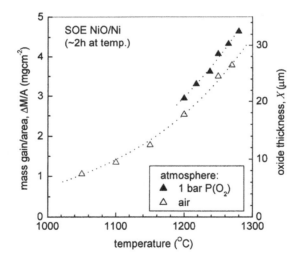

FIG. 2. Mass gain/area for NiO growth on textured Ni after ~2h oxidation in air and in pure oxygen (1 bar) atmosphere.

High Temperature Superconductors

1b) Oxygen Oxidation Studies at Imperial College: In accordance to our previous studies[12] oxidation of RABiTS pure Ni was done for a constant time of 2 or 3 hours at different temperatures (1000-1300°C) in flowing oxygen. Single phase (100)NiO was found to be formed in a narrow temperature range near 1250°C. Having achieved this 'optimum temperature', isothermal oxidation at various oxidation times were performed. Formation of single phase (100)NiO was seen at a very short oxidation time of 0.2-10 min (oxide thickness < 6μm). We lose the cube oriented NiO when the oxide thickness is 10-30 μm. Cube oriented NiO formed again after the thickness of the scales are > 30μm i.e. after oxidation for > 150 min. A plot of I(100)/I(111)NiO versus oxidation temperature is shown in Fig. 3 with an inset of in plane texture measurements (pole figures) for NiO formed at 2 min and at 150 min. The reasons for the predominance (100)NiO texture at short oxidation times and then increasing (111) texture for intermediate times followed by increasing (100)NiO plane for much longer times is: for very short oxidation times, the (100)NiO planes has the closest lattice match with the (100)Ni surface and therefore readily forms[13]. For immediate times, the oxidation is dominated by the growth of the grains, namely the close packed (111) grains. For longer times, the flux of Ni cations moving through the scales of NiO is reduced considerably because the oxide is far too thick. Then, the growth and lateral spread of grains at the surface takes over namely growth of (100)NiO grains. These grow laterally along fast <111> directions, consuming (111)NiO grains.

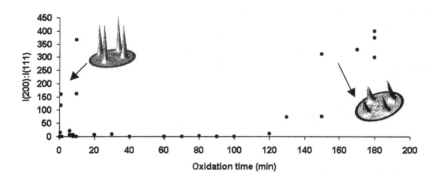

FIG. 3. I(200)/I(111)NiO versus oxidation time (min). Inset is (111) NiO pole figure showing sharp in-plane texture.

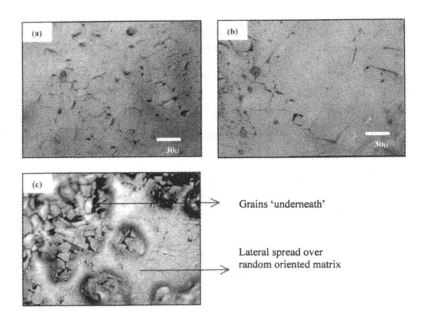

Grains 'underneath'

Lateral spread over
random oriented matrix

FIG. 4. SEM micrographs of (a) 2 min (b) 180 min (c) 180 min oxidised Ni showing abnormal grains

Surface microstructures of the textured oxide are shown in Fig. 4. Fig. 4-c shows the lateral oxide growth confirming our conclusion on the mechanism of texture evolution of the NiO. For a well textured NiO normally, large flat grains growing in a matrix of randomly oriented grains (for short oxidation time oxide) or square grains which coalesce together to form flat surface for oxide formed after 150 min of oxidation. The root mean square (rms), roughness of the oxide is roughly 150nm for the flat surface oxide. The formation of the large highly oriented grains at a very narrow temperature range and for specific time is concluded to result from the formation of abnormal grains growth of surface grains.

2. SOE of RABiTS Ni-alloys: NiO is always the first oxide to be formed on Ni-Cr alloys[14] thermal oxidation of Ni-Cr at high pO_2 (> 10^{-10} atm at 1000°C). On Ni-Fe, mixed oxides are formed at high temperature oxidation mainly $NiFe_2O_4$.

2a) SOE of Ni-10% and 13% Cr: For Ni-10%Cr and 13%Cr, oxidation was done isothermally at 1050°C as per earlier Ni-superalloy oxidation studies[15]

for different times. The out of plane texture was similar to that shown in Fig. 3 for pure Ni. For both alloys, the optimum out-of-plane texture of cube oriented NiO was obtained for shortest oxidation time of few minutes, although a strong texture still remained after longer time. For an oxidation times of 1 min < t < 40 min, the NiO layer showed both $0°$ and $45°$ in-plane orientation with respect to the underlying oxide. For an optimum oxidation time, the NiO thickness was ~ $5\mu m$ and was sharply textured both in and out-of-plane, as shown in Fig. 5. Formation of Cr_2O_3 plays an important role in the scale behaviour on Ni-Cr alloys. NiO always forms the outer oxide layer in agreement with Ahmad[14]. A fully connected Cr_2O_3 layer formed at the interface of NiO and Ni acted as a diffusion barrier to suppress further oxidation. The textured NiO is rather smooth (rms ~ 200nm) when formed at short oxidation time. A typical surface microstructure of the high quality NiO is shown in Fig. 6-a and 6-b is the cross section of oxides on Ni-Cr.

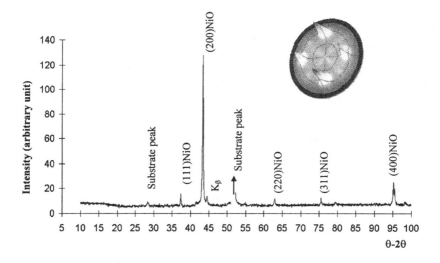

FIG. 5. XRD pattern for NiO formed on Ni10%Cr after 40 min of oxidation. Inset shows (111) pole figure of single in-plane textured NiO.

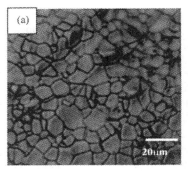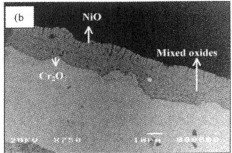

FIG. 6. (a) Optical surface micrograph of NiO formed on Ni-10%Cr. (b) Cross section of NiO on Ni-Cr.

2b) SOE of NiFe 50/50 Tape: Fig. 7 shows x-ray traces for the bare Ni-Fe substrate and a typical sample oxidised at around 850°C for 60 minutes. The oxide is cubic and the lattice parameter is calculated as $a=(8.352\pm0.010)$ Å. The important thing to note about the texture of the oxide is that it is a combination of (100) and (111), rather than being randomly oriented. The (311) and (440) peaks which have high intensities in the powder pattern do not appear in the θ-2θ scan. It is likely that the oxide is neither entirely $NiFe_2O_4$ nor Fe_3O_4, but has some intermediate composition, best described as $(Ni_xFe_{1-x})^{2+}Fe^{3+}_2O_4$. This oxide has a spinel structure. A further feature of Fig. 7 is the fact that after oxidation, a large shoulder is produced on the high angle side of the (200) NiFe peak. This indicates that there is an excess of Ni over Fe in the remaining alloy, consistent with the formation of the iron-rich oxide detailed above. Although a mixed (111) and (100) layer is evident, the degree of (100) nature varies with temperature, with no (100) oriented material at low temperatures, a maximum around 800°C, and then less at higher temperatures.

In order to study a layered structure of the oxide, a sample was intentionally made to spall after the oxidation. This is done by rapidly quenching the oxidising tape by turning off the heating current suddenly. Complete spalling happens reliably only if the oxide layer is sufficiently thick, which is probably associated with the elastic strain produced in the film due to a large volume change. As the oxide spalls in large pieces, it is possible to x-Ray both the oxide flakes and the underlying tape. These θ-2θ scans are shown in figure 1a and b and clearly show that the oxide which was removed has a (111) texture whilst the remaining tape is covered with an oxide which is (100) oriented.

This fact that a thin (100) layer remains on the surface may be utilised to consistently produce cube textured oxides on the surface of the Ni-Fe tape. If the tapes are further oxidised they maintain the cube texture, and the rocking curve of Fig. 8 demonstrates the excellent alignment.

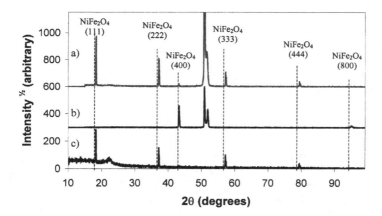

FIG. 7. The (111) oxide (c) may be completely removed from the oxidised tape (a) to leave behind a NiFe tape which has a cube textured oxide layer (b).

The atomic force microscope images of Fig. 9. indicate that the oxide layer is rough, which could be a problem for subsequent superconductor deposition. The (111) layer which is produced initially is very rough, but when removed, it leaves a much smoother cube textured oxide layer. Unfortunately, the re-oxidation process results in a rough surface again.

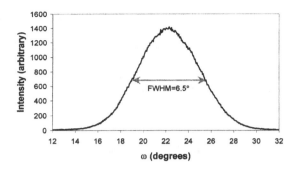

FIG. 8. Figure 2 {400} rocking curve showing excellent alignment after re-oxidation of Ni-Fe tape

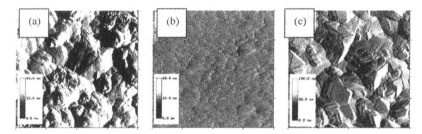

FIG. 9. Atomic force micrographs of a 10 x 10 μm² region of the surfaces of the oxide layers. a) is the (111) oxide, b) the surface (1000) oxide and (c) the thicker (100) oxide

High Temperature Superconductors

3a. Liquid phase processing of Nd₂₋ₓCeₓCuO₄ on Ni / NiO: *3a. Liquid phase processing of $Nd_{2-x}Ce_xCuO_4$ on Ni / NiO:* For the film growth of $Nd_{2-x}Ce_xCuO_4$ a precursor with a composition of 50 wt% CuO and 50 wt% (Nd_2O_3 + CeO_2) was used. The material was printed on the oxidized Ni tape and heated up to a temperature of 1300°C in order to melt this composition completely according to the phase diagram[16] .The samples were kept on this temperature for about 10 min and than fast cooled down to a temperature of about 1240°C to initiate the nucleation. After keeping it for several minutes at these temperature the sample was further cooled down slowly with a rate between 5 and 10°/min leading to further growth of $Nd_{2-x}Ce_xCuO_4$ on the substrate. Optical microscopy showed large grains with a smooth surface (Fig. 10). X-ray diffraction measurement revealed a almost completely c-axis growth of the $Nd_{2-x}Ce_xCuO_4$ and a decrease of the out-of-plane alignment to a value of about 3° FWHM in comparison to 5° of the NiO layer. However texture measurements showed a variety of in-plane orientations resulting from large grains with single crystalline quality rotated around the substrate normal. These results combined with those in section 3b below indicate that the control of the nucleation process is strongly dependent on lattice mismatch between the NiO and the $Nd_{2-x}Ce_xCuO_4$ layers.

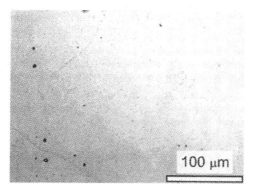

FIG. 10. Optical image of a $Nd_{2-x}Ce_xCuO_4$ film grown by liquid phase epitaxy on surface oxidized biaxial textured Ni tapes showing a flat surface

3b. Liquid phase processing of Nd₂₋ₓCeₓCuO₄ on Ni / NiO: *3b. Liquid phase processing of $Nd_{2-x}Ce_xCuO_4$ on Ni / NiO:* The $Nd_{2-x}Ce_xCuO_4$ (x=0-0.15) buffer layer on top of NiO was grown from liquid phase using a number of liquid-phase processing (LPP) methods. The first method used was the traditional top dipping LPE growth. The composition of the high temperature solution for the LPE growth was 20%Nd_2O_3:80%CuO. The liquid was prepared at 1200°C and the LPE growth was carried out at about 1150 °C. In a second method, a precursor with composition around 30%Nd_2O_3:70%CuO was screen-printed on the substrates using an organic binder and then heated up to 1200 °C to reach a single liquid state. This temperature was kept for about 20 minutes and then cooled down to room temperature at 300 °C/hr. Because of the large surface area to volume ratio and the faster evaporation of CuO than Nd_2O_3, the liquid became supersaturated rapidly, leading to the nucleation and growth of Nd_2CuO_4 on the substrates.

In the third method, a growth system similar to the traditional zone-refining furnace was used. A hydraulically pressed and sintered source rod of 40%Nd_2O_3:60%CuO was fed into the focus of an infrared beam to produce a molten zone at one end. When the increasing size of the molten zone exceeded the limit of the surface tension, the liquid dropped down onto the substrate below, which was maintained at a suitable temperature (*ca* 1100-1200 °C) for the epitaxial growth induced from the highly textured substrate. Because of the extremely good wetting property of the high-temperature cuprate solution, the liquid drops quickly spread out on the substrates.

A XRD pole figure of $Nd_{2-x}Ce_xCuO_4$ (x=0 for this sample) grown on NiO/Ni is shown in Fig.11, which shows an every good biaxial texture with FWHM=5°. There was a 45° angle between the *a/b*-axes of the Nd_2CuO_4 and NiO, indicating that the epitaxial relation between this two compounds was [100]//[110]. The lattice parameter of the *a/b*-axis was about 0.394nm for Nd_2CuO_4 and 0.417 nm for NiO, therefore, 3 unit cells of Nd_2CuO_4 could grow along two units of the [110] axis of NiO with lattice mismatch as small as 0.3%. A cross-sectional SEM of the Nd_2CuO_4/NiO/Ni structure is shown in Fig.12. The interface between the Nd_2CuO_4 and NiO was quite smooth and Nd_2CuO_4 could grow into all the groves between NiO grains. Therefore, LPP is a good method to grow $Nd_{2-x}Ce_xCuO_4$ on the relatively rough surface of NiO.

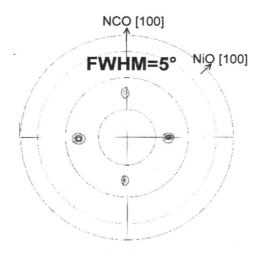

FIG. 11 Fig.11 (103) pole figure of Nd_2CuO_4 on NiO/Ni

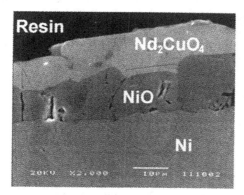

FIG. 12. SEM image of cross-section

Conclusions

We report on basic studies of a new coated conductor architectures: $Ni/NiO/Nd_{2-x}Ce_xCuO_4/YBCO$. Thin, highly textured and smooth NiO layers could be formed on pure Ni, Ni-Cr and Ni-Fe alloys. A smooth dense $Nd_{2-x}Ce_xCuO_4$ was formed on the surface of the NiO.

Acknowledgements

This research was funded by the EPSRC.

References

[1] X. Qi and J.L. MacManus-Driscoll; Journal of Crystal Growth, 213,(2000), 312

[2] X. Qi, M. Soorie, Z. Lockman, J. L. MacManus-Driscoll, Journal of Materials Research 17 (1) (2002) 1-4.

[3] E D Specht, A Goyal, D F Lee, F A List, D M Kroeger, M Paranthaman, R K Williams and D K Christen, Supercond. Sci. Technol. 11 No 10, (1998) 945-949

[4] W. Goldacker and R.Nast, private communication, (2001)

[5] A. Tuissi and E. Villa, CNR-TEMPE, Corso Promessi Sposi 29, 23900 Lecco, Italy

[6] J. Knauf, R. Semerad, W. Prusseit, B. DeBoer, J. Eickemeyer, IEEE Trans. on App. Supercon. 11, (2001), 2885 -2888

[7] Z. Lockman, X. Qi, A. Berenov, W. Goldacker, R. Nast, B. deBoer, B. Holpfapzel, & J. L. MacManus-Driscoll, Surface Oxidation of Cube-textured Ni-Cr for the formation of a NiO Buffer Layer for Superconducting Coated Conductor, (2002) to be published in Physica C

[8] P. Kofstad, in Nonstoichiometry, Diffusion, and Electrical Conductivity in Binary metal Oxides, Wiley-Interscience, N. York 1972.

[9] Z. Lockman, W. Goldacker, R. Nast, B. de Boer and J.L. MacManus-Driscoll, IEEE Trans. Appl. Supercond. 11, (2001), 3325.

[10] A. Kursumovic, R. Tomov, R. Hühne, B.A. Glowacki, J.E. Evetts, A. Tuissi, E. Villa and B. Holzapfel submitted to Physica C

[11] K. Matsumoto, S-B. Kim J-G. Wen and I. Hyrabayashi, IEEE Trans Appl. Supercon., 9, (1999), 1539.

[12] Z. Lockman, W. Goldacker, R. Nast, B. de Boer and J.L. MacManus-Driscoll, Physica C, 351, 1, (2001), 34-37

[13] N.N. Khoi, W.W. Smeltzer and J.D. Embury, " Growth and Structure of NiO on Ni Crystal Faces", Journal of Electrochemical Society, 122 (11) (1975) 1495-1502.

[14] B. Ahmad and P. Fox, Oxidation of Metals 52 (1/2) (1999) 113 – 138 and G. Calvarin, R. Molins, and A.M. Huntz, "Oxidation Mechanism of Ni-20Cr Foils and Its Relation to the Oxide- Scale Microstructure", Oxidation of Metals 53 (1/2) (2000) 25-48.

High Temperature Superconductors

[15] G. C. Wood and T. Hodgkiess, "Characteristic Scales on Pure Nickel-Chromium Alloys at 800-1200°C, Journal of the Electrochemical Society, 113 no. 4 (1966) 319-327

[16] Tanaka, I., Komai, N. and Kojima, H., Physica C, 190 (1990), 112-113

ION TEXTURING OF AMORPHOUS YTTRIA-STABILIZED ZIRCONIA TO FORM A TEMPLATE FOR YBa$_2$Cu$_3$O$_7$ DEPOSITION

Paul Berdahl, Jinping Liu, Ronald P. Reade, and Richard E. Russo
Environmental Energy Technologies Division
E. O. Lawrence Berkeley National Laboratory
Berkeley, CA 94720

ABSTRACT
We are investigating a novel technique for texturing of buffer layers, such as yttria-stabilized zirconia (YSZ), that we term ITEX (Ion TEXturing). The goal is to achieve the high-quality texture now routinely produced by oblique IBAD (Ion Beam Assisted Deposition), in a method that is faster, simpler, and less expensive. In the new ITEX method, an amorphous film is synthesized first, and subsequently bombarded with an oblique ion beam, causing the surface region of the film to crystallize. YBa$_2$Cu$_3$O$_7$ deposition on the YSZ ITEX film yields a highly c-axis oriented superconducting film that also has in-plane orientation (critical current density, 0.25 MA cm^{-2}). It is proposed that a future challenge for molecular dynamics computations is to simulate the ITEX process, in which atom or ion bombardment causes fully oriented crystals to form in an amorphous film.

INTRODUCTION

The synthesis of high-current YBa$_2$Cu$_3$O$_7$ superconducting films on metal tape substrates requires a fully oriented template/buffer layer to enable epitaxial growth of the superconducting film. In 1992 it became clear that such structures can be fabricated,[1,2] using IBAD of YSZ, as critical current density values (J_c) reached 0.6 MA cm^{-2} (77 K, 0T) at that time. Since then the YSZ IBAD process has been refined and improved at a number of laboratories, with $J_c > 1$ MA cm^{-2} now routinely achieved. While the YSZ IBAD process is now quite advanced technically, it is still not completely understood and, more importantly, appears too slow to enable the low-cost manufacturing of high-current superconducting tapes. The Achilles heel of the process is that the desired texture in the YSZ film develops gradually, by competitive grain growth, during the deposition of a layer that is thousands of unit cells thick. Still, the YSZ IBAD process has been important historically because it demonstrated that there was at least one way to fabricate high-current YBa$_2$Cu$_3$O$_7$ conductors.

In the search for better processes for coated conductor fabrication a number of innovations have been pursued. Noteworthy strategies under investigation include rolling assisted texturing of metal tapes (of Ni, and Ni alloys) followed by several epitaxial (epi) buffer layers,[3] and inclined substrate deposition (ISD).[4-6] Of particular note here is the oblique IBAD of MgO on an amorphous substrate.[7,8] In contrast with IBAD of YSZ, a high degree of texture develops in MgO layers that are only a few unit cells thick.

The work we are reporting here is derived from earlier research on YSZ IBAD[2]. It is similar to present-day MgO IBAD work since the layer directly textured by the ion beam is quite thin. We first deposit an amorphous YSZ film, and then subsequently bombard the film surface with an oblique ion beam.[9] A capping CeO_2 layer can be used or can be omitted.

EXPERIMENT

The first step in an ITEX process is the deposition of a non-crystalline, preferably amorphous, buffer layer. We used mechanically polished Haynes alloy 230 for the metal substrate and deposited an amorphous YSZ (a-YSZ) layer by pulsed laser deposition (PLD).[11] After oblique ion bombardment an epi $YBa_2Cu_3O_7$ layer was deposited by standard PLD techniques, either directly on the ITEX-YSZ or on a capping CeO_2 layer.

Some experiment details are as follows. The a-YSZ layer was synthesized under conditions of a nominal vacuum ($<10^{-6}$ torr) at room temperature. The ion bombardment step was performed with a 3 cm Kaufman ion source using 300 eV Ar ions, at 5 mA for 1.5 min. The bombardment angle was $55°$ from the surface normal. The chamber pressure was 0.1 Pa (0.8 mtorr), half Ar and half O_2. The substrate heater block temperature was 800 °C. The $YBa_2Cu_3O_7$ layer then could be deposited immediately after turning off the ion source and adjusting the O_2 pressure for standard PLD deposition.

A Staib Instruments differentially pumped 35 keV RHEED system was used to evaluate the crystallinity of the YSZ surface, which appeared comparable to images obtained after IBAD processing[9,10]. When the sample was rotated about the surface normal, the RHEED pattern changed with a fourfold symmetry, directly indicating preferential in-plane orientation. *Ex situ* AFM images showed a smooth surface for unbombarded samples, while the ion-treated surface shows grain boundary grooving, with a grain size of roughly 30 nm.

X-ray diffraction results on an overlying $YBa_2Cu_3O_7$ layer are shown in Figs. 1 and 2. In this case, no CeO_2 interlayer was employed. The YBCO layer is highly c-axis oriented, while the (103) φ scan shows weak but definite in-plane

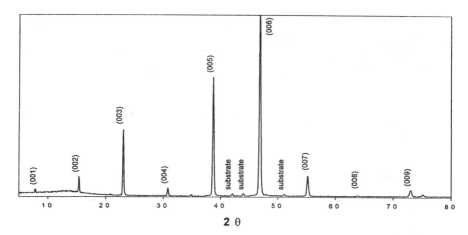

Fig. 1. X-ray diffraction from YBCO film on ITEX YSZ buffer on metal substrate, showing strong c-axis texture.

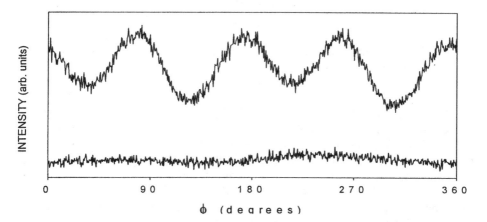

Fig. 2. (103) YBCO phi scans on ITEX YSZ (upper curve) and amorphous YSZ without ITEX (lower curve).

orientation. A sample prepared in the same way but with the ion bombardment step omitted did not show any in-plane alignment. Electrical measurements on a 50 μm bridge of the ITEX sample, as shown in Fig. 3, yielded a critical current density of 0.25 MA cm^{-2} (77K, 0T).

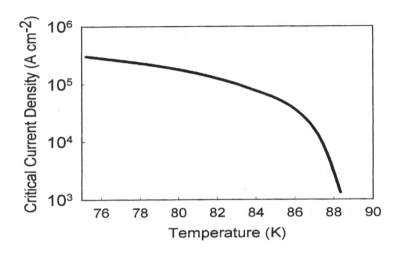

Fig. 3. Critical current density vs. temperature for YBCO film on ITEX YSZ buffer.

Since epitaxial YBa$_2$Cu$_3$O$_7$ layers on YSZ often have some misaligned grains, rotated about the surface normal by 45°, it is common to employ a PLD CeO$_2$ capping layer over the YSZ. Figure 4 shows high-resolution TEM images of the CeO$_2$/YSZ interface, demonstrating directly that excellent epitaxy is achieved, at deposition temperatures of 500 °C and above. During the ion processing of the YSZ, the penetration depth of the 300 eV Ar ions is expected to only be 1 to 2 nm. These images do not show any obvious ion damage in the interface region, which may be a consequence of annealing.

High Temperature Superconductors

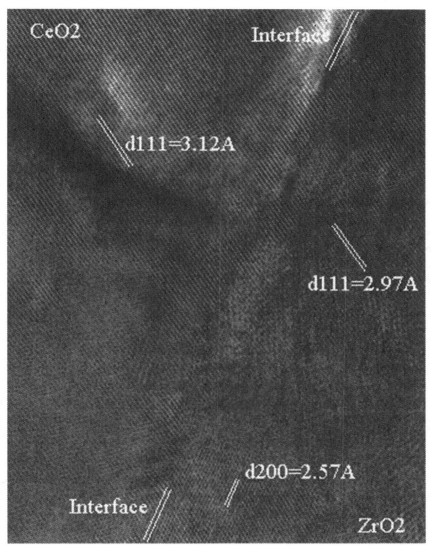

Fig. 4. High-resolution transmission electron micrograph of the interface region between the ITEX-treated YSZ (ZrO₂) buffer layer and an overlying epitaxial layer of CeO₂.

AMORPHOUS YSZ FILM SYNTHESIS

Amorphous YSZ films have not yet been widely synthesized. It is likely that all of the key experimental parameters have not yet been identified or adequately controlled. Film density, stoichiometry and deposition temperature are examples of parameters that may be critical. The chemistry at the YSZ/metal interface may also be a factor.

A number of a-YSZ films have been synthesized by PLD at room temperature. The use of oxygen partial pressure above 10^{-4} torr leads to crystallinity in the films, as judged by $\theta/2\theta$ x-ray diffraction patterns.[11] Films deposited in the 10^{-6} torr range and below are usually amorphous, but we observe some variation. That is, these films may be wholly amorphous, as judged by x-ray diffraction, or they may contain small quantities of crystallites. As an illustration, in some films a small (111) peak occurs in the precursor films. In these films, during ITEX processing (ion bombardment) at 800 °C, the (111) crystallites grow by solid-state reaction and prevent the desired (002) texture from forming. Thus, further investigation is needed to characterize a-YSZ film synthesis in more detail, to ascertain the best conditions for ITEX precursor fabrication.

PROSPECTS FOR MOLECULAR DYNAMICS SIMULATIONS

Prior molecular dynamics simulations have been performed to assess the nature of ion beam assisted deposition.[12] Also, computations have also addressed accumulation of ion damage in and the amorphization of crystalline materials.[13] Further, we note that simulations of ion beam induced epitaxial crystallization have been performed.[14] To our knowledge, however, simulations of an ITEX-type process have not yet been performed to describe the introduction of crystals into the surface region of an amorphous matrix by means of low-energy atom or ion bombardment. This is not a trivial problem. Suitable interaction potentials must be chosen so that the amorphous and crystalline phases are suitably stable. Furthermore, thermal annealing (finite temperature) effects must be incorporated into the computation, so that bombardment-induced defects can be ameliorated. Two obvious target systems are YSZ and SiO_2.[15]

If ion texturing can be simulated by molecular dynamics, we can then expect to obtain crystallite orientation distribution functions for comparison with experiment.

SUMMARY

The novel ITEX process for forming textured template layers has been used to fabricate a simple $YBa_2Cu_3O_7$/YSZ/metal structure with a critical current density of 0.25 MA cm^{-2}. While these results are encouraging, further improvements in texture are needed to eliminate more high-angle grain boundaries and thus raise the critical current. One important route to improving the technology at this point is to understand and improve the synthesis of the amorphous precursor films so that no undesired texture is obtained. Future molecular dynamics computations may prove valuable in gaining a more fundamental understanding of ion-bombardment induced crystallization, and thereby lead to further process improvements.

ACKNOWLEDGMENTS

We would like to thank Dr. Xiangyun Song for performing the electron microscopy measurements. This work was supported by the U. S. Department of Energy under Contract No. DE-AC03-76SF00098, and by a cooperative research and development agreement with American Superconductor Corp.

REFERENCES
[1] Y. Iijima, N. Tanabe, O. Kohno, and Y. Ikeno, "In-Plane Aligned YBa$_2$Cu$_3$O$_{7-x}$ Thin Films Deposited on Polycrystalline Metallic Substrates," *Appl. Phys. Lett.* **60**, 769-771 (1992).

[2] R. P. Reade, P. Berdahl, S. M. Garrison, and R. E. Russo, "Laser Deposition of Biaxially Textured Yttria-Stabilized Zirconia Buffer Layers on Polycrystalline Metallic Alloys for High Critical Current Y-Ba-Cu-O Thin Films," *Appl. Phys. Lett.* **61**, 2231-2233 (1992).

[3] C. Park, D. P. Norton, D. K. Christen, D. T. Verebelyi, R. Feenstra, J. D. Budai, A. Goyal, D. F. Lee, E. D. Specht, D. M. Kroeger, and M. Paranthaman, *IEEE Trans. Appl. Supercond.* **9**, 2276-2278 (1999).

[4] K. Hasegawa, H. Mukai, M. Konishi, J. Fujikami, K. Ohmatsu, K. Hayashi, K. Sato, S. Honjo, H. Ishii, Y. Sato, and Y. Iwata, in *Advances in Superconductivity X*, p. 607, edited by K. Osamura and I. Hirabayashi, Springer, Tokyo, 1998.

[5] M. Bauer, R. Semerad, and H. Kinder, "YBCO Films on Metal Substrates with Biaxially Aligned MgO Buffer Layers," *IEEE Trans. Appl. Supercond.* **9**, 1502-1505 (1999).

[6] B. Ma, M. Li, Y. A. Jee, R. E. Koritala, B. L. Fisher, U. Balachandran, "Inclined-Substrate Deposition of Biaxially Textured Magnesium Oxide Thin Films for YBCO Coated Conductors," *Physica C* **366**, 270-276 (2002).

[7] C. P. Wang, K. B. Do, M. R. Beasley, T. H. Geballe, and R. H. Hammond, *Appl. Phys. Lett.* **71**, 2955-2957 (1997).

[8] J. R. Groves, P. N. Arendt, S. R. Foltyn, Q. X. Jia, T. G. Holesinger, H. Kung, E. J. Peterson, R. F. DePaula, P. C. Dowden, L. Stan, and L. A.

Emmert, "High critical current density $YBa_2Cu_3O_{7-\delta}$ thick films using ion beam assisted deposition MgO bi-axially oriented template layers on nickel-based superalloy substrates," *J. Mat. Res.* **16,** 2175-2178 (2001).

[9]R. P. Reade, P. Berdahl, and R. E. Russo, "Ion-beam nanotexturing of buffer layers for near-single-crystal thin-film deposition: Application to $YBa_2Cu_3O_{7-\delta}$ superconducting films," *Appl. Phys. Lett.* **80,** 1352-1354 (2002).

[10]V. Betz, B. Holzapfel, and L. Schultz "Growth of biaxially aligned buffer layers for YBCO tapes by ion-beam assisted laser deposition and in situ RHEED texture analysis," *IEEE Trans. Appl. Supercond.* **7,** (1997).

[11]R. P. Reade, X. L. Mao, and R. E. Russo, "Characterization of Y-Ba-Cu-O thin films and yttria-stabilized zirconia intermediate layers on metal alloys grown by pulsed laser deposition," *Appl. Phys. Lett.* **59,** 739-741 (1991).

[12]L. Dong, L. A. Zepeda-Ruiz, and D. J. Srolovitz, "Sputtering and in-plane texture control during the deposition of MgO," *J. Appl. Phys* **89,** 4105-4112 (2001).

[13]H. Hensel and H. M. Urbassek, "Implantation and damage under low-energy Si self-bombardment," *Phys. Rev. B* **57,** 4756-4763 (1998).

[14]B. Weber, D. M. Stock, and K. Gartner, " MD simulations of ion beam induced epitaxial crystallization at a-Si/c-Si interfaces: interface structure and elementary processes of crystallization," *Nucl. Instr. & Meth. in Phys. Res. B* **148,** 375-380 (1999).

[15]T. Mizutani, *J. Non-Cryst. Solids* **181,** "Compositional and structural modifications of amorphous SiO_2 by low-energy ion and neutral beam irradiation," 123-134 (1995).

YBCO/YSZ/HASTELLOY SUPERCONDUCTING TAPES BY IBAD MAGNETRON DEPOSITION

S. Gnanarajan and N. Savvides
CSIRO Telecommunications and Industrial Physics
Bradield Road, WestLindfield NSW 2070
Australia

ABSRACT

Superconducting YBCO/YSZ/hastelloy tapes were fabricated by depositing epitaxial YBCO films on biaxially aligned YSZ layers on polished hastelloy substrates. YSZ buffer layers were deposited by ion beam assisted magnetron deposition. The degree of biaxial alignment in YSZ and YBCO films was determined for x-ray pole figures and phiscans. YSZ layers of different thicknesses reveal biaxial alignment develops at thickness as small as 100 nm and enhanced biaxial alignment was observed in homoepitaxial YSZ layers. The best YBCO films had biaxial alignment phiscan FWHM of 9°–10° and critical current density, $J_c(77 \text{ K}, 0 \text{ T}) = (0.9 - 1.2) \text{ MA cm}^{-2}$.

INTRODUCTION

Ion beam assisted deposition (IBAD) is well known to influence the growth and properties of thin films. Recently IBAD techniques have been used to deposit biaxially aligned oxide films on technical substrates to be used as lattice matched substrates or templates for subsequent epitaxial growth [1,2]. A major driving force for this work is to develop commercially viable and scalable process to fabricate long lengths of high-current superconductor tapes that use heat-resistant metal substrates[3-6]. One such process makes use of biaxially aligned oxide buffer layers on hastelloy metal tape to fabricate $YBa_2Cu_3O_7$ (YBCO) tapes. Suitable oxide buffer layers include yttria stabilized zirconia (YSZ) [1-3, 7], cerium oxide (CeO_2) [5, 8], magnesium oxide (MgO) [6] and praseodymium oxide Pr_6O_{11} [9]. The buffer layer serves both as a template and as a barrier to diffusion metal species during the high-temperature deposition.

The critical current density, J_c, of YBCO tapes is strongly influenced by the degree of the biaxial alignment of the buffer layer. In general the crystalline quality of the buffer layer must be similar to a single crystal so that the buffer layer has pure [001] axis alignment (normal to the substrate plane) and excellent in-plane alignment ([010] and [100] axes in the substrate plane). The in-plane alignment is usually assessed by determining the full width at half maximum (FWHM) $\Delta\phi$ of x-ray ϕ-scans [1-4]. Previous studies have shown short lengths of YBCO tapes having a high J_c (~10^6 A/cm^2) suitable for commercial applications can be produced provided the biaxially aligned oxide buffer layer is highly aligned with $\Delta\phi < 14°$ [2].

The basic IBAD technique involves a physical vapor source which provide the depositing species and a Kaufman type ion beam source which is directed towards the growing film at a specific angle of 55° to the normal of the film plane [1,2, 7-12]. The ion bombardment causes [001] axis growth normal to the film plane and also induces in plane alignment of the [010] and [100] axes.

Iijima.et.al [11] and Freyhardt et.al[12] have investigated the development of the biaxial alignment of the IBAD YSZ with increasing film thickness and showed that it is difficult to achieve $\Delta\phi < 10°$ in relatively thin buffer layers by simple IBAD techniques. For YBCO/YSZ/hastelloy tapes the YBCO film is grown epitaxially over the YSZ buffer layer. Thus the degree of biaxial alignment imparted to the YBCO film is determined primarily by the biaxial alignment at the surface of the YSZ buffer layer. We have investigated the development of the biaxial alignment of IBAD YSZ buffer layer as a function of thickness and developed a novel scheme whereby an additional YSZ layer is grown epitaxially over the IBAD-YSZ layer to form a bi-layer buffer architecture (i.e. epi-YSZ/IBAD-YSZ) on hastelly substrates. Thin films of YBCO were grown onto the buffered substrates to form YBCO tapes. Here we report measurements of the degree of biaxial alignment as function of the thickness of the IBAD and epi YSZ buffer layers and show that the epi-YSZ/IBAD-YSZ/hastelloy architecture leads to YBCO tapes that have enhanced biaxial alignment with $\Delta\phi < 14°$.

EXPERIMENTAL

The IBAD system consisted of an UHV chamber fitted with an unbalanced magnetron sputter gun to provide the physical vapour source and a 30 mm diameter Kaufman-type ion beam source to bombard the growing YSZ film with energetic Ar$^+$ ion [7, 14]. The ion beam source was operated using pure Ar gas (2 sccm) and its collimated beam was directed at an angle of 55° to the normal of the substrate plane to bombard the film with Ar$^+$. The ion energy and arrival rate ratio of ions to atoms were varied in the ranges 300-400 eVand 0.6–0.8 to achieve optimum film properties.

The planar magnetron was fitted with crystalline YSZ target and operated at 200 W rf power (13.56 MHz). The discharge gas species was Ar/O2 mixture (mass flow ratio 50:1) at a pressure of 0.1 Pa. The hastelloy substrates were held onto a high-temperature metal block substrate heater, which located opposite the target at a distance of 50 mm.

Both IBAD and epi YSZ layers were deposited under the same magnetron discharge conditions. The epi-YSZ layers were grown at a rate of 12 nm/min. During IBAD a significant amount of condensable material is re-sputtered by the ion beam. Consequently the deposition rate for the IBAD-YSZ layers was reduced to approximately 6 nm/min. Detailed investigations to assess the quality of YSZ layers and their suitability for YBCO tape production established that YSZ films free of micro cracks and pin holes that retained their integrity during subsequent processing into YBCO tapes could be produced at temperatures in the range of 300 – 500°C. In this study we deposited the IBAD-YSZ at 300°C and epi-YSZ over layer at 500°C. Note that during the epi-YSZ growth the ion beam is switched off. To enhance the adatom mobility of the condensing species we applied a small negative dc bias (20-50 V) to the hastelloy substrate.

The superconducting YBCO films (300 nm thick) were deposited onto the buffered hastelloy substrates at 730°C in a separate chamber. We used a non ion assisting unbalanced magnetron to sputter stoichiometric YBCO target in an Ar/O2 gas mixture at a pressure of 50 Pa [14]. This method generally yields YBCO films on single crystal MgO(100) and SrTiO3(100) substrates which have Jc = (2-4)x 10^6 A/cm^2 [14].

The crystalline structure and orientation of all the films were determined by x-ray θ-2θ diffraction and their in-plane biaxial alignment was determined by X-ray phi scan measurements using Philips X'pert diffraction system.

RESULTS AND DISCUSSION
IBAD YSZ films

The biaxial alignment of the IBAD YSZ films was investigated with increasing film thickness. The thickness of the films varied from 100 to 900 nm. The in-plane biaxial alignment of the films was studied using YSZ(111) phi -scans (Fig.1). The phi-scan of the film with thickness of 100 nm indicates it has biaxial alignment with FWHM of about 37.8°. Fig.1 shows the biaxial alignment increases rapidly with increasing thickness up to about 300 nm to 17° FWHM and above 300 nm, biaxial alignment of the films increases slowly to 15.4° FWHM at 900 nm. The variation of FWHM with thickness found in our experiments were similar to the results reported by Iijima.et.al [11] and Freyhadt.et.al [12], however in our films the biaxial

Figure I. YSZ(111) φ - scans with increasing IBAD YSZ film thickness

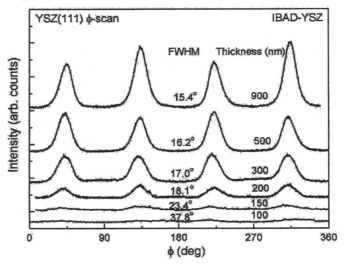

Figure II. YBCO(103) φ - scans with increasing IBAD YSZ film thickness

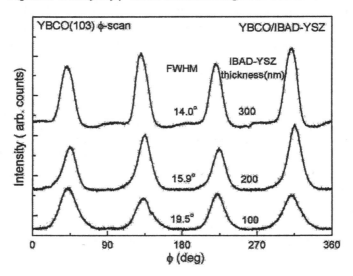

High Temperature Superconductors

alignment increases more rapidly with increasing thickness below 300 nm. The phi-scan measurements mentioned above can only be considered as the mean biaxial alignment over the whole YSZ film rather than the biaxial alignment on the surface of the film.

To determine the biaxial alignment on the surface the IBAD YSZ films, we deposited YBCO films on these substrates. The biaxial texture of the YBCO film is considered as the biaxial texture of the YSZ film on the surface. The biaxial alignment of the YBCO films were determined using YBCO(103) phi-scans. The YBCO(103) phi-scans of the YBCO films on the 100nm, 200nm and 300nm IBAD YSZ films were shown in Fig.2. The YBCO film on 100 nm thick YSZ film had YBCO(103) phi-scan FWHM of 19.5°. The biaxial alignment of the YBCO films on the IBAD YSZ films increases with increasing thickness of the IBAD YSZ films up to about 300 nm to FWHM of 14° and saturates for further increase in thickness. It is clear that there is little improvement in texture above 300 nm. Hence for films above 300nm thickness IBAD YSZ is essentially growing epitaxially.

Homoepitaxial YSZ/IBAD YSZ films

The epi YSZ layers were grown on 300 nm IBAD YSZ films. The biaxial alignment of the epi YSZ/IBAD YSZ films was investigated as a function of epi layer thickness of 50 – 400 nm. For the 50nm epi YSZ film on IBAD YSZ film, YSZ(111) phi-scan FWHM was about 16.6° (Fig.3). The biaxial alignment improved as the thickness of the epi layer increased up to 200 nm and no further improvement was observed above 200 nm. The biaxial alignment of the 200 nm epi YSZ/300nm IBAD YSZ layers have FWHM of 13.4° which is slightly better than the FWHM of the YBCO film (14°) of the thickest IBAD YSZ layer.

The YBCO films were deposited on the epi YSZ layers to determine the biaxial alignment on the surface of the epi YSZ films. Again the biaxial alignment of the YBCO films were determined by the YBCO(103) phiscans (Fig.4). As expected the biaxial alignment of the YBCO film on 50 nm epi YSZ layer is as good as the biaxial alignment of the YBCO film on the thickest IBAD YSZ film. The YBCO layer on the 200 nm epi YSZ film has FWHM of less than 9.1°. The YBCO films with biaxial alignment phiscan FWHM of 9°–10° had critical current density, $J_c(77 \text{ K}, 0 \text{ T}) = (0.9 - 1.2)$ MA cm^{-2}.

Figure III. YSZ(111) φ - scans with inceasing epi YSZ film thickness

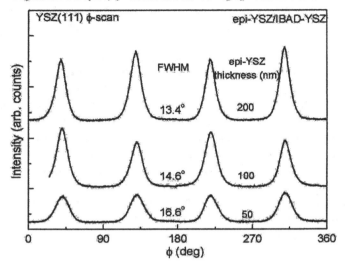

Figure IV. YBCO(103) φ - scans for increasing epi-YSZ film thickness

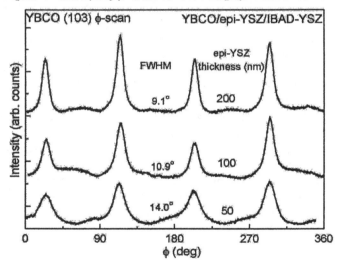

High Temperature Superconductors

Fig.5 shows FWHMs of the all the films plotted against the YSZ film thickness. It is clear there is significant increase in the biaxial alignment on the surface of the typical epi YSZ layer (FWHM = 9.1°) in comparison to the typical surface of the IBAD YSZ layers (FWHM = 14°. This process not only increases the critical current density in the YBCO film by a factor of two but also decreases the time needed to deposit the buffer layer.

Various mechanisms have been proposed to explain IBAD induced biaxial alignment in thin films. Ressler.et.al [15] mechanism is based on ion induced damage to the top layers of the film due to Ar^+ ion penetration into the top few unit cells. According to Iijima.et.al [11] after the early stage of growth the orientations in the directions of [010] and [100] develop by collaboration of selective growth of top layers and homoepitaxy on grains in the film which were already biaxially aligned. The enhancement in biaxial alignment observed in our results due to homoepitaxial growth appears to be consistent with the above models. The homoepitaxy seems to occur on the good biaxially aligned grains and it also heals the damage done to the top layer of the film by the Ar^+ ions, which leads to the enhanced biaxial alignment for the YBCO films grown on the homoepitaxial YSZ layer.

Figure V. YSZ thickness vs φ - scan FWHM for IBAD YSZ, epi YSZ and YBCO films

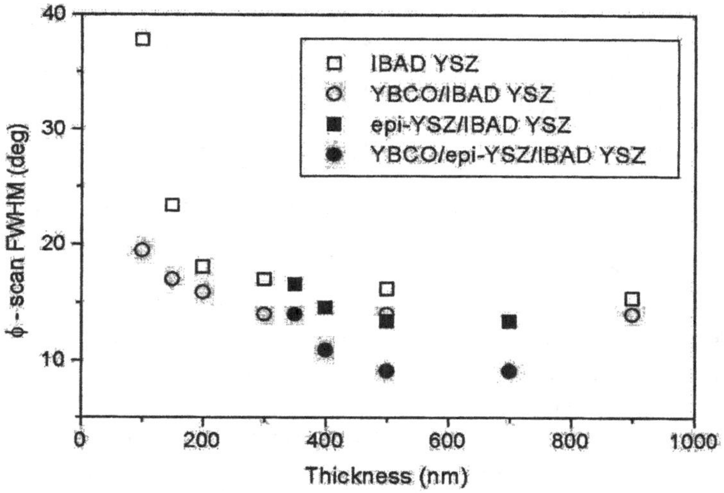

CONCLUSIONS

Biaxial alignment of ion beam assisted magnetron deposited YSZ films were investigated as a function thickness. The biaxial alignment of the YBCO film on the 100 nm thick IBAD YSZ film had FWHM of 19.5°. The improvement in the biaxial alignment of the YBCO film on IBAD YSZ film saturates to a value of about 14° at about 300 nm IBAD YSZ thickness

YBCO films grown on epi-YSZ/IBAD YSZ films develop better biaxial alignment than the YBCO films grown on IBAD YSZ films. The enhancement increases with increasing homoepitaxial layer thickness up to about 200 nm with FWHM of about 9°. The YBCO films with biaxial alignment phiscan FWHM of 9°–10° had critical current density, J_c(77 K, 0 T) = (0.9 – 1.2) MA cm^{-2}.

REFERENCES

[1] Y. Iijima, N. Tanabe,O. Kohno and Y. Ikeno, Appl. Phys. Lett. **60** (1992) 769.

[2] X. D. Wu, S. R. Foltyn, P. Arendt, J. Townsend, C. Adams, I. H. Campell, P. Tiwari, Y. Coulter and D. E. Peterson, Appl. Phys. Lett. **65** (1994) 1961.

[3] N. Savvides, S. Gnanarajan, J. Herrmann, A. Thorley, A. Katsaros and A. Molodyk, Mat. Res. Soc. Symp. Proc. **616** (2000) 199.

[4] Y. Iijima and K.Matsumoto, Supercond. Sci. Technol. **13** (2000) 68.

[5] J. E. Mathis, A. Goyal, D. F. Lee, F. A. List, M. Paranthaman, D. K. Christen, E. D. Spect, D. M. Kroger and P. M. Martin, Jpn. J. Appl. Phys **37** (1998) 1379.9

[6] T. G. Holesinger, S. R. Foltyn, P. N. Arendt, H. Kung, Q. X. Jia, R. M. Dickerson, P. C. Dowden, R. F. DePaula, J. R. Groves and J. Y. Coulter, J. Mater. Res. **15** (2000) 1110

[7] S. Gnanarjan, A. Katsaros and N. Savvides, Appl. Phys. Lett. **70** (1997) 2816.

[8] S. Gnanarajan and N. Savvides, Thin Solid Films **350** (1999) 124.

[9] C. P Wang, K. B. Do, M. R. Beasley, T. H. Geeballe and R. H Hammond, Appl. Phys. Lett. **71** (1997) 2995.

[10] V. Betz, B. Holzapfel, D. Raouser and L. Shultz, Appl. Phys. Lett. **71** (1997) 2952.

[11] Y. Iijima, M. Hosaka, N. Tanabe, N. Sadakata, T. Saitoh and O. Kohno, J. Mater. Res. **13** (1998) 3106.

[12] J. wiesmann, J. Dzick, J. Hoffmann, K. Heinemann and H. C. Freyhardt, J. Mater. Res. **13** (1998) 3149.

[13] N. Savvides and A. Katsaros, Appl. Phys. Lett. **62** (1993) 528.

[14] N. Savvides and A. Katsaros, Thin Solid Films **228** (1993) 182.

[15] K. G. Ressler, N. Sonnenberg, M. J. Cima, J. Am Ceram. Soc. **80** (1997) 2637.

RESIDUAL STRESS MEASUREMENT IN YBCO THIN FILMS*

Jae Hong Cheon and J. P. Singh

Energy Technology Division

Argonne National Laboratory

Argonne, IL 60439

ABSTRACT

Residual stress in YBCO films on Ag and Hastelloy C substrates was determined by using 3-D optical interferometry and laser scanning to measure the change in curvature radius before and after film deposition. The residual stress was obtained by appropriate analysis of curvature measurements. Consistent with residual thermal stress calculations based on the thermal expansion coefficient mismatch between the substrates and YBCO film, the measured residual stress in the YBCO film on Hastelloy C substrate was tensile, while it was compressive on the Ag substrate. The stress values measured by the two techniques were generally in good agreement, suggesting that optical interferometry and laser scanning have promise for measuring residual stresses in thin films.

INTRODUCTION

In recent years, substantial effort has been concentrated on processing Y-Ba-Cu-O (YBCO)-based coated conductors by various techniques because of their potential to provide improved electrical properties in magnetic fields [1,2]. Typical processing approaches include ion-beam-assisted deposition, rolling-assisted biaxially textured substrate method, pulsed laser deposition (PLD), e-beam co-evaporation, and metal organic deposition [3-7]. In a coated conductor system, a metallic or ceramic substrate is coated with a ceramic superconductor (i.e., YBCO) with buffer layers in between that act as diffusion barriers between the substrate and YBCO film. Various substrates such as $LaAlO_3$, Hastelloy C, nickel, and silver (Ag) have been used to process coated conductors. Because the various layers in coated conductors have different expansion coefficients, residual strains (and hence stresses) develop in different layers during processing. These stresses may cause microcracking and damage of the superconductor layer, resulting in degradation of superconducting properties and reduction in service life. Therefore, it is critical to evaluate residual stresses in coated conductors and understand the role of various processing parameters that influence these stresses.

Several techniques can be used to measure residual stresses in various thin layers (films) of coated conductors. X-ray diffraction [8] has been used to measure residual stresses in films by determining the change in lattice spacing due to the stresses. In addition, the curvature measurement technique [9] has been used to measure residual stresses in films by evaluating the change of the radius of curvature of the film as a result of the stresses. In this study, we determined the residual stresses in YBCO films on Ag and Hastelloy C substrates by a curvature measurement technique using 3-D optical interferometry and laser scanning.

MATERIALS AND METHODS

Coated conductors using YBCO films on Ag and Hastelloy C substrates were processed by the PLD technique. The substrates (1.5 cm x 1.2 cm) were polished to an approximately 1 μm finish on both sides and were subsequently subjected to an ultrasonic cleaning to obtain flat and clean surfaces. The substrate thickness (ranging from 30 μm to 200 μm) was measured by scanning electron microscopy.

The PLD process was carried out with a Lambda Physik COMPex 201 Excimer laser having a $Kr-F_2$ gas mixture as the laser medium [10]. The substrates were placed on a rotating inclined sample holder. The samples were glued to the holder with a silver paste and were heated to 700-800°C during the deposition process. A commercial YBCO target (1 inch in diameter and 0.25 inch thick) was used and the distance from target to substrate was maintained at 4-8 cm. The oxygen partial pressure in the deposition chamber was maintained in the range of 100-300 mTorr by flowing ultra-high-purity oxygen through the chamber. The energy density for deposition was estimated to be ≈1-3 J/cm². The thickness of the deposited YBCO films was measured to be ≈300 nm. Four PLD specimens were prepared for each substrate. The details of the deposition process are described in Reference [10].

Residual stresses in the YBCO films on Ag and Hastelloy C substrates were evaluated by measuring the change in the radius of curvature of the substrates before and after the film deposition. It was assumed that the curvature change in substrate also represents the change in curvature of the film in the absence of film delamination. The radius of curvature was measured by 3-D optical interferometric surface mapping [11] and laser scanning [12]. For both the optical and laser scanning methods, three to five curvature measurements in the x- and y-directions were made for each specimen (Fig. 1).

In the optical interferometric technique, the curvature of the film was determined by counting the number of interference fringes [13]. In the laser scanning technique, a laser beam was reflected off the curved surfaces of the coated substrate specimens. A rotating mirror guided the reflected laser beam to scan the curved surface in the desired direction. A position-sensitive photodetector determined the intensity change in the reflected beams from different locations resulting from the surface curvature [13]. The radius of curvature (R) was obtained from the gradient (κ) of the plot representing laser detection intensity as a function of mirror position (Fig. 2). With the measured value of κ, the radius of curvature was calculated from Equation 1 [12].

$$R = 2/(0.36442\kappa) \qquad (1)$$

Subsequently, the residual stress (σ) in the YBCO film was estimated from Equation 2 [14].

$$\sigma = E_f t_s^2 / (6\Delta R t_f) \qquad (2)$$

In Equation 2, $E_f = E/(1 - \nu)$ is the bi-axial Young's modulus of elasticity, where E is the Young's modulus of elasticity, and ν is Poisson's ratio. Also, ΔR is the difference in the radius of curvature of the film (or the substrate) before and after the film deposition, t_s is the substrate thickness, and t_f is the film thickness. The thicknesses of the substrate and film were measured by scanning electron microscopy, and the elastic moduli were taken from the literature (Table 1) [15, 16].

RESULTS AND DISCUSSION

The optical interferometric images in Fig. 3 show the curvature of the two substrates before and after YBCO deposition. These images were constructed by computer based on an interference fringe analysis and were subsequently used to determine the radius of curvature of different substrates before and after the YBCO film deposition. Table 2 shows the radii of curvature measured by optical inferometric and laser scanning techniques for Hastelloy C and Ag substrates before and after deposition of YBCO film. From Table 2, the change in the radius of curvature (ΔR) after film deposition was derived for the different substrates and techniques. The residual stresses in the YBCO film were then estimated by substituting the measured values of ΔR, t_s, and t_f into Equation 2. These measured stresses are plotted in Fig. 4. We also estimated the residual stresses (σ) due to the thermal expansion mismatch by analytical approach, using Equation 3 [17].

$$\sigma = \frac{E}{(1-\nu)}\Delta\alpha\Delta T \qquad (3)$$

where E and ν are the elastic modulus and Poisson's ratio of the film, respectively; $\Delta\alpha$ is the difference in thermal expansion coefficients of the substrate and the film; and ΔT is the difference between the room and processing temperature. For comparison, these calculated values of the residual stresses are shown in Figure 4.

As seen in Figure 4, there is a reasonably good agreement between the stress values measured by the optical and laser techniques. The residual stresses in the YBCO film on Hastelloy C substrate were measured to be 47 and 76 MPa by the optical and laser methods, respectively. The corresponding stresses in the YBCO film on Ag substrate were measured to be 1024 and 1078 MPa. However, these measured values differed from the residual stresses calculated from the thermal expansion coefficient mismatch with Equation 3: 160 MPa for Hastelloy C substrate and 882 MPa for Ag substrate.

We reached three important conclusions from our results. First, the residual stresses in YBCO film on Hastelloy C substrate are tensile, while the stresses are compressive for the Ag substrate. This is inferred from the expansion coefficient of YBCO being larger than that of Hastelloy C and smaller than that of Ag (Table 1). Second, the measured tensile stress in the YBCO film on Hastelloy C substrate is slightly smaller than the calculated value, whereas the measured compressive stress in the film on Ag substrate is slightly larger than the calculated value (Fig. 4). This difference is believed to be due to the presence of an internal compressive stress that developed during the PLD processing due to Ar ion-peening effects. Verification of this hypothesis is underway. Third, the residual stresses measured by the two techniques with the same substrate are the same order of magnitude (Fig. 4), indicating that these techniques show considerable promise for stress measurements in thin YBCO films.

ACKNOWLEDGMENT

This work supported by the U.S. Department of Energy (DOE), Energy Efficiency and Renewable Energy, as a part of a DOE program to develop electric power technology, under Contract W-31-109-Eng-38. We thank Meiya Li and Beihai Ma for providing coated-conductor specimens with YBCO film.

REFERENCES

1. A. Goyal et al., Appl. Phys. Lett. 69 (1996) 1795.

High Temperature Superconductors

2. D. P. Norton et al., Science 274 (1996) 755.

3. Y. Iijima, N. Tanabe, O. Kono, and Y. Ikeno, Appl. Phys. Lett. 60 (1992) 759.

4. H. Qing, D. K. Christen, J. D. Budai, E. D. Specht, D. F. Lee, A. Goyal, D. P. Norton, M. Paranthaman, F. A. List, and D. M. Kroeger, Physica C 275 (1997) 155.

5. K. Fujino, N. Yoshida, N. Hayashi, S. Okuda, T. Hara, and H. Ishii, Advances in Superconductivity VI, Springer, Tokyo, 1994, p. 763.

6. H. Kim, J. Yoo, K. Jung, J. Lee, S. Oh, and D. Youm, Supercond. Sci. Technol. 13 (2000) 995.

7. P. C. McIntyre, M. J . Cima, and A. Roshkon, J. Appl. Phys. 77 (1995) 5263.

8. H. P. Klug and L. E. Alexander, X-ray Diffraction Procedures for Polycrystalline and Amorphous Materials, 2nd edn, Wiley, New York , 1974, ch. 11.

9. P. A. Flinn, MRS Proceedings 130 (1989) 41.

10. B. Ma, M. Li, Y. A. Jee, R. E. Koritala, B. L. Fisher, and U. Balachandran Physica C 366 (2002) 270.

11. R. E. Cuthrell, D.M. Mattox, C. R. Peeples, P.L. Dreike, and K.P. Lamppa, J. Vac. Sci. Technol., A 6 (1988) 2914.

12. P. A. Flinn, D. S. Gardner, and W. D. Nix, IEEE Transactions on Electronic Devices 34 (1987) 689.

13. W. D. Nix, Metallurgical Transactions A 20A (1989) 2217.

14. G. G. Stoney, Proceedings of the Royal Society, London, Series A 82 (1909) 172.

15. Goodfellow Cambridge Limited, Huntingdon, England, "Technical Data : Ag and Hastelloy C" www.goodfellow.com (2002).

16. A. S. Raynes, S. W. Freiman, F. W. Gayle, and D. L. Kaiser, Journal of Applied Physics 70 (1991) 5254.

17. D. Burgreen, Elements of Thermal Stress Analysis, C. P. Press, New York, 1971, p. 462.

Table 1. Mechanical and physical properties of Hastelloy and Ag substrates and YBCO

	Hastelloy C[15]	Ag[15]	YBCO[16]
Melting Temperature (°C)	1300	961.9	1270
Thermal Expansion Coefficient (10^{-6} K^{-1})	12.5	19.4	13.4
Elastic Modulus (GPa)	170-220	100	135-157

Table 2. Radius of curvature measured by optical interferometric and laser scanning methods for Hastelloy and Ag substrates before and after deposition of YBCO film.

	Optical Interferometry		Laser Scanning	
	Hastelloy C	Ag	Hastelloy C	Ag
Radius of Curvature before Deposition (μm)	2.07e+05	4.92e+05	2.18e+05	5.12e+05
Radius of Curvature after Deposition (μm)	2.06e+05	6.02e+05	2.16e+05	6.46e+05

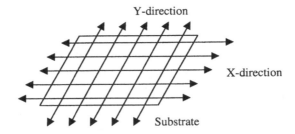

Y-direction

X-direction

Substrate

Fig. 1. Schematic showing x- and y-directions on the measured curved substrates.

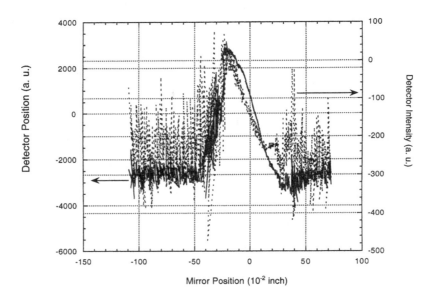

Fig. 2. Typical variation of laser detection intensity as a function of mirror position.

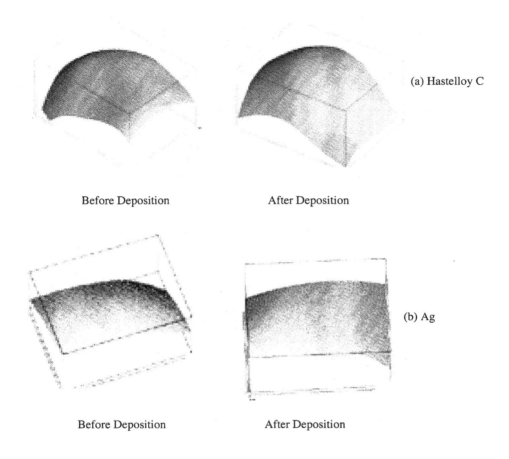

(a) Hastelloy C

Before Deposition   After Deposition

(b) Ag

Before Deposition   After Deposition

Fig. 3. Optical interferometric images showing the curvature of substrates before and after YBCO deposition on (a) Hastelloy C and (b) Ag substrates.

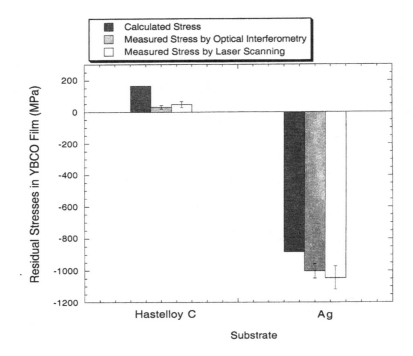

Fig. 4. The measured and calculated (based on thermal mismatch) residual stresses in YBCO films on Hastelloy C and Ag substrates.

High Temperature Superconductors

GROWTH KINETICS AND TEXTURE OF $(Nd,Ce)_2CuO_4$ / NiO BUFFERS ON Ni-BASED RABITS

Z. Lockman, X. Qi and
J.L. MacManus-Driscoll
Department of Materials
Imperial College of Science Technology
and Medicine
Prince Consort Road
London, SW7 2BP, UK

W. Goldacker and R. Nast
Forschungszentrum Karlsruhe
Technik und Umwelt
Hermann-v-Helmholtz-Platz 1
76344 Eggenstein-
Leopoldshafen
Karlsruhe, Germany

B. deBoer
Institut für Festkörper-und
Werkstoffforschung, Helmholtzstraße 20
Dresden, Germany

A. Kursumovic, R. Tomov[*],
R. Hühne, J.E. Evetts and
B.A. Glowacki
Department of Materials Science
and Metallurgy
University of Cambridge
Pembroke Street
Cambridge CB2 3QZ, UK

R. Major
Carpenter Technology (UK) Ltd,
Napier Way,
Crawley, RH10 2RB, UK

ABSTRACT

We report on basic studies of a new coated conductor architecture: $Ni/NiO/Nd_{2-x}Ce_xCuO_4/YBCO$. Various Ni compositions were explored including pure Ni, Ni-Cr, and Ni-Fe, and a new liquid phase processing (LPP) technique to make $Nd_{2-x}Ce_xCuO_4$ (NCCO) buffer layers was utilized. Fundamental studies of surface oxidation epitaxy of Ni and Ni-alloys were undertaken. It was shown that a very thin NiO layer can be formed on pure Ni with excellent texture and low

[*] corresponding author: rit21@cam.ac.uk

surface roughness (~80nm). For Ni-Cr, ~5micron thick NiO surface layer with excellent texture and surface finish (roughness ~200nm) can be produced. For Ni-Fe, a smooth, thin Ni_2FeO_4 was formed underneath a surface (111) oxide, which was then successfully removed. The (NCCO) layer was fabricated by LPP to have a surface roughness of a few 10's of nm.

INTRODUCTION

Since the discovery of high T_c superconductors in 1986, there has been an enormous effort worldwide to establish a scaleable method for the fabrication of coated conductors on metallic tapes. Although short length samples with J_c over 1 MA/cm^2 have been prepared by a number of vapor deposition methods, e.g. pulsed laser deposition (PLD), they are so far failed to produce long length conductors at a reasonable cost. An alternative route to vapor deposition methods is LPE, which has been widely used for the growth of semiconductor and garnet thin films during 1970s and 1980s. The advantages of LPE compared are the very fast growth rate, typically 1-10 μm/min, and the capability to grow thick films without degrading the structural perfection[i]. However, LPE growth of REBCO thick films on metallic substrate is much more difficult. The high growth temperature largely increases the risk of reaction between the substrate and the film. The solubility of RE's in the BaO:CuO solution is very low, therefore nucleation and growth is more difficult than the traditional LPE of semiconductors and garnets. Compared to PLD and other vapor deposition methods, LPE has a much lower available supersaturation and hence a much smaller driving force for nucleation. As a consequence, a close lattice matched substrate is required. At present, most of the buffer layers were developed for the use in the vapor deposition methods and therefore are not suitable for LPE growth. Nd_2CuO_4, a tetragonal crystal with a=b≈0.394 nm, is a good potential buffer for LPE. As a part of the RE-Ba-Cu-O system[2] Nd_2CuO_4 neither poisons superconductivity nor reacts strongly with the barium cuprate solution at high temperature. It has a fairly close lattice match and thermal expansion match to REBCO and NiO. In this paper, we report on the growth of $Nd_{2-x}Ce_xCuO_4$ thick films on surface oxidised Ni-alloy substrates by scaleable methods. We also report on the basics of the surface oxidation of different Ni-alloys.

EXPERIMENTAL

100μm thick pure nickel, Ni-10%Cr, Ni-13%Cr and Ni-Fe were prepared by RABiTS method[3]. A summary of all Ni and Ni-alloys used for the SOE is shown in Table I. The tape surface was finished by a standard electro-polishing or chemical etching method (5%HF). The foils were then ultrasonically cleaned in acetone and methanol.

Table I Summary of Ni and Ni-based tapes and SOE conditions

Foils	thickness (μm)	anneal.temp. (°C)	anneal.time (min)	atmosphere	heating source	
Ni	80-100	1000-1300	0.2-180	O_2/air	radiant	[4, 5]
Ni-10%Cr	100	1050	0.2-180	air	radiant	[4, 8]
Ni-13%Cr	100	1050	0.2-180	air	radiant	[6]
Ni-50%Fe	13-25	500-1100	0.2-60	Ar/O_2	ohmic	[7, 9]

The oxidation conditions of different Ni and Ni-based alloys are given in Table I. The progress of oxygenation was monitored by mass and thickness increase. Mass was measured by a precise microbalance after each partial oxidation. Thickness was measured both by a micrometer and using secondary electron microscope (SEM) or optical microscope. The texture analysis was carried out by X-ray diffraction techniques: θ-2θ scans, rocking curves and pole figures. Optical microscopy was used to monitor the development of microstructure during film growth.

In order to systematically study high temperature oxidation process of Ni-Fe, a number of Ni-Fe tapes were oxidized using O_2 and Ar/O_2 mixtures in a UHV chamber, tapes being heated resistively using a direct current. There are two methods of carrying out the oxidation, which turn out to show important differences. One way is to preheat the tape in vacuum, then introduce the Ar/O_2 gas mixture, or alternatively to introduce the required pressure of gas into the chamber, then ramp up the current. Below around 950°C a continuous oxide is grown, which is dark gray in appearance at sufficient thickness. At higher temperatures however, macroscopic oxide grains become visible on the surface due to secondary re-crystallization and abnormal grain growth.[9]

RESULTS AND DISCUSSIONS

Pure Nickel: At low temperatures <1000°C, NiO film growth is influenced by the grain boundaries in the "parent" Ni-phase which has an average grain size of about 30 μm. The growth is faster around grain boundaries creating humps on the NiO surface. The grain size of the oxide film is in the order of 1 μm. At higher temperatures, the grain size of the NiO film increases being >20 μm at 1250°C after 2h on textured Ni. The NiO growth has a specific flat surface that is a consequence of predominantly volume diffusion of growing species. At very high temperatures, >1300°C, highly concave growth around grain boundaries makes the film very rough[10].

Evolution of texture with temperature: A working window for SOE NiO/Ni was established from the ratio of XRD intensities I(200)/I(111)NiO. For samples oxidized for approximately two hours in air at different temperatures the results are shown in Fig. 1-a. As can be seen, there is a pronounced peak around 1260°C, at somewhat higher temperature than found in oxygen oxidation presumably due to using air rather than 100% oxygen[11]. As expected, a sharp in-plane alignment is achieved at the "peak" of the working window, Fig.1-b. In Fig.1-c the pole figure shows that an acceptable in-plane orientation can be achieved in the range of $\Delta T < 20°$. It should be noted, that at temperatures >1270°C, increased (220) peak intensities are present that add-up to overall deterioration of the NiO/Ni epitaxy.

The sample mass per unit area, $\Delta m/A$, for NiO/Ni films grown for two hours, both in pure oxygen and in air, is shown in Fig. 2. The sample mass increases due to formation of NiO film in the same manner as oxide thickness. These data correspond to data on epitaxial development shown in Fig. 1. The oxide film thickness was calculated here, but corresponds closely to experimentally measured values that were performed on some of the samples. As can be seen, an almost perfect epitaxy that was achieved at around 1260°C corresponds to ~25 µm thick NiO . The kinetics of NiO film growth in air at different temperatures versus time was investigated in more detail in Ref[12]. This kinetics was found to obey expected parabolic ($\sqrt{t}$) law. The activation energy calculated from the kinetic data was 220 kJ/mol for temperatures above 1100°C, corresponding to that for volume diffusion of Ni in NiO in excellent agreement with literature data. There are three distinct temperature regimes affecting the growth. At lower temperatures (<1000°C), due to boundary dominated diffusion a predominant (111)NiO orientation was found. At higher temperatures (1150-1270°C) a flat oxide surface was obtained on both textured and as-rolled Ni tapes, indicating uniform volume diffusion during the growth. However, both (111) and (100) oriented NiO grains are nucleated initially and continue to grow competing with each other. Thus lateral growth occurs by competitive growth of the (001) and (111) NiO grains.

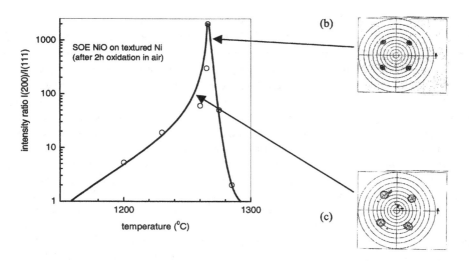

FIG. 1. (a) Effective working window for SOE NiO/Ni in air; (b) pole figure for a narrow temperature range of excellent in-plane texture, I(200)/I(111)>1000, and (c) satisfactory in-plane texture , with I(200)/I(111)>50.

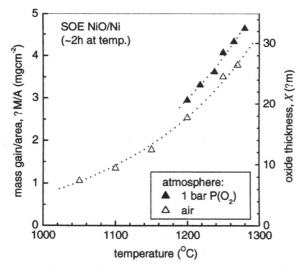

FIG. 2. Mass gain/area for NiO growth on textured Ni after ~2h oxidation in air and in pure oxygen (1 bar) atmosphere.

Evolution of texture with time: Formation of single phase (100)NiO was seen at a very short oxidation time of 0.2-10 min (oxide thickness < 6μm). We lose the cube oriented NiO when the oxide thickness is 10-30 μm. Cube oriented NiO formed again after the thickness of the scales are > 30μm i.e. after oxidation for > 150 min. A plot of I(100)/I(111)NiO versus oxidation temperature is shown in Fig. 3 with an inset of in plane texture measurements (pole figures) for NiO formed at 2 min and at 150 min. The reasons for the predominance (100)NiO texture at short oxidation times and then increasing (111) texture for intermediate times followed by increasing (100)NiO plane for much longer times is: (i) for very short oxidation times, the (100)NiO planes has the closest lattice match with the (100)Ni surface and therefore readily forms[13]; (ii) for intermediate times, the oxidation is dominated by the growth of the grains, namely the stable (100) and (111) grains; (iii) for longer times, the flux of Ni cations through the scales of NiO is reduced considerably because the oxide thickness. Then, the growth and lateral spread of grains at the surface takes over, namely growth of (100)NiO grains. These grow laterally along fast <111> directions, consuming (111)NiO grains.

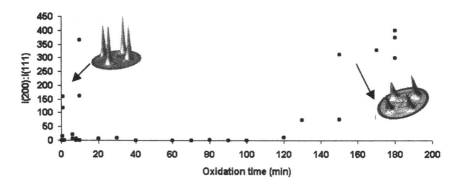

FIG. 3. I(200)/I(111)NiO versus oxidation time (min). Inset is (111) NiO pole figure showing sharp in-plane texture.

Surface microstructures of the textured oxide grown for very shor (2 mni) and long times (180 min) are shown in Fig. 4a and b respectively. Fig. 4-c shows regions of the lateral oxide growth confirming our conclusion about the mechanism of texture evolution in the NiO. The root mean square (rms), roughness of the oxide is roughly 150nm for the flat surface oxide. The formation of the large highly oriented grains at a very narrow temperature range and for specific time is concluded to result from the competitive growth of the surface grains.

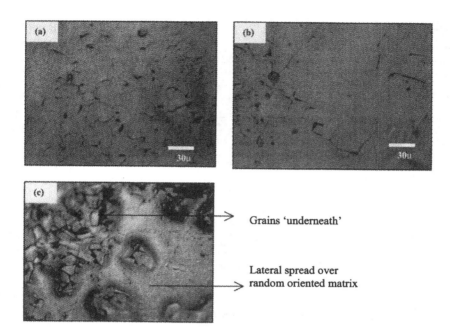

Grains 'underneath'

Lateral spread over
random oriented matrix

FIG. 4 SEM.micrographs of oxidized Ni (a) 2 min, (b) 180 min showing disrupted growth..

Ni-10%Cr and Ni-13%Cr: For Ni-10%Cr and 13%Cr, oxidation was done isothermally at 1050°C as per earlier Ni-superalloy oxidation studies[14]. The out-of-plane texture was similar to that shown in Fig. 3 for pure Ni. For both alloys, the optimum out-of-plane texture of cube oriented NiO was obtained for the shortest oxidation time of a few minutes, although a strong texture still remained after longer times. For an oxidation times of 1 min< t <40 min, the NiO layer showed both 0° and 45° in-plane orientations with respect to the substrate. For an optimum oxidation time, ~ 5μm thick NiO sharply textured both in and out-of-plane was produced, as shown in Fig. 5. Cr_2O_3 plays an important role in the scale behaviour on Ni-Cr alloys. NiO always forms the outer oxide layer in agreement with Ahmad[15]. A fully connected Cr_2O_3 layer formed at the interface of NiO and Ni acted as a diffusion barrier to suppress further oxidation. The textured NiO is rather smooth (rms ~ 200nm) when formed for short oxidation times. A typical surface microstructure of the high quality NiO is shown in Fig. 6-a, and 6-b is the cross section of oxides on Ni-Cr.

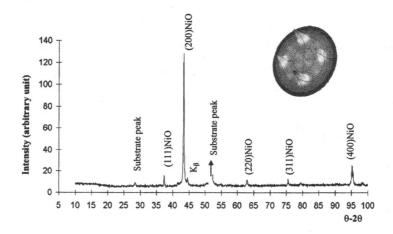

FIG. 5. XRD pattern for NiO formed on Ni10%Cr after 40 min of oxidation

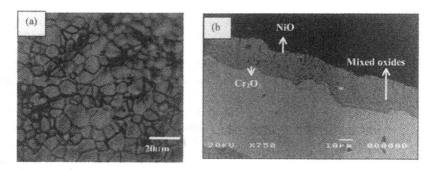

FIG. 6 (a) Optical surface micrograph of NiO formed on Ni-10%Cr. (b) Cross-section of NiO on Ni-Cr.

Ni-50%Fe tape: Fig. 7 shows x-ray traces for the bare Ni-Fe substrate and a typical sample oxidised at around 850°C for 60 minutes. The oxide is cubic and the lattice parameter is calculated as $a= 8.352$ Å. The important thing to note about the texture of the oxide is that it is a combination of (100) and (111), rather than being randomly oriented. The (311) and (440) peaks, which have high

High Temperature Superconductors

intensities in the powder pattern, do not appear in the θ-2θ scan. It is likely that the oxide is neither entirely $NiFe_2O_4$ nor Fe_3O_4, but has some intermediate composition, best described as $(Ni_xFe_{1-x})^{2+}Fe^{3+}_2O_4$. This oxide has a spinel structure. A further feature of Fig. 7 is the fact that after oxidation, a large shoulder is produced on the high angle side of the (200) NiFe peak. This indicates that there is an excess of Ni over Fe in the remaining alloy, consistent with the formation of the iron -ich oxide detailed above. Although a mixed (111) and (100) layer is evident, the degree of (100) nature varies with temperature, with no (100) oriented material at low temperatures, a maximum around 800°C, and then less at higher temperatures.

In order to study a layered structure of the oxide, a sample was intentionally made to spall after the oxidation. This is done by rapidly quenching the oxidizing tape by turning off the heating current suddenly. Complete spalling happens reliably only if the oxide layer is sufficiently thick, which is probably associated with the elastic strain produced in the film due to a large volume change. As the oxide spalled in large pieces, it was possible to take x-ray spectra from both the oxide flakes and the underlying tape. These θ-2θ scans are shown in figure 1b and c and clearly show that the oxide which was removed has a (111) texture whilst the remaining tape is covered with an oxide which is (100) oriented.

FIG. 7 X-ray diffraction of surface oxidized Ni-50%Fe tapes. The (111) oxide (c)

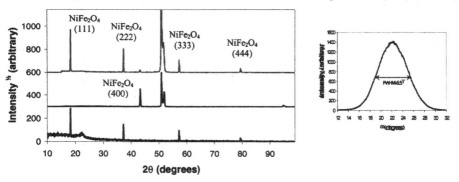

may be completely removed from the oxidized tape (a) to leave behind a NiFe tape, which has a cube textured oxide layer (b). (The inset on the right shows {400}-rocking curve with excellent grain alignment after re-oxidation of Ni-Fe tape).

This fact that a thin (100) oriented layer remains on the surface may be utilized to consistently produce cube-textured oxides on the surface of the Ni-Fe tape. If the tapes are further oxidized they maintain the cube texture, and the rocking curve of Fig. 7 demonstrates the excellent alignment. The atomic force microscope images of Fig. 8 indicate that the oxide layer is rough. The (111) layer, which is produced initially, is very rough, but when removed, it leaves a much smoother cube textured oxide layer. Unfortunately, the re-oxidation process results in a rough surface again.

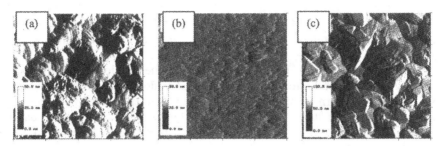

FIG. 8. Atomic force micrographs of a 10 x 10 μm^2 region of the surfaces of the oxide layers: a) the (111) oxide, b) the surface (100) oxide and (c) the thicker (100) oxide obtained after re-oxidation.

Liquid phase processing of $Nd_{2-x}Ce_xCuO_4$ on Ni / NiO: For the film growth of $Nd_{2-x}Ce_xCuO_4$, a precursor with a composition of 50 wt% CuO and 50 wt% (Nd_2O_3 + CeO_2) was used. The material was printed on the oxidized Ni tape and heated up to a temperature of 1300°C in order to melt this composition completely. The samples were kept at this temperature for about 10 min and than fast cooled down to a temperature of about 1240°C to initiate the nucleation. After keeping it for several minutes the samples were further cooled down slowly with a rate between 5 and 10°/min leading to the growth of $Nd_{2-x}Ce_xCuO_4$ on the substrate. Optical microscopy showed large grains with a smooth surface (Fig. 9). X-ray diffraction measurements revealed almost complete c-axis oriented growth of the $Nd_{2-x}Ce_xCuO_4$ and an improvement in the out-of-plane alignment - 3° FWHM in comparison to 5° of the NiO layer. However texture measurements showed a variety of in-plane orientations resulting from the large grains with single crystalline quality rotated around the substrate normal. These results combined with those in next section indicate that the control of the nucleation process is strongly dependent on lattice mismatch between the (100)NiO and the $Nd_{2-x}Ce_xCuO_4$ layers.

High Temperature Superconductors

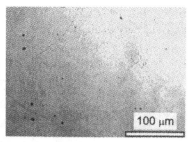

FIG. 9. Optical image of a Nd$_{2-x}$Ce$_x$CuO$_4$ film grown by liquid phase epitaxy on surface oxidized biaxial textured Ni tapes showing a flat surface

Liquid phase processing of Nd$_2$CuO$_4$ on Ni/NiO: The Nd$_2$CuO$_4$ buffer layer on top of NiO was grown from liquid phase using a number of liquid-phase processing (LPP) methods. The *first method* used was the traditional top dipping LPE growth. The composition of the high temperature solution for the LPE growth was 20%Nd$_2$O$_3$: 80%CuO. The liquid was prepared at 1200°C and the LPE growth was carried out at about 1150 °C. In a *second method*, a precursor with composition around 30%Nd$_2$O$_3$: 70%CuO was screen-printed on the substrates using an organic binder and then heated up to 1200 °C to reach a single liquid state. This temperature was kept for about 20 minutes and then cooled down to room temperature at 300 °C/hr. Because of the large surface area to volume ratio and the faster evaporation of CuO than Nd$_2$O$_3$, the liquid became supersaturated rapidly, leading to the nucleation and growth of Nd$_2$CuO$_4$ on the substrates.In the *third method*, a growth system similar to the traditional zone-refining furnace was used. A sintered source rod of 40%Nd$_2$O$_3$: 60%CuO was fed into the focus of an infrared beam to produce a molten zone at one end. When the increasing size of the molten zone exceeded the limit of the surface tension, the liquid dropped down onto the substrate below, which was maintained at a suitable temperature (1100-1200°C) for the epitaxial growth induced from the highly textured substrate. Because of the extremely good wetting property of the high-temperature cuprate solution, the liquid drops quickly spread out on the substrates.

An XRD pole figure of Nd$_2$CuO$_4$ (x=0 for this sample) grown on NiO/Ni is shown in Fig.10. In the absence of Ce the in-plane texture of the Nd$_2$CuO$_4$ layer appeared to be significantly improved. There is a 45°-angle rotation between the *a/b*-axes of the Nd$_2$CuO$_4$ and NiO, indicating that the epitaxial relation between these two compounds was [100]//[110]. The measured lattice parameter a=0.394nm for Nd$_2$CuO$_4$ and 0.417 nm for NiO. Therefore, 3 unit cells of Nd$_2$CuO$_4$ align with two units of the [110] axis of NiO. The lattice mismatch is only 0.3%. A cross-sectional SEM of the Nd$_2$.CuO$_4$/NiO/Ni structure is shown in

Fig.11. The interface between the Nd$_2$CuO$_4$ and NiO was quite smooth and Nd$_2$CuO$_4$ grew into all the groves between NiO grains.

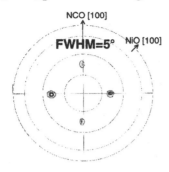

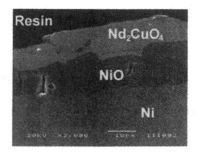

FIG. 10 (103) pole figure of Nd$_2$CuO$_4$ on NiO/Ni

FIG. 11 SEM cross-section image of Ni/NiO/Nd$_{2-x}$Ce$_x$CuO$_4$ architecture.

CONCLUSIONS

We report on basic studies of a new coated conductor architectures: Ni/NiO/Nd$_{2-x}$Ce$_x$CuO$_4$/YBCO. Thin, highly textured and relatively smooth NiO layers could be formed on pure Ni, Ni-Cr and Ni-Fe alloys. A smooth dense well-textured Nd$_{2-x}$Ce$_x$CuO$_4$ (x=0-0.15) was formed on the surface of the NiO by LPP methods.

ACKNOWLEDGEMENTS

Presented research was conducted jointly between Department of Materials Science and Metallurgy at University of Cambridge and Department of Materials in Imperial College in London and was funded by the EPSRC. The support from ABB-Research, Zurich is also gratefully acknowledged.

REFERENCES:

[1] X. Qi and J.L. MacManus-Driscoll, "Liquid-phase epitaxial growth of REBa2Cu3O7-d (RE=Y,Yb,Er) thick films at reduced temperatures", *Journal of Crystal Growth*, **213** 312-318 (2000).

[2] X. Qi, M. Soorie, Z. Lockman and J.L. MacManus-Driscoll, "Rapid growth of $Nd_{2-x}Ce_xCuO_4$ thick films as a buffer for the growth of rare-earth barium cuprate-coated conductors", *Journal of Materials Research*, **17** [1] 1-4 (2002).

[3] E.D Specht, A. Goyal, D.F Lee, F.A List, D.M Kroeger, M. Paranthaman, R. K Williams and D.K Christen, "Cube-textured nickel substrates for high-temperature superconductors ", *Supercond. Sci. Technol.* **11** (10) 945-949 (1998).

[4] R. Nast, W. Goldacker B. Obst and G.Linker, "Recrystallisation kinetics of biaxially textured Ni substrates for YBCO coated conductors," *Inst. Phys. Conf. Ser.,* **167** 411-414 (2000).

[5] A. Tuissi and E. Villa, CNR-TEMPE, Corso Promessi Sposi 29 23900 Lecco, Italy

[6] J. Knauf, R. Semerad, W. Prusseit, B. DeBoer and J. Eickemeyer, "YBaCuO – deposition on metal tape substrates", *IEEE Trans. on App. Supercon.*, **11** 2885 - 2888 (2001).

[7] B.A. Glowacki, M. Vickers, N. Rutter, E. Maher, F. Pasotti, A. Baldini and R. Major "Texture development in long lengths of NiFe tapes for superconducting coated conductor', *Journal of Materials Science*, **37** 157- (2002).

[8] Z. Lockman, X. Qi, A. Berenov, W. Goldacker, R. Nast, B. deBoer, B. Holpfapzel, and J. L. MacManus-Driscoll, "Surface oxidation of cube-textured Ni-Cr for the formation of a NiO Buffer Layer for Superconducting Coated Conductor", to be published in *Physica C.*

[9] N.A. Rutter, B.A. Glowacki, J.H. Durrell, J.E. Evetts, H. te Lintelo, R. De Gryse and J. Denul, "Formation of native cube textured oxide on a flexible NiFe tape substrate for coated conductor application," *Inst. Phys. Conf. Ser.,* **167** 407-410 (2000).

[10] P. Kofstad, in Nonstoichiometry, "Diffusion, and electrical conductivity in binary metal oxides", Wiley-Interscience, N. York 1972.

[11] Z. Lockman, W. Goldacker, R. Nast, B. de Boer and J.L. MacManus-Driscoll, "Study of thermal oxidation of NiO on pure Ni, Ni-10%Cr and Ni-V9% tapes", *IEEE Trans. Appl. Supercond.* **11** 3325 (2001).

[12] A.Kursumovic, R. Tomov, R. Hühne, B.A. Glowacki, J.E. Evetts, A. Tuissi, E. Villa and B. Holzapfel , submitted to *Physica C*

[13] N.N. Khoi, W.W. Smeltzer and J.D. Embury, " Growth and structure of NiO on Ni crystal faces", *Journal of Electrochemical Society*, **122** (11) 1495-1502 (1975).

[14] G.C.Wood and T.Hodgkiess, "Characteristic scales on pure Nickel-Chromium Alloys at 800-1200°C", *Journal of the Electrochemical Society*, **113**(4) 319-327 (1966).

[15] B.Ahmad and P.Fox, "STEM analysis of the transient oxidation of a Ni-20Cr alloy at high temperatures ", *Oxidation of metals* **52**(1/2) 113-138 (1999).

CA DOPING OF YBCO THIN FILMS

A. Berenov, J.L. MacManus-Driscoll, D. MacPhail
aDepartment of Materials, Imperial College, Prince Consort Road, London, SW7 2BP, UK

S. Foltyn
Superconductivity Technology Center, Los Alamos National Laboratory, Los Alamos, New Mexico 87545, USA

ABSTRACT

Ca-gels were applied to PLD YBCO thick films followed by high temperature post-annealing. Parameters of Ca diffusion in YBCO (diffusion coefficients, activation energy, etc) as a function of temperature were determined by depth profiling SIMS. Fast Ca diffusion along the grain boundaries was observed. The effect of Ca-doping on T_c and J_c was studied by VSM. Ca distribution in the samples, induced by preferential diffusion along the grain boundaries, did not appear to change superconducting properties within the grains but showed increased J_C across the grain boundaries.

INTRODUCTION

It is well known that the properties of grain boundaries in YBCO differ drastically from those of bulk material. It has been shown both experimentally and theoretically that the grain boundaries in YBCO superconductors are depleted of carriers as compared to the bulk [1]. This depletion leads to weak intergrain links and limit critical currents in superconductor. Recently, the promise for improved grain boundaries has been shown in YBCO thin films which have been doped throughout with Ca [2, 3]. In this study the effect of Ca-doping of grain boundaries in YBCO has been investigated in an attempt to increase hole concentration on the grain boundaries and attainable J_C.

EXPERIMENTAL

Starting materials were $YBa_2Cu_3O_{7-\delta}$ thin films grown by PLD on (100) $SrTiO_3$ [4]. A dilute (0.05M) solution of $Ca(NO_3)_2$ (99.99+%) was dropped on the surface of the specimens. Samples were annealed in air at different preparation conditions (Table I). All the films annealed at 870°C were later oxygenated in flowing air at 500°C for 2 days. Several samples were not doped and used as reference.

The Ca concentration profiles were measured by SIMS (Atomika 6500). The depth of the crater was measured by an optical surface profilometer. The raw data was converted into the Ca diffusion profiles, by calculating the isotopic fraction from the raw intensities and converting the sputter time to depth assuming a constant sputtering rate. Diffusional profiles were fitted using the following solution of Fick's second law:

$$C(x,t)/C_0 = \exp\left(-\frac{x^2}{4D_{bulk}t}\right) \qquad (1)$$

where $C(x,t)$ is the concentration of Ca as a function of distance from the surface, x, for an annealing time, t. D_{bulk} is the bulk diffusion coefficient, and C_0 is the concentration at the surface. In several diffusional profiles, short-circuiting tails were observed which were attributed to the fast Ca diffusion along the grain boundaries. The product of grain boundary diffusion coefficient and grain boundary width was calculated by the method proposed by Chung and Wuensch [5].

Samples were characterized by XRD, SEM. Magnetization was measured by VSM. Critical current densities were calculated using Bean model [6]. Tc was determined from magnetic permeability measurements.

RESULTS AND DISCUSSIONS

XRD scans of starting PLD films showed only highly intensive (00l) peaks with c-axis oriented perpendicular to the film surface. Pole figures on (111)

Table I. Preparation conditions and results of measurements

Sample	Ca-doped	Doping conditions	"c" parameter, Å	D_{bulk}, cm^{2}s^{-1}	$D_{g.b.}\delta$,cm^3 s^{-1}	T_C, K
YE102	yes	870°C, 30 min		NA	NA	89.5
YE103	yes	870°C, 20 hrs	11.698	NA	NA	92.0
YE104	no	870°C, 20 hrs	11.693	NA	NA	92.8
YE105	yes	870°C, 6 hrs	11.681	NA	NA	91.6
YE202	yes	870°C, 2 hrs	11.778	6.2×10^{-14}	4.2×10^{-18}	92.6
YE203	no	870°C, 2 hrs	11.752	NA	NA	NA
YE204	yes	798°C, 6 hrs	11.753	2.7×10^{-14}	3.1×10^{-18}	NA
YE205	yes	700°C, 24 hrs	11.767	7.9×10^{-15}	NA	NA
YE501	yes	700°C, 45 min		1.3×10^{-14}	NA	NA

YBCO peak showed in-plane alignment of the film. No secondary phases were observed in starting thin films. After doping peaks belong to CaO and BaO were observed. The intensity and presumably the amount of BaO increased with the duration of anneal. Interestingly, no Y-rich phases were observed in Ca doped samples and no BaO was present in undoped samples. Therefore it is possible to conclude that some Ca is substitute onto the Ba site in $YBa_2Cu_3O_{7-\delta}$. This observation is supported by the structural study of $Y_{1-x}Ca_xBa_2Cu_3O_{7-\delta}$ which showed that Ca substituted onto Y site at low Ca concentrations (<11%), whereas at higher concentrations, Ca also substituted onto Ba site [7]. Values of the "c" lattice parameters are given in Table 1. No clear correlation of the duration of doping with the "c" parameter was observed.

The values of the Ca chemical diffusion coefficient as well as the grain boundary diffusion coefficient are given in Table I. The Ca-diffusion coefficient in PLD films is about 20 times lower than in the bulk YBCO [3]. This discrepancy is attributed to the anisotropy of Ca diffusion along the a-b planes and c- direction in $YBa_2Cu_3O_{7-\delta}$. Analogous anisotropy in the ion diffusion was observed for oxygen [8] and barium diffusion [9] in $YBa_2Cu_3O_{7-\delta}$. The activation energy for Ca chemical diffusion along c-direction is ~ 113 kJ/mole. This value is smaller than the activation energies of tracer diffusion of Ba and Y in $YBa_2Cu_3O_{7-\delta}$, but

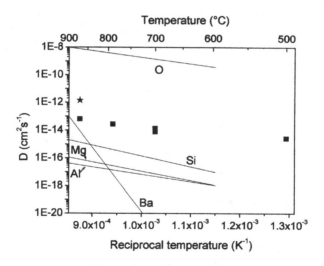

Fig. 1. Arrhenius plot of Ca diffusion coefficient in $YBa_2Cu_3O_{7-x}$ thin films (■). Solid lines are literature data on ion diffusion in YBCO [8,9].

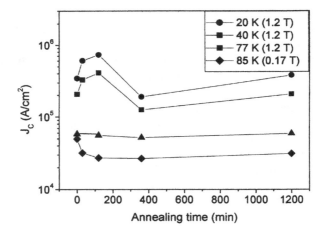

Figure 2. Effect of annealing time on the critical current densities of Ca
doped films in annealed air at 870°C.

compares reasonably well with the activation energy for the chemical diffusion of
Ag and Co in $YBa_2Cu_3O_{7-\delta}$ [9].

The grain boundary diffusion coefficient was calculated assuming the width of
the grain boundary to be of the order of 1 nm [10]. The value was 3.1×10^{-11} cm^2s^{-1}
for the film annealed at 798°C. Diffusion enhancement factors ($D_{bulk}/D_{g.b.}$) are
roughly estimated to be of the order of 10^3. Thus Ca diffuses much faster along
the grain boundaries in $YBa_2Cu_3O_{7-\delta}$ than through the bulk. Consequently,
preferential doping of the grain boundaries appears to be possible.

No correlation of T_C was observed with the duration of the Ca anneal implying
that no significant doping of grain occurs. However pronounced effect of the Ca
annealing time on Jc was observed (Fig. 2). Film doped with Ca at 870°C for 2
hrs, showed an increase of J_C by a factor of two at 40 K. The decrease of the J_C at
prolonged annealing times is thought to be related to the overdoping of the grains
with Ca supplied from the grain boundaries leading to a decrease of T_C in the
grains close to the grain boundary.

The success of the surface gel doping technique depends strongly on the
microstructure of the sample to be coated. Any extended defects other than grain
boundaries could contribute to the Ca diffusion and result in the degradation of
the superconducting properties. As an example YBCO film was cut in three parts.
Two resulting films were doped with Ca (YE607 and YE606) and one left as a
reference (YE605). All films were annealed at 500°C in oxygen. As expected J_C

High Temperature Superconductors

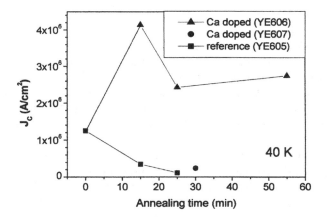

Figure 3. Effect of annealing time on the critical current densities of films annealed in oxygen at 500°C.

for YE607 film increased with annealing and later decreased with the optimal annealing time of 15 min (Fig. 3). At the same time J_C for another film (YE606) decreased with annealing and followed the data of the reference film (YE605). Short annealing time required to optimally dope the film at 500°C is surprising, as it required 2 hrs to achieve analogous effect at 870°C where much faster diffusion rates are expected. It is thought that the pores observed by SEM on YE607 caused higher Ca doping rate due to gel penetration in the pores and/or faster diffusion along the pores as compared to the grain boundaries. The decrease of J_C in YE605 and YE606 films after annealing in oxygen is currently unclear.

CONCLUSIONS

Ca chemical diffusion was studied in $YBa_2Cu_3O_{7-\delta}$. Fast diffusion along the grain boundaries was observed in c-oriented thin films. The optimum duration of Ca annealing increased J_C by a factor of 2 in epitaxial $YBa_2Cu_3O_{7-\delta}$ thin films at 40 K. The doping /diffusion technique developed in this work allows preferential doping of grain boundaries in $YBa_2Cu_3O_{7-\delta}$.

ACKNOWLEDGEMENTS
The authors are grateful EPSRC for financial support.

High Temperature Superconductors

REFERENCES

[1] H. Hilgenkamp, C.W. Scheider, R.R. Schulz, B. Goetz, A. Schmehl, H. Bielefeldt and J. Mannhart, "Modifying electronic properties of interfaces in high-T_c superconductors by doping", *Physica C*, **326-327** 7-11 (1999).

[2] A. Schmehl, B. Goetz, R.R. Schulz, C.W. Schneider, H. Bielefeldt, H. Hilgenkamp and J. Mannhart, "Doping-induced enhancement of the critical currents of grain boundaries in YBa$_2$Cu$_3$O$_{7-\delta}$", *Europhysics Letters*, **47** [1] 110-115 (1999).

[3] A.V. Berenov, R. Marriott, S.R. Foltyn, J.L. MacManus-Driscoll, "Effect of Ca-doping on grain boundaries and superconducting properties of YBa$_2$Cu$_3$O$_{7-\delta}$", *IEEE Transactions on Applied Superconductivity*, **11** [1] pt 3 3780-3783 (2001).

[4] N. Malde, L.F. Cohen, G. Gibson, S.R. Foltyn, J. Nelstrop, J.J.Wells, A. Berenov, J.L. MacManus-Driscoll, "Raman and X-ray characterisation of structural disorder in YBCO thick films grown at high rates for IBAD conductors", *Institute of Physics Conference Series*, **167** 415-418 (1999).

[5] Yong-Chae Chung and B.J. Wuensch, "An improved method, based on Whipple's exact solution, for obtaining accurate grain-boundary diffusion coefficients from shallow solute concentration gradients", *Journal of Applied Physics*, **79** [11] 8323-8229 (1996).

[6] C.P. Bean, *Physical Review Letters*, **8** 250 (1962).

[7] G.Böttger, H. Schwer, E. Kaldis, K. Bente, "Ca doping of YBa$_2$Cu$_3$O$_{7-\delta}$ single crystals: structural aspects", *Physica C*, **275** 198–204 (1997).

[8] J.L. Routbort, N. Chen, K.C. Goretta, S.J. Rothman, p. 569 in *High-Temperature Superconductors: Materials Aspects. Procedings of the ICMC'90 Topical Conference on Materials of High-Temperature Superconductors*. Edited by H.C. Freyhardt, R. Flükiger, M. Peuckert. Informationsgesellschaft, Verlag, 1991.

[9] N. Chen, S.J. Rothman, J.L. Routbort, K.C. Goretta, "Tracer diffusion of Ba and Y in YBa$_2$Cu$_3$O$_x$", *Journal of Material Research*, **7** [9] 2308-2316 (1992).

[10] J.R. Farver, R.A. Yund, D.C. Rubie, "Magnesium grain boundary diffusion in forsterite aggregates at 1000 ° -1300 °C and 0.1 MPa to 10 GPa", *Journal of Geophysical Research*, **99** [B10] 19809-19819 (1994).

DEMONSTRATION OF HIGH CURRENT DENSITY YBCO FILMS ON ALL SOLUTION BUFFERS

M. Parans Paranthaman, S. Sathyamurthy, H.Y. Zhai, H.M. Christen, S. Kang, and A. Goyal
Oak Ridge National Laboratory, Oak Ridge, TN 37831, USA

ABSTRACT

Chemical solution deposition has been emerged as a viable, low-cost, non-vacuum process for fabricating long lengths of $YBa_2Cu_3O_{7-\delta}$ (YBCO) coated conductors. Single, epitaxial buffer layers of $La_2Zr_2O_7$ (LZO) have been grown directly on biaxially textured and strengthened Ni-W (3 at.%) substrates by a solution process. Smooth, epitaxial, and crack-free buffer layers with desired thickness were produced using multiple coating. YBCO films grown on 60 nm thick LZO-buffered Ni-W substrates using pulsed laser deposition yielded very high critical current densities of 1.9 MA/cm^2 at 77 K and self-field. These results are comparable to those typically obtained on identical substrates using our standard architecture of $CeO_2/YSZ/Y_2O_3/Ni$. The dependence of critical current density, J_c on thick YBCO films grown on 60 nm thick LZO layers is investigated.

INTRODUCTION

Since the discovery of high temperature superconductivity in cuprate based superconductors, much progress has already been made, from the near-term commercialization of the first-generation bismuth strontium calcium copper oxide (BSCCO) superconductor tapes to the continuing advancement in second-generation conductors coated with yttrium barium copper oxide (YBCO). The development of second-generation coated conductors continues to show the steady improvement toward the long-length processing capabilities. The U.S. Department of Energy's vision is, *"Low-cost, high-performance YBCO Coated Conductors will be available in 2005 in kilometer lengths. For applications in liquid nitrogen, the wire cost will be less than $ 50/kA-m, while for applications requiring cooling to temperatures of 20-60 K the cost will be less than $ 30/kA-*

m. " One of the important critical needs to achieve this goal is to develop a simple low-cost buffer layer technology on strengthened Ni-alloy substrates.

Metal-organic decomposition (MOD) based solution precursor approaches have been examined as a potential low-cost process for manufacturing YBCO coated conductors. This is mainly due to the *speed/cost* advantage of this process. Recently, ORNL and American Superconductor Corporation have reported critical currents of 118 A with a standard deviation of less than 3% over a 1-m length of centimeter-wide RABiTS tape with the architecture of YBCO/CeO$_2$/YSZ/Y$_2$O$_3$/Ni/Ni-W using solution-based metal-organic decomposition (MOD).[1] Although there has been significant progress in the development of low-cost manufacturing technologies for select components of the YBCO coated conductors, key deposition technologies must still be improved. Even though the oxide buffer layers are much thinner than the YBCO layer, the buffer deposition accounts for over 20% of the total conductor cost – the single highest projected cost element in the entire conductor. Hence, the coated conductor technology dictates that a considerable and concentrated effort must be devoted to the development of higher rate, high quality buffer layer deposition processes as alternatives to the already established methods.

The main focus of this work is to develop a chemical solution deposition process to grow high quality lanthanum zirconium oxide, La$_2$Zr$_2$O$_7$ (LZO) buffer layers on strengthened Ni-W substrates. LZO has a cubic pyrochlore structure with a close lattice match with YBCO (less than 1% lattice mismatch). Recently, we have demonstrated the growth of high current density YBCO films on both solution LZO seed layers[2,3] and all solution LZO layers[4] grown on biaxially textured Ni substrates. Here we report the results of using a single LZO buffer layer deposited on strengthened Ni-W (3 at. %) substrates with reduced magnetism for YBCO coated conductors.

EXPERIMENTAL PROCEDURE

The LZO precursor solution with 0.25 M cation concentration was prepared by refluxing a stoichiometric mixture of lanthanum isopropoxide and zirconium n-propoxide in 2-methoxyethanol in a Schlenk-type apparatus. As-rolled Ni-W (3 at. %) tapes were cleaned by ultrasonification in iso-propanol. The tapes were then annealed in either reducing atmosphere (Ar-H$_2$ 4%) or in high vacuum at 1300 °C to get the desired cube texture. The metal substrates were coated with LZO film by spin coating the precursor solution at 2000 rpm for 30 s, followed by annealing at 1100 °C for 1 h in a flow of 1 atm Ar-H$_2$ 4% gas mixture. The process was repeated three times to get 60 nm thick LZO layers. On the LZO-buffered Ni-W substrates, pulsed laser deposition was used for YBCO deposition, conducted at 780 °C and pO$_2$ of 120 mTorr. After deposition, the samples were first cooled to 500 °C at a rate of 5 °C/min; then the O$_2$ pressure was increased to

550 Torr, and the samples were cooled to room temperature. YBCO thickness was varied from 200 nm to 1500 nm.

The films were analyzed by X-ray diffraction techniques. SEM micrographs were taken using a Hitachi S-4100 field emission microscope. The thickness of both buffer layers and YBCO were determined by Rutherford backscattering spectroscopy. The resistivity and transport critical current density, J_c, were measured using a standard four-probe technique. The voltage contact spacing was 0.4 cm. Values of J_c were calculated using a 1 μV/cm criterion.

RESULTS AND DISCUSSION

Typical θ-2θ scans for 200 nm thick YBCO films grown on 60 nm thick solution LZO buffered Ni-W substrates are shown in Figure 1. These scans indicate the presence of c-axis aligned films. It is interesting to note that there is

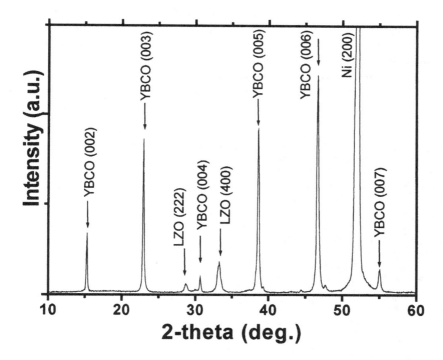

FIGURE 1. A typical θ-2θ scan for a 200 nm thick YBCO on solution LZO buffered Ni-W substrates. The YBCO film has a preferred c-axis orientation.

no NiO present in the film. However, small amounts of (222) LZO peaks were present. In certain cases, (222) LZO peaks were absent. Detailed RHEED studies are being conducted to determine the texture of the surface layers and the results will be published later. Detailed XRD results from ω and ϕ scans on YBCO (200 nm)/LZO (60 nm)/Ni-W revealed good epitaxial texturing. The full width at half maximum (FWHM) values for Ni (002) and YBCO (006) are 5.7° and 6.8°, and those of Ni (111) and YBCO (103) are 7.7° and 9.3°, respectively. The YBCO (103) pole figure revealed the presence of a single four-fold cube texture. SEM micrograph for a 200 nm thick PLD YBCO film on 60 nm thick LZO-buffered Ni-W substrate is shown in Figure 2. The microstructure of the YBCO film looks like a typical dense PLD film. These films had zero resistance temperature, T_c close to 90 K. The field dependence of J_c for the same film is shown in Figure 3. The zero-field J_c was close to 1.9 MA/cm^2 at 77 K. J_c at 0.5 T was about 25 % of the zero-field J_c. This indicates the presence of a strongly linked YBCO film. The dependence of critical current, I_c and J_c on YBCO thickness is shown in Figure 4. As the YBCO thickness increased, the I_c increased steadily but the J_c decreased significantly. A high I_c of over 70 A/cm-width was measured for 1.1 μm thick YBCO film grown on just 60 nm thick LZO-buffered Ni-W substrate. Optimization of thick YBCO film growth is in progress. Efforts are also being made to demonstrate the growth of solution MOD YBCO layers on all solution buffers.

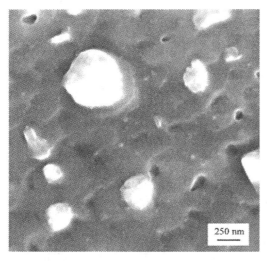

FIGURE 2. SEM micrograph for a 200 nm thick YBCO film on 60 nm thick solution LZO-buffered Ni-W substrate.

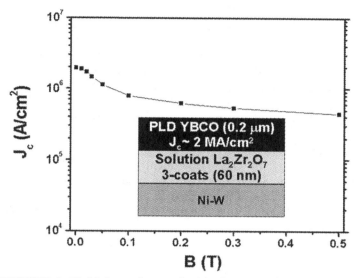

FIGURE 3. Field dependence of critical current density, J_c, for 200 nm thick PLD YBCO film on 3-coats of solution LZO buffered-Ni-W substrate.

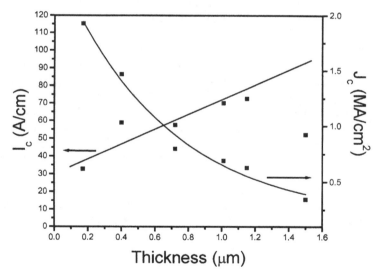

FIGURE 4. Variation of YBCO film thickness on 60 nm thick solution LZO-buffered Ni-W and critical current, I_c and critical current density, J_c.

SUMMARY

We have demonstrated that YBCO films with good superconducting properties can be grown on 3-coats of 60 nm thick solution LZO buffered Ni-W substrate. This proves that $La_2Zr_2O_7$ can be an excellent diffusion barrier for Ni. This also eliminates the need for both Ni protective layers and seed layers. For a 1.1 μm thick YBCO film on just 60 nm thick LZO/Ni-W, an I_c of over 70 A/cm-width was measured. This promises a route for producing long lengths of YBCO coated conductors using a completely non-vacuum buffer layer technology.

ACKNOWLEDGEMENTS

This work was supported by the U.S. Department of Energy, Division of Materials Sciences, Office of Science, Office of Energy Efficiency and Renewable Energy. The research was performed at the Oak Ridge National Laboratory, managed by U.T.-Battelle, LLC for the USDOE under contract DE-AC05-00OR22725.

REFERENCES

[1] D.T. Verebelyi, U. Schoop, C. Thieme, S. Annavarapu, S. Cui, X. Li, W. Zhang, T. Kodenkandath, L. Fritzemeier, Q. Li, A.P. Malozemoff, N. Nguyen, D. Buczek, J. Lynch, J. Scudiere, M. Rupich, A. Goyal, L. Heatherly and M. Paranthaman' "Uniform performance of meter long YBCO on textured Ni alloy substrates," submitted to Physica C (2002).

[2] T.G. Chirayil, M. Paranthaman, D.B. Beach, D.F. Lee, A. Goyal, R.K. Williams, X. Cui, D.M. Kroeger, R. Feenstra, D.T. Verebelyi, and D.K. Christen, "Epitaxial growth of $La_2Zr_2O_7$ thin films on rolled-Ni substrates by sol-gel process for high T_c superconducting tapes," *Physica C* **336**, 63-69 (2000).

[3] S. Sathyamurthy, M. Paranthaman, T. Aytug, B.W. Kang, P.M. Martin, A. Goyal, D.M. Kroeger, and D.K. Christen, "Chemical solution deposition of lanthanum zirconate buffer layers on biaxially textured Ni-1.7% Fe-3% W alloy substrates for coated-conductor fabrication," *J. Mater. Res.* **17**, pp. 1543-1549 (2002).

[4] S. Sathyamurthy, M. Paranthaman, H-Y. Zhai, H.M. Christen, P.M. Martin, and A. Goyal, "Lanthanum zirconate: A single buffer layer processed by solution deposition for coated conductor fabrication," *J. Mater. Res.* **17**, pp.2181-2184 (2002).

Bulk Processing

PROCESSING AND PROPERTIES OF MELT PROCESSED Y-Ba-Cu-O CONTAINING DEPLETED URANIUM OXIDE

D. A. Cardwell, N. Hari Babu and M. Kambara
IRC in Superconductivity
University of Cambridge
Madingley Road, Cambridge,
CB3 0HE, UK.

C. D. Tarrant and K. R. Schneider
Urenco plc,
Capenhurst,
Cheshire,
CH1 6ER, UK.

ABSTRACT

Large, single domain Y-Ba-Cu-O (YBCO) containing up to 0.8 wt. % depleted and natural UO_2 has been fabricated successfully by top seeded melt growth (TSMG) processing. The effect of Pt doping and the addition of Y_2BaCuO_5 (Y-211) and Y_2O_3 to the starting composition on the sample microstructure has been investigated and the bulk processing developed to yield samples with high critical current densities of 45×10^3 A/cm^2 at 1 T at 77K. Depleted U-doped YBCO samples of 2cm in diameter have been fabricated without added Pt samples and observed to contain a fine distribution of second phase, uranium-containing inclusions. These samples trap a magnetic field of almost 0.4T at 77K.

INTRODUCTION

Bulk Y-Ba-Cu-O (YBCO) processed in the form of large, high critical current density, J_c, grains has significant potential for engineering applications due to its ability to trap large magnetic fields[1]. It has long been established that values of J_c in YBCO fabricated by the top seeded melt growth technique (TSMG) can be improved by introducing fine Y_2BaCuO_5 (Y-211) normal phase inclusions within the superconducting $YBa_2Cu_3O_{7-\delta}$ (Y-123) matrix[2]. These inclusions, or features associated with them such as dislocations, form effective magnetic flux pinning sites within the bulk material, which is associated directly with increased J_c. Addition of small amounts of Pt or CeO_2 to the YBCO precursor powder mixture (i.e. Y-123 and Y-211) has been reported to be particularly effective in refining the Y-211 particle size in the bulk matrix[3]. More recently, large Pt-free YBCO single grains, which contain a fine distribution of Y-211 inclusions and exhibit improved field trapping performance, have been fabricated by an infiltration and growth process[4]. Elimination of Pt in the YBCO microstructure is desirable from the point of view of its cost and the lack of general availability of soluble platinum salts, which are essential for the production of precursor powders via a solution route. Despite these developments, however, the average Y-211 particle size in the melt processed microstructure remains typically at around 1 μm.

Various irradiation techniques have been employed in an attempt to further improve flux pinning in YBCO, including fast neutron[5], high energy heavy ion[6] and electron bombardment[7]. Although these methods are successful in increasing the dislocation and local defect density, which also act as effective flux pinning sites in this material, they tend to be both inconvenient and expensive and therefore unsuitable for routine medium-scale production of samples for practical applications. An alternative and more feasible method for introducing local non-superconducting regions into the bulk YBCO microstructure has been suggested by Fleisher et al[8] in polycrystalline YBCO and by Weinstein et al[9] in melt processed single grain YBCO superconductors. This method consists of adding small amounts of fissile uranium (U^{235}) to the superconductor during processing. The fully processed sample is then irradiated with thermal neutrons to induce nuclear fission and create so-called "U/n" fission track pinning centres throughout its bulk[10]. The relatively large amount of energy released from the fission of U^{235} results in the formation of an amorphous YBCO phase over a number of Y-123 unit cells. This process creates locally non-superconducting regions in the superconducting matrix of dimensions of typically 15 nm. In order to obtain the greatest pinning energy per pinning centre, the size of these defect areas should be comparable to the diameter of the flux line core, 2ξ, where ξ is the coherence length (~1.5 nm at 0K in YBCO). The size and geometry of the U/n pinning sites have been reported to satisfy this criterion[10]. Weinstein et al[11] have also reported the presence of fine $(U_{0.6}Pt_{0.4})YBa_2O_6$ (UPYBO) second phase deposits in melt textured YBCO, which modify the chemistry of phase formation and form effective, so-called "U-Chem" pinning sites. A clear advantage of U-Chem pinning sites is that they avoid the need to irradiate the sample following melt processing. More recently Weinstein et al[12,13] have reported even greater flux pinning in YBCO samples doped with two foreign elements, one of which may be uranium. These pinning centres are again associated with the formation of a fine second phase composition analogous to UPYBO, which forms a double perovskite structure, and with refinement of the Y-211 phase. Significantly, Weinstein *et al* speculate that Pt can be substituted by Zr in these samples without loss of pinning strength. As noted above, the elimination of Pt as a Y-211 refining agent in the YBCO bulk microstructure is extremely desirable.

In this paper we report the processing of bulk YBCO superconductors containing depleted uranium in the form of large single grains by TSMG. We report refinement of the microstructure for single grain samples containing various amounts of Y-211, Y_2O_3 and Pt added to the parent YBCO composition. We also report current density and trapped field measurements for the most promising large grain sample of the limited composition range investigated.

EXPERIMENTAL

Precursor powders were prepared to investigate the effect of the addition of depleted uranium (DU) oxide, Pt and Y_2O_3 on the microstructure and critical current density of melt processed YBCO. This involved mixing thoroughly using a mortar and pestle high purity ceramic powders of Y-123 (Pi-Kem), Y-211 (Pi-Kem), Y_2O_3, Pt and depleted uranium oxide in the required ratios to give the following compositions;

(a) To investigate the effect of DU variation with excess Y-211 but no Pt;
Y-123 + 30mol% Y-211;
Y-123 + 30mol% Y-211 + 0.1wt% DU;
Y-123 + 30mol% Y-211 + 0.4wt% DU;
Y-123 + 30mol% Y-211 + 0.8wt% DU.

(b) To investigate the effect of DU variation with excess Y-211 and Pt;
Y-123 + 30mol% Y-211;
Y-123 + 30mol% Y-211 + 0.1wt% DU + 0.1wt% Pt;
Y-123 + 30mol% Y-211 + 0.4wt% DU + 0.1wt% Pt;
Y-123 + 30mol% Y-211 + 0.8wt% DU + 0.1wt% Pt.

(c) To investigate the effect of DU variation with excess Y_2O_3 and Pt;
Y-123 + 30mol% Y_2O_3 + 0.1wt% Pt;
Y-123 + 30mol% Y_2O_3+ 0.1wt% DU + 0.1wt% Pt;
Y-123 + 30mol% Y_2O_3+ 0.4wt% DU + 0.1wt% Pt.

(d) To investigate the effect of DU variation with excess Y_2O_3 and no Pt;
Y-123 + 30mol% Y_2O_3;
Y-123 + 30mol% Y_2O_3 + 0.1wt% DU;
Y-123 + 30mol% Y_2O_3+ 0.4wt% DU.

High Temperature Superconductors

259

Differential thermal analysis (DTA) of the precursor powders showed that addition of DU does not affect significantly the peritectic temperature of the system. Cylindrical pellets of 2.5 cm diameter and 1 cm thickness were subsequently prepared from the powders listed above and melt processed using small Nd-Ba-Cu-O (NdBCO) melt textured pseudo-crystals to seed the large grain growth. A box furnace with a cold finger and controllable thermal gradients was used for this purpose, as described in reference[14]. The microstructural features of polished surfaces of the large grain samples were investigated using polarised optical microscopy. These were then annealed in flowing O_2 at 450°C for 70 hours to obtain the superconducting orthorhombic phase. Samples of approximate geometry 1mm × 1mm × 0.5mm were cut from the single grain samples and their magnetic moment hysteresis loops measured at different temperatures using a Quantum Design superconducting quantum interference device (SQUID) magnetometer with fields applied parallel to their crystallographic c-axes. Values of J_c were subsequently calculated from the hysteresis loop width using the Bean critical state model. The magnetic field trapping capability of each sample at 77K was measured by scanning a Hall probe sensor approximately 0.5mm above the sample surface following magnetisation by a pulsed copper coil solenoid with a peak field output of ~ 0.5T.

RESULTS AND DISCUSSION

Figure 1 shows optical micrographs of a single grain of Y-123 + 30mol% Y-211 *fabricated without platinum* and containing (a) 0 wt% DU, (b) 0.1wt% DU and (c) 0.8wt% DU. The area selected for the optical micrograph image in each case is ~1 cm away from the centre of the grain, along the a/b direction of the specimen. It can be seen that the Y-211 (grey in contrast) phase particles are distributed uniformly within the Y-123 matrix. The microstructure of the grain shown in figure 1(a), which contains no DU, is characterised by a coarse distribution of Y-211 inclusions of average size around 3-4 μm (with a maximum diameter of up to 10 μm). These particles become trapped in the Y-123 matrix during the peritectic solidification process on cooling. Figure 1(b) illustrates the change in morphology, size and distribution of the second phase inclusions associated with the addition of depleted uranium oxide to the precursor powder. It can be seen that the inclusions are dispersed more finely in the Y-123 matrix compared with the un-doped specimen. In addition, two different distinguishable sizes of second phase particles are clearly present in the DU containing samples. The larger of these are similar in size and morphology to those present in figure 1(a), and have been confirmed by electron probe microanalysis (EPMA) to be Y-211. The smaller, more spherical particles in figure 1(b), on the other hand, are associated with another phase in the Pt-free, U-doped YBCO microstructure.

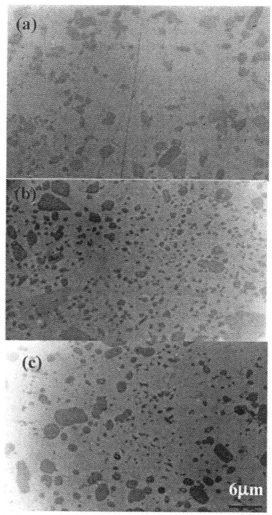

Fig. 1. Optical micrograph of TSMG YBCO with initial composition (a) Y-123 + 30 mol% Y-211 (b) Y-123 + 30 mol% Y-211 + 0.1 wt% DU and (c) Y-123 + 30 mol% Y-211 + 0.8wt % DU. These micrographs were taken at approximately the same position from the grain centre.

EPMA has confirmed these to be uranium containing. Chen *et al*[15] observed similar fine sized particles in an earlier study and identified their chemical composition as $U_2Y_3Ba_5O_x$. The reduced size of the particles observed in the present study suggests a reduced interfacial energy with the liquid in the peritectic state, which would inhibit Ostwald ripening at elevated temperature[16]. Furthermore, increasing the concentration of DU up to 0.8wt%, does not appear to increase the number of smaller particles present in the microstructre significantly, as is evident in figure 1(c). The addition of an optimum amount of Pt of 0.1 wt% to undoped YBCO has been observed to refine the size of Y-211 phase particles within the bulk material[17]. Further reduction of the Y-211 particle size was observed by adding this optimum Pt concentration to the uranium doped samples (i.e. to form the compositions identified in (b) in the Experimental section). Of these, the sample containing 0.1 wt% DU was found overall to contain the finest, most uniform distribution of second phase particles, with approximate average size 1 μm, as illustrated in figure 2 (a). This suggests that the presence of both DU

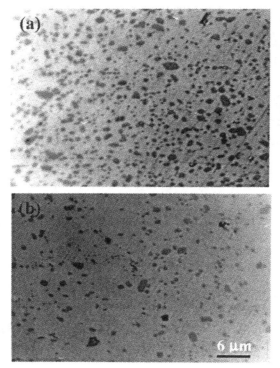

Fig. 2 Optical micrograph of TSMG YBCO with initial composition Y-123 + 30 mol% Y-211 + 0.1 wt% Pt with (a) 0.1 wt% DU and (b) 0.8 wt% DU.

and Pt contributes to the refinement of second phase particles in the YBCO microstructure via a relatively complex, interactive mechanism. Weinstein *et al*[9] identified the composition of the fine sized second phase particles, in the case of

Pt containing samples, to be $U_3Pt_2Y_5Ba_{12}O_x$. The sample containing 0.8 wt% DU in the present study (shown in figure 2(b)), contains similar sized larger second phase particles (i.e. Y-211) as the sample processed with 0.1wt% DU. Increasing the DU content of the sample containing 30 mol% Y-211 and 0.1wt% Pt beyond 0.1wt%, therefore, does not yield further refinement of the Y-211 phase for the range of compositions studied here. Further refinement of the YBCO microstructure has been reported for samples containing higher concentrations of Pt (up to 0.5wt%) and uranium (up to 0.8wt%)[18]. The aim of the present study, however, is to investigate minimising the amount of Pt in bulk YBCO without reducing its field trapping potential, and the specific effect of Y_2O_3 enrichment on the sample microstructure. As a result this range of compositions is not investigated here.

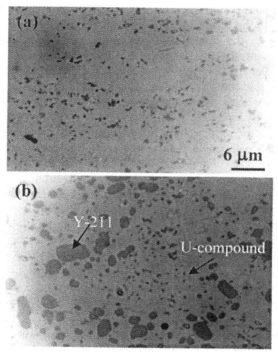

Fig. 3. Optical micrographs of TSMG YBCO with initial composition Y-123 + 30 mol% Y-211 + 0.8 wt% DU grown isothermally at 1000°C for 10h and then cooled slowly at 1°C/h to 990°C before rapid cooling to room temperature (a) close to seed crystal (effectively grown isothermally) and (b) a few mm away from the seed (slow cooling).

The distribution of the two types of second phase particles within the TSMG microstructure was investigated further by changing the undercooling during melt growth. This was achieved by processing a pellet of initial Pt-free composition Y-123 + 30mol% Y-211 + 0.8wt% DU isothermally at 1000°C for 10 hours and then slow cooling at 1°C/h to 990°C before furnace cooling to room temperature. Interesting variations in the spatial distribution of second phase particles along the a/b plane were observed in this sample (i.e. in the plane of the sample surface, perpendicular to its thickness). It can be seen that very small second phase particles are trapped during initial grain growth near the centre of the grain, as shown in Figure 3(a). These particles are not distributed randomly, but tend to form a regular pattern in the vicinity of the seed[19]. Two different sized particles can be observed in the large grain microstructure a few millimetres from the seed position when the sample is grown at constant temperature and cooled at 1°C/h. One of these is smaller in size (<1 μm) and one larger (3-4 μm), as shown in Figure 3(b). This indicates that, although the addition of DU promotes the formation of fine sized second phase U-compound precipitates[9], the presence of large Y-211 inclusions in the composition prior to melt processing (from 'standard' Y-211 enrichment) hinders the general refinement of the second phase particles. This suggests that ripening of second phase particles may be inhibited by reducing the Y-211/liquid interfacial energy by adding Pt or by employing a Y-rich compound[20], rather than the addition of excess Y-211 to the precursor powders.

Figure 4(a) shows an optical micrograph of a melt processed YBCO sample containing Y-123 + 30 mol% Y_2O_3 + 0.1 wt% DU + 0.1 wt% Pt. The distribution of the second phase particles is uniform throughout the specimen, with a size of around 200-300nm, which is extremely small. Preliminary EPMA studies indicate that these inclusions at least contain uranium, platinum and copper (U-Pt-Cu) and, therefore, are of a different composition to that of $(U_{0.6}Pt_{0.4})YBa_2O_6$, reported by Weinstein et al[11]. No change in the average size of the fine particles was observed as the amount of DU was increased from 0.1wt% to 0.4wt%. In addition, the morphology of the majority of the U-Pt-Cu particles is spherical. This suggests that uranium plays a critical role in reducing the interfacial energy with the liquid phase at elevated temperature, as discussed above, and is directly responsible for the observed fine distribution of secondary phase particles in the bulk superconductor matrix. It is believed that small spherical shaped particles generate a higher density of defects in the Y-123 matrix surrounding the normal phase particle, which form particularly effective flux pinning sites. Figure 4(b) shows the microstructure of a sample containing Y-123 + 30 mol% Y_2O_3 + 0.1 wt% DU to illustrate the effect of the absence of Pt from the DU doped YBCO microstructure. The size and morphology of the second phase particles is nearly same as in series (c) described in the Experimental section, indicating that addition of Pt is not required to refine the fine, second phase particles for large DU doped YBCO grains.

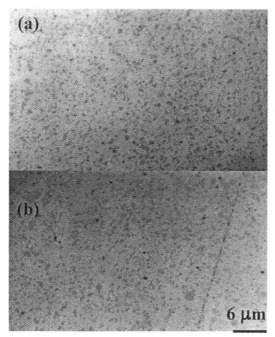

Fig. 4. Optical micrograph of TSMG YBCO with initial composition Y123 + 30 mol% Y_2O_3 + 0.1wt% DU (a) with 0.1 wt% Pt and (b) with no Pt.

In order to investigate the formation of these fine sized second phase particles further, we have studied the microstructure of a variety of bulk samples after quenching from the peritectically decomposed state (i.e. from the pro-peritectic stage of melt processing). The following compositions were investigated for this purpose;

(i)   Y-123 + 0.8wt% DU;
(ii)   Y-123 + 30mol% Y-211 + 0.8wt% DU;
(iii)   Y-123 + 30mol% Y-211 + 0.8wt% DU + 0.1wt% Pt;
(iv)   Y-123 + 30mol% Y-211 + 0.8wt% DU + 0.5wt% Pt;
(v)   Y-123 + 30mol% Y_2O_3 + 0.1wt% DU + 0.1wt% Pt.

Bulk pellets of these compositions were heated to 1045 °C, held for 30 minutes and quenched in liquid nitrogen. Figure 5 shows the microstructure of the samples with the above initial compositions. The presence of large Y-211 particles (again, grey in contrast), which are formed when Y-123 decomposes into Y-211 and the $Ba_3Cu_5O_8$ liquid phase, dominates the microstructure of composition (i), as shown in figure 5(a). The average size of these large Y-211 particles in the liquid is around 8μm, which is very close to that observed in the un-doped Y-123 system[21.]

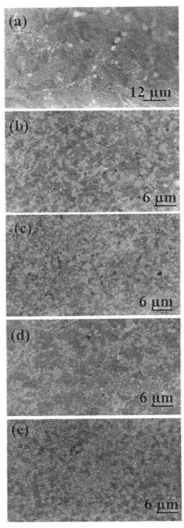

Fig. 5. Optical micrographs of bulk YBCO samples after quenching from the peritectically decomposed state in liquid nitrogen (a) Y-123 + 0.8wt% DU, (b) Y-123 + 30mol% Y-211 + 0.8wt% DU, (c) Y-123 + 30mol% Y-211 + 0.8wt% DU + 0.1wt% Pt, (d) Y-123 + 30mol% Y-211 + 0.8wt% DU + 0.5wt% Pt and (e) Y-123 + 30mol% Y_2O_3 + 0.1wt% DU + 0.1wt% Pt.

Athur *et al*[21] reported that these particles coarsen significantly to around 18μm after the sample is held at 1075 °C for 30 hours. However, a large number of sub-micron sized particles are also present in composition (i) (Y-123 + 0.8wt% DU), which could be of the $U_2Y_3Ba_5O_x$ composition reported by Chen *et al*[15], in

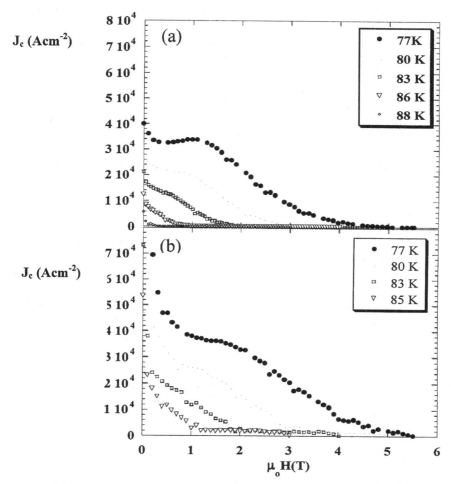

Fig. 6. Critical current density, J_c, of YBCO containing (a) Y-123 + 30 mol% Y-211 + 0.1 wt% DU + 0.1wt% Pt and (b) Y-123 + 30 mol% Y_2O_3 + 0.1 wt % DU + 0.1wt% Pt at different temperatures.

addition to the larger sized Y-211 particles. By enriching this composition with 30mol% Y-211, the sub-micron sized uranium compound precipitates are more predominant in the liquid and the average size of the Y-211 phase particles is reduced to 2-3 μm, as shown in figure 5(b). This microstructure is comparable to that of the textured grain (shown in figure 1(c)) where Y-211 particles and small U-compound precipitates coexist in the Y-123 matrix. The large Y-211 particles in the liquid (evident in figure 5(b)) ripen during the peritectic reaction, leading to the formation of few 4-5 μm sized Y-211 particles in the fully melt processed YBCO matrix (figure 1(c)). The few but relatively large Y-211 particles in the liquid may be refined by the addition of 0.1wt% Pt to composition (ii), as shown in figure 5(c). This behaviour is similar to that observed in the YBCO system without the addition of uranium. Addition of Pt prevents the coarsening of Y-211 particles in the pro-peritectic state by effectively reducing the interfacial energy between particle and liquid. As a result, the largest size of the Y-211 phase particles in the liquid in the peritectically decomposed state (figure 5(c)) remains the same as that in the YBCO matrix after solidification under slow cooling [figure 2(b)]. No significant reduction in either the size of the uranium compound or the size of the Y-211 phase particles is apparent for the pro-peritectic sample with higher Pt content (composition (iv), figure 5(d)) compared with lower Pt content. By enriching the initial composition with Y_2O_3 rather than Y-211 (Y-123 + 30mol% Y_2O_3 + 0.1wt% DU + 0.1wt% Pt), however, very uniform and smaller second phase particles are observed in the liquid in the pro-peritectic state, as shown in Figure 5(e). Even though this sample contains less DU than compositions (i) to (iv), most of the particles are of a sub-micron size and are distributed uniformly in the Ba-Cu-O liquid phase. In addition, no significant coarsening of second phase particles was observed for this composition after peritectic solidification under slow cooling, as illustrated by the very fine microstructure in the YBCO single grain shown in figure 4(a).

The critical current densities as a function of applied magnetic field at different temperatures are shown in figure 6 for the samples with initial composition (a) Y-123 + 30mol% Y-211 + 0.1% DU + 0.1wt% Pt and (b) Y-123 + 30mol% Y_2O_3 + 0.1% DU + 0.1wt% Pt, respectively. J_c is increased at medium fields for both samples resulting in the emergence of a peak in $J_c(H)$. A relatively high J_c of 35×10^3 A/cm^2 is observed at 77 K and 1 T for the DU doped sample containing excess Y-211. J_c decreases, however, as the applied field is increased further. A significantly higher J_c of 80×10^3 A/cm^2 at 0.1 T at 77K and 45×10^3 A/cm^2 at 1 T at 77K is observed for the sample prepared with excess Y_2O_3, as shown in figure 6(b). Furthermore, J_c remains relatively constant at 45×10^3 A/cm^2 over an applied field range of 0.8 T < B < 2 T and decreases relatively slowly at higher fields compared to the sample containing excess Y-211. The

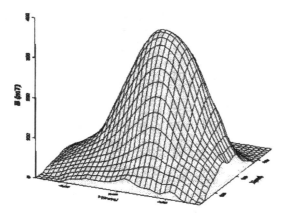

Fig. 7. Trapped flux profile of a 14 mm × 14 mm sized grain with the composition Y-123 + 30mol% Y_2O_3 + 0.1% DU + 0.1wt% Pt.

occurrence of a peak in J_c at relatively high field (2 T) in this sample is very encouraging from an applications point of view. The trapped flux profile of a 14 mm × 14 mm sized grain with the composition Y-123 + 30mol% Y_2O_3 + 0.1% DU + 0.1wt% Pt is shown in figure 7. A maximum trapped field of 0.38 T is obtained in this sample. The relatively high critical current density and high trapped field of U-doped YBCO containing excess Y_2O_3 is direct evidence of its potential for high magnetic field engineering applications.

CONCLUSIONS

Microstructural investigations of YBCO single grain superconductors containing various amounts of excess Y-211, Y_2O_3, Pt and depleted uranium have confirmeded that addition of uranium is effective in yielding very fine sized secondary phase particles containing U and/or Pt in YBCO prepared with excess Y-211. By enriching an initial composition of Y-123 + 0.1wt% DU + 0.1wt% Pt with 30mol% Y_2O_3 rather than 30mol% Y-211, as is common practice, large grain bulk YBCO superconductors have been fabricated with very fine sized, uranium containing second phase particles of ~ 200-300nm in diameter which exhibit high critical current densities, even at relatively high applied magnetic fields. Small sized particles are still observed without addition of Pt to the precursor powder for uranium doped YBCO prepared with Y_2O_3. This is an important development in

High Temperature Superconductors

reducing the cost of bulk melt processed material which exhibits high J_c and good field trapping capability. A relatively large trapped field of ~0.38 T at 77K for a 14 mm x 14 mm sized grain for YBCO prepared from Y-123 + 0.1wt% DU + 0.1wt% Pt with 30mol% Y_2O_3 indicates that uranium doping has considerable potential for high magnetic field engineering applications.

ACKNOWLEDGEMENTS

We are grateful to Prof. Roy Weinstein of the Texas Center for Superconductivity, Houston, for a number of valuable and productive discussions during the preparation of this manuscript. Hari Babu is is a Leverhulme Fellow supported by the Leverhulme Trust.

REFERENCES

[1] D. A. Cardwell, "Processing and properties of large grain (RE)BCO," *J. Mat. Sci. Eng. B*, **B53** 1-10 (1998).

[2] M. Murakami, M. Morita, K. Doi, and K. Miyamoto, A new process with the promise of high J_c in oxide superconductors," *Jpn. J. Appl. Phys.* **28** 1189-1194 (1989).

[3] N. Ogawa, I. Hirabayashi, and S. Tanaka, "Preparation of a high-J_c YBCO bulk superconductor by the platinum doped melt growth method," *Physica C*, **177** 101-105 (1991).

[4] N. Hari Babu, M. Kambara, P. J. Smith, D. A. Cardwell and Y. Shi, "Fabrication of large single-grain Y-Ba-Cu-O through infiltration and seeded growth processing," *J. Mater. Res.,* **15** 1235-1238 (2000).

[5] A.Umezawa, G. W. Crabtree, J. Z. Liu, H. W. Weber, W. K. Kwok, L. H. Nunez, T. J. Moran, C. H. Sowers and H. Claus, "Enhanced critical magnetization currents due to fast neutron irradiation in single-crystal $YBa_2Cu_3O_{7-\delta}$," *Phys. Rev. B* **36** 7151-7154 (1987).

[6] L. Civale, A. D. Marwick, T. K Worthington, M. A. Kirk, J. R. Thompson, L. Krusin-Elbaum, Y. R. Sun, J. R. Clem, and F. Holtzberg, "Vortex confinement by columnar defects in $YBa_2Cu_3O_7$ Crystals: Enhanced pinning at high fields and temperatures," *Phys. Rev. Lett.* **67**, 648-651 (1991).

[7] J. Giapintzakis, W. C. Lee, J. P. Rice, D. M. Ginsberg, and I. M. Robsertson, R. Wheeler and M. A. Kirk, M.-O. Ruault, "Production and identification of flux-pinning defects by electron irradiation in $YBa_2Cu_3O_{7-x}$ single crystals" *Phys. Rev. B* **45** 10677-10683 (1992).

[8] R. A. Fleisher, H. R. Hart, Jr., K. W. Lay, and F. E. Luborsky, "Increased flux pinning upon thermal-neutron irradiation of uranium-doped $YBa_2Cu_3O_7$," *Phys. Rev. B* **40** 2163-2169 (1989).

[9] R. Weinstein, R. Sawh, Y. Ren and D. Parks, "The role of uranium, with and without irradiation, in the achievement of $J_c \sim 300\ 000$ A cm^{-2} at 77 K in large grain melt-textured Y123" *Mater. Sci. Eng B,* **53**, 38-44 (1998).

[10] R. Weinstein, Y. Ren, R. Sawh, A. Gandini, W. Hennig, M. Murakami, T. Mochida, N. Chikumoto, N. Sakai, G. Krabbes, W. Bieger, D. Milliken, S. X. Dou, S. Tönie, M. Eisterer and H. W. Weber, "Properties of HTS for Successful U/n Processing" *Physica C,* **341-348**, 1415-1418 (2000).

[11] Y. Ren, R. Weinstein and R. Sawh, "New Chemical Pinning Centre from Uranium Compound in Melt Textured YBCO", *Physica C,* **282-287**, 2275-2276 (1997).

[12] R. Weinstein and R. Sawh, "A Class of Chemical Pinning Centres including Two Elements Foreign to HTS", Submitted to *Supercond. Sci. Technol.* (2002).

[13] R. Sawh, R. Weinstein, D. Parks, A. Gandini, Y. Ren and I. Rusakova, "Tungsten and Molybdenum Double Perovskites as Pinning Centres in Melt Textured Y123", Submitted to *Physica C* (2002).

[14] D. A. Cardwell, W. Lo, H. D. E Thorpe and A. Roberts, *J. Mater. Sci. Letts.,* **14**, 1444 (1995).

[15] I-G Chen, J. Liu, R. Weinstein and R. Sawh, p 657 in *Advances in Superconductivity IX* edited by S. Nakajima and M. Murakami (Tokyo: Springer-Verlag) (1997).

[16] G. H. Greenwood, *Acta Metall.,* **4**, 243 (1956).

[17] W. Lo, D. A. Cardwell, C. D. Dewhurst, H.-T. Leung, J. C. L. Chow and Y. H. Shi, "Controlled Prtoessing and Properrties of Large Pt doped YBCO Pseudo-crystals for Electromagnetic Applications", *J. Mater. Res.,* **12**, 2889, (1997).

[18] R-P Sawh, Y. Ren, R Weinstein, W. Hennig, and T. Nemoto, "Uranium chemistry and pinning centers in high temperature superconductor," *Physica C* **305** 159-166 (1998).

[19] N. Hari Babu, M. Kambara, Y-H. Shi, D. A. Cardwell, C. D. Tarrant and K. R. Schneider, "Processing and microstructure of single grain, uranium-doped Y-Ba-Cu-O superconductor," *Supercond. Sci. Tech.* **15** 104-110 (2002).

[20] G. Krabess, P. Schatzle, W. Bieger, U. Wiesner, G. Stover, M. Wu, T. Strasser, D. Litzkendorf, K. Fischer, P. Gornert, "Modified melt texturing process for YBCO based on the polythermic section $YO_{1.5}$-Ba$_{0.4}$Cu$_{0.6}$O in the Y-Ba-Cu-O phase-diagram at 0.21 bar oxygen pressure,"*Physica C,* **244**, 145-152 (1995).

[21] S. P. Athur, V. Selvamanickam, U. Balachandran, and K. Salama, " Study of growth kinetics in melt-textured $YBa_2Cu_3O_{7-x}$," J. Mater. Res., **11**, No12, 2976-2989 (1996).

APPLICATION OF RE123-BULK SUPERCONDUCTORS AS A PERMANENT MAGNET IN MAGNETRON SPUTTERING FILM DEPOSITION APPARATUS

Uichiro Mizutani
Department of Crystalline Materials
Science, Nagoya University
Furo-cho, Chikusa-ku, Nagoya
Japan 464-8603

Takashi Matsuda
DIAX Co., Ltd.
44 Chikoin Shimoyashiki-cho,
Kasugai,
Japan 486-0906

Yousuke Yanagi and Yoshitaka Itoh
IMRA MATERIAL R&D Co., Ltd.
5-50 Hachiken-cho, Kariya,
Japan 448-0021

Hiroshi Ikuta
Center for Integrated Research in Science
and Engineering, Nagoya University
Furo-cho, Chikusa-ku, Nagoya,
Japan 464-8603

Tetsuo Oka
AISIN SEIKI, Co., Ltd.
2-1 Asahi-machi, Kariya, Japan 448-8650

ABSTRACT

In the first stage of the present work, we constructed the prototype magnetron sputtering apparatus by replacing an ordinary Nd-Fe-B magnet with a superconducting permanent magnet capable of producing more than 8 Tesla immediately above its surface. The Sm123 bulk superconductor, 36 mm in diameter, was magnetized at low temperatures around 30 K by a static or pulsed field. Through the measurement of the dependence of the deposition rate on the radial magnetic flux density $B_{//}^{max}$, we were led to conclude that a deposition rate higher than 0.17 nm/s (100 Å/min) for films like Cu will be possible even under the Ar gas pressure less than 2.7×10^{-2} Pa (2×10^{-4} Torr), provided that the value of $B_{//}^{max}$ can be increased to 1 Tesla above the surface of the target. Encouraged by this conclusion, we have newly constructed a more practical magnetron sputtering system, which allows us to mount a bulk superconductor with a diameter up to 60 mm. The Ar gas pressure dependence of the discharge current was measured in this system by using the 36 mmϕ-Sm123 bulk superconductor. The data turn out to be promising enough to achieve our goal in very near future, provided that the present superconductor is replaced by the one with the diameter of 60 mm.

INTRODUCTION

The $REBa_2Cu_3O_{7-\delta}$ (RE=Sm, Nd, Gd etc.) (hereafter abbreviated as RE123) bulk superconductors have received much attention from the viewpoint of practical applications. A number of ideas have been proposed to make use of magnetic field trapped by the superconductor and to use it as a very strong permanent magnet. Included are a superconducting motor [1], a magnetic separator [2] and a magnetic field generator [3]. For example, a superconducting motor has been successfully constructed by simultaneously magnetizing twenty Y123 bulk superconductors cooled to 77 K by feeding a pulsed current through solenoid coils wound around them in series [1].

Magnetron sputtering has developed very rapidly over the last decade to facilitate the deposition of a wide range of industrially important coatings in response to an increasing need for high-quality functional films. In the basic sputtering process, a target or cathode plate is bombarded by energetic ions generated in glow discharge plasma produced immediately above the target. The bombardment process consists of the removal of target atoms with their subsequent condensation onto a substrate as a film. Secondary electrons are also emitted from the target surface as a result of the ion bombardment and serve as maintaining the plasma. In the basic sputtering mode, several drawbacks exist like its low deposition rates, low ionization efficiencies in the plasma and high substrate heating effects. These difficulties have been largely swept away by the development of magnetron sputtering [4].

The magnets are arranged in such a way that one pole is positioned at the central axis of the target and the second one forms a ring of soft magnetic material or permanent magnets around the outer edge of the target. Magnetrons make use of the fact that a closed loop of magnetic flux starting from the center of the target and ending near its periphery can constrain secondary electron motion to the vicinity of the target due to the Lorentz force. Trapping the electrons near the target substantially enhances the frequency of electron-atom collisions. The enhanced ionization efficiency results in a dense plasma in the vicinity of the target. This leads to an increase in ion bombardments against the target, resulting in higher sputtering rates and higher deposition rates at the substrate. In addition, an increased ionization efficiency achieved in the magnetron mode allows the glow discharge to be maintained at lower operating pressures of typically 10^{-1} Pa compared to 1 Pa in the basic mode. The Nd-Fe-B permanent magnet has been most frequently employed to confine secondary electrons near the target. The magnetic field component parallel to the target surface, $B_{//}$, is generally 0.03-0.05 Tesla and, at most, 0.2 Tesla, even when the magnetic circuit is carefully designed.

In the present work, we first report on the construction and performance of the prototype magnetron sputtering apparatus and, subsequently, that of a more practical system by employing a magnetized bulk superconductor as a permanent magnet in place of ordinary Nd-Fe-B magnet. The practical deposition system was designed to install a bulk superconductor with the diameter up to 60 mm with the aim at achieving as high magnetic flux density $B_{//}$ as possible. We showed that a deposition rate exceeding 0.17 nm/s (100 Å/min) for metal films like Cu will be made possible even at an Ar gas pressure less than 2.7×10^{-2} Pa in very near future, provided that $B_{//}$ is increased to 1 Tesla above the target.

SELECTION OF A BULK SUPERCONDUCTOR

A c-axis oriented Sm123 bulk superconductor was employed in the present magnetron sputtering apparatus. It was synthesized by the melt-texturing method under the Ar-gas flow atmosphere for a mixture of Sm123 and Sm_2BaCuO_5 (Sm211) powders with the molar ratio of Sm123 to Sm211 equal to 3:1 [5,6]. 15 wt.% of Ag_2O powder was added to increase its mechanical strength. The growth of 211 particles during the melt processing was suppressed by adding 0.5 wt.% Pt, while a Nd123 (001) crystal was used as a seed to grow the c-axis oriented single-domain Sm123 crystal. The final product after post-annealing in oxygen-gas flow was 36 mm in diameter and 14 mm in thickness.

CONSTRUCTION OF THE PROTOTYPE MAGNETRON SPUTTERING APPARATUS

As shown in Fig.1, the film deposition apparatus is composed of two separable housings. The top housing represents a main vacuum chamber, in which the cathode target is located at its bottom and a substrate stage 70 mm above the target. The bottom housing consists of a cylindrical head, in which the bulk superconductor, 36 mm in diameter, is mounted on a soft iron block cooled by the Gifford-McMahon (G-M) refrigerator, and a ring of soft iron yoke around the superconductor.

The bottom housing is initially detached from the top one, when the bulk superconductor is magnetized by either a static or pulsed field. The soft iron ring is also removed. In the case of static magnetization, the cylindrical head is inserted into the bore, 100 mm in inner diameter, of a 10 Tesla superconducting solenoid magnet and the magnetic flux of 6 Tesla was trapped at the surface of the bulk superconductor by cooling to 30 K in the field cooling mode [5,6]. The magnetization was alternatively performed by inserting the cylindrical head into a Cu solenoid coil immersed in liquid nitrogen in the pulsed field magnetization mode [7-9]. The trapped magnetic field could be increased to 4 Tesla at the surface of the superconductor at 30 K by applying the so called IMRA technique [7, 8].

After magnetization, the cylindrical head is placed at the center of the ring of the soft iron yoke, 64 mm in inner diameter, 84 mm in outer diameter and 45 mm in height (see Fig.1). The distribution of the trapped flux density was measured by scanning a three-axial Hall sensor (AREPOC Ltd., AXIS-3) over the top surface of the cylindrical head. Its center is taken as origin. The radial component $B_{//}$ is plotted in Fig.2 as functions of r and z along directions parallel and perpendicular to its surface, respectively, when the trapped field of 6 Tesla was produced at the

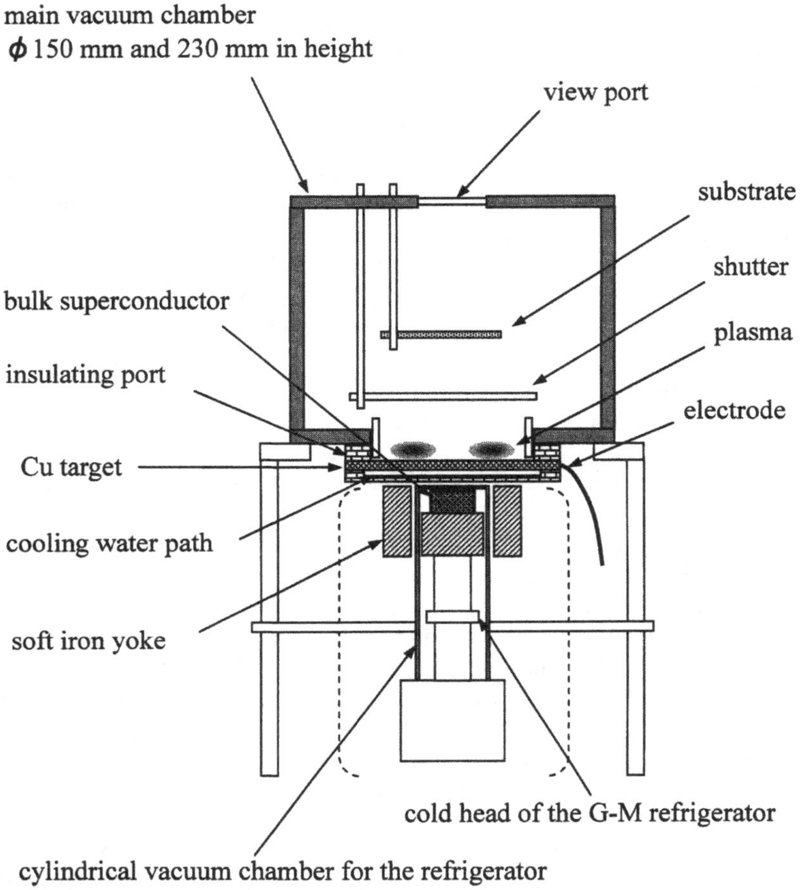

main vacuum chamber
ϕ 150 mm and 230 mm in height

view port

substrate

shutter

plasma

electrode

bulk superconductor

insulating port

Cu target

cooling water path

soft iron yoke

cold head of the G-M refrigerator

cylindrical vacuum chamber for the refrigerator

Fig.1 Schematic illustration of the prototype magnetron sputtering apparatus equipped with a superconducting permanent magnet, 36 mm in diameter, cooled by a refrigerator to 30K. Upon magnetization, the housing enclosed by a dashed line is detached.

surface of the superconductor magnetized by the field cooling mode. A dashed line drawn 5 mm above the cylinder surface corresponds to the top surface of the target. Chain lines indicate the position where a magnetic field vector is parallel to the surface of the target, i.e., $B_z = 0$. We found that the value of a horizontally aligned magnetic field, which is hereafter referred to as $B_{//}^{max}$, reaches 0.45 Tesla at r=25 mm from the center on the target. This magnetic field is one order of magnitude higher than that obtained by a conventional Nd-Fe-B permanent magnet.

High Temperature Superconductors

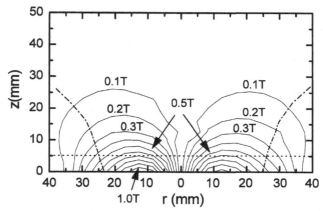

Fig. 2 Radial magnetic flux density $B_{//}$ as functions of r and z along directions parallel and perpendicular to the surface of the cylindrical head in the prototype apparatus, respectively. A field of 6 Tesla was trapped at the surface of the superconductor. A dashed line refers to the top surface of the target and the chain lines to the position where a magnetic field vector becomes parallel to the surface of the target, i.e., $B_z = 0$.

DISCHARGE CHARACTERISTICS OF THE PROTOTYPE MAGNETRON SPUTTERING APPARATUS

The discharge current is plotted in Fig.3 as a function of the target voltage

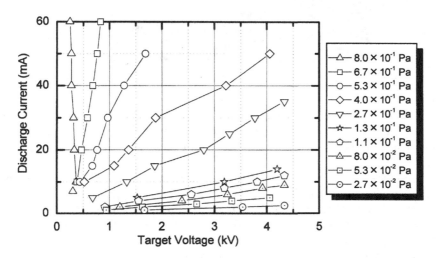

Fig. 3 Discharge current as a function of the target voltage under various Ar gas pressures in the presence of $B_{//}^{max} = 0.45$ Tesla obtained by static field magnetization in the prototype apparatus.

under various Ar gas pressures in the presence of $B_{//}^{max}$= 0.45 Tesla obtained by static field magnetization discussed above. Note that we studied the *I-V* characteristics only in the low Ar gas pressure range, though the discharge current could far exceed 100 mA when the Ar gas pressure is increased beyond the value of 6.7×10^{-1} Pa. As shown in Fig.3, we could confirm the persistence of discharging down to an Ar gas pressure of 2.7×10^{-2} Pa. The data in the low pressure range follow an approximate linear relation $I=\alpha(V-V_0)$, where V_0 is about 0.25 kV. This behavior is different from the relation $I=\beta(V-V_0)^2$ reported by Westwood et al. [10] in the planar magnetron at the discharge current above about 500 mA. Figure 4 shows a photograph of the discharge observed from the viewport at a pressure of 2.7×10^{-2} Pa.

Fig. 4 Glow discharge plasma observed through viewport in the prototype magnetron sputtering apparatus.

The Ar gas pressure dependence of the discharge current is plotted in Fig.5 under three different radial magnetic flux densities $B_{//}^{max}$ of 0.18, 0.38 and 0.45 Tesla. One can clearly see that the discharge current consistently increases with increasing $B_{//}^{max}$. To show the effect of the magnetic field more explicitly, the deposition rate, when pure Cu is used as a target, is plotted in the units of nm/s in Fig.6 as a function of $B_{//}^{max}$ under three different Ar gas pressures. We find that the deposition rate increases almost linearly with increasing $B_{//}^{max}$. In particular, an extrapolation of the data at 2.7×10^{-2} Pa to $B_{//}^{max}$ =1.0 Tesla indicates that the practically important deposition rate of 0.17 nm/s would be comfortably realized even at such low gas pressure, at least, for metal films like Cu.

High Temperature Superconductors

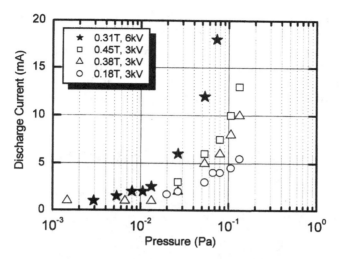

Fig. 5 Discharge current as a function of Ar gas pressure under different radial magnetic flux densities $B_{//}^{max}$ of 0.18, 0.38 and 0.45 Tesla in the prototype apparatus and 0.31 Tesla in the newly constructed system (see the next Section). The target voltage was fixed at 3 and 6 kV, respectively.

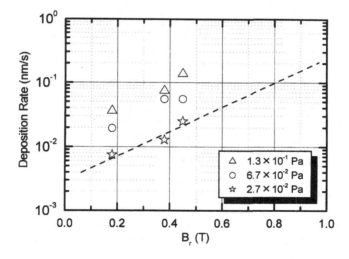

Fig. 6 Deposition rate as a function of $B_{//}^{max}$ under three different Ar gas pressures in the prototype apparatus. An extrapolated dashed line drawn through the data for the pressure of 2.7×10^{-2} Pa exceeds a practically important deposition rate of 0.17 nm/s or 100 Å/min at $B_{//}^{max}$=1.0 Tesla for metal films like Cu.

High Temperature Superconductors

CONSTRUCTION OF A MORE PRACTICAL MAGNETRON SPUTTERING SYSTEM

Encouraged by the performance of the prototype apparatus, we have newly constructed a more practical magnetron sputtering system, which can accommodate a bulk superconductor with a diameter up to 60 mm and is again separable into the top and bottom housings. A photograph of the system is shown in

Fig. 7 A whole assembly of our newly constructed magnetron sputtering system equipped with a superconducting permanent magnet.

Fig.7. In the top housing, the planar magnetron cathode, 125 mm in diameter, is installed in a stainless steel vacuum chamber, 500x500 mm in length and 600 mm in height. The oxygen-free highly conductive (OFHC) Cu plate, 3 mm in thickness, was used as a target in the present work. Deionized water is circulated underneath the target to suppress excessive heating during sputtering.

The Sm123 bulk superconductor, 36 mm in diameter and 15 mm in height, was again mounted on a cold stage located at the center of a stainless steel vacuum chamber to protect it from freezing the moisture. The distance from the top surface of the cylindrical head to that of the target was 8 mm in the present system. The magnetization of the bulk superconductor can again be achieved either by static or pulsed field in the same way as the prototype discussed above. We show below the discharge characteristics, when the bulk superconductor was magnetized at 35 K by the pulsed field-IMRA method [7-9].

After magnetizing the bulk superconductor, the bottom housing was connected to the bottom of the main vacuum chamber to perform magnetron sputtering of OFHC-Cu. The superconductor now serves as a permanent magnet and produces magnetic field above the target. The maximum radial magnetic flux density of 0.31 Tesla was produced immediately above the target surface.

The discharge current was measured as a function of the Ar gas pressure, when the target voltage is kept at 6 kV. The results are incorporated into Fig.5, where the data obtained using the prototype apparatus with the target voltage of 3 kV have already been discussed. One can clearly see that, in spite of a decrease in $B_{//}^{max}$ to 0.31 Tesla, the discharge current increased about twice as compared to that in the prototype one with $B_{//}^{max}$=0.45 Tesla. This enhancement is apparently attributed to an increase in the target voltage to 6 kV.

As mentioned above, our final objective is to achieve the deposition rate more than 0.17 nm/s for films like Cu under the Ar gas pressure less than 2.7×10^{-2} Pa by increasing $B_{//}^{max}$ up to 1 Tesla. The following experiments are under way to reach this goal.

(1) The replacement of the Sm123 bulk superconductor, 36 mm in diameter, by a RE123 bulk superconductor, 60 mm in diameter, is critical. A RE123 bulk superconductor of this size will be able to produce $B_{//}^{max}$ exceeding 1 Tesla above the target. The synthesis of Sm123 bulk superconductors, 60 mm in diameter, is now in progress.

(2) The RE123 superconducting permanent magnet is surrounded by a ring of soft iron yokes to enhance $B_{//}^{max}$. An optimization of the magnetic circuit including the shape and location of the soft-iron yokes will be made by performing computer simulations, which will allow us to calculate the magnetic flux distribution produced from a given magnetic circuit.

CONCLUSION

The prototype magnetron sputtering apparatus was constructed by replacing ordinary Nd-Fe-B magnet with a 36 mm-diameter Sm123 bulk superconductor, which was magnetized at 30 K by a static or pulsed field. It could produce the radial component of magnetic field of 0.45 Tesla above the target. The deposition rate higher than 0.17 nm/s (100 Å/min) for films like Cu will be possible under the Ar gas pressure less than 2.7×10^{-2} Pa (2×10^{-4} Torr), if the value of $B_{//}^{max}$ can be increased to 1 Tesla. Encouraged by the results above, we have newly constructed a more practical magnetron sputtering system, which can accommodate a bulk superconductor with a diameter up to 60 mm. The Ar gas pressure dependence of the discharge current was so far measured in this system by using the Sm123 bulk superconductor with 36 mm in diameter. The data turn out to be promising enough to achieve our goal, i.e., a discharge current higher than 10 mA under the vacuum of 2×10^{-4} Torr by replacing the present superconductor with a 60 mm-diameter one free from weak links.

ACKNOWLEDGMENTS

The authors have benefited from many fruitful discussions with Dr.A.Imai, Nagoya Industrial Science Research Institute, and Mr.K.Sakurai, Shizuoka ANELVA Co., Ltd., during the course of the present research work. We are most thankful for this invaluable cooperation and help. This work was supported by a Grant-in-Aid for the Development of Innovative Technology from the Ministry of Education, Sports, Culture, Science and Technology.

REFERENCES

1. Y.Itoh, Y.Yanagi, M.Yoshikawa, T.Oka, S.Harada, T.Sakakibara, Y.Yamada and U.Mizutani, "High-Temperature Superconducting Motor Using Y-Ba-Cu-O Bulk Magnets", *Jpn.J.Appl.Phys.* **34** 5574-78 (1995).
2. N.Saho, presented at the *American Ceramic Society's 104th Annual Meeting* (St.Louis, U.S.A., 2002).
3. T.Oka, Y.Itoh, Y.Yanagi, M.Yoshikawa, H.Ikuta and U.Mizutani, "Construction of a 2-5 T class superconducting magnetic field generator with use of an Sm123 bulk superconductor and its application to high-magnetic field demanding devices", *Physica* C **335** 101-6 (2000).
4. P.J.Kelly and R.D.Arnell, "Magnetron sputtering: a review of recent developments and applications", *Vacuum* **56** 159-72 (2000).
5. H. Ikuta, A. Mase, Y. Yanagi, M. Yoshikawa, Y. Itoh, T. Oka and U. Mizutani, "Melt-Processed Sm-Ba-Cu-O Superconductors Trapping Strong Magnetic Field", *Supercond. Sci. Technol.* **11** 1345-7 (1998).
6. U.Mizutani, A.Mase, H.Ikuta, Y.Yanagi, M.Yoshikawa, Y.Itoh and T.Oka, "Synthesis of c-axis oriented single-domain Sm123 superconductors capable of trapping 9 Tesla at 25 K and its application to a strong magnetic field generator", Mat.Sci.Eng. **B65** 66-8 (1999).
7. U.Mizutani, T.Oka, Y.Itoh, Y.Yanagi, M.Yoshikawa and H.Ikuta, "Pulsed-field Magnetization applied to high-Tc superconductors", *Applied Superconductivity* **6** 235-46 (1998).
8. Y.Yanagi, Y.Itoh, M.Yoshikawa, T.Oka, T.Hosokawa, H.Ishihara, H.Ikuta and U.Mizutani, "Trapped field distribution on Sm-Ba-Cu-O Bulk superconductor by pulsed-field magnetization", Proc. of the 12th Int.Symp. on Superconductivity (ISS'99) (October 17-19, 1999, Morioka), 470-2.
9. H.Ikuta, H.Ishihara, T.Hosokawa, Y.Yanagi, Y.Itoh, M.Yoshikawa, T.Oka and

U.Mizutani, "Pulse field magnetization of melt-processed Sm-Ba-Cu-O", Supercond.Sci.Technol. **13** 846-9 (2000).
10. W.D.Westwood, S.Maniv and P.J.Scanion, J.Appl.Phys. **54** 6841-6 (1983)

TAILORING DISLOCATION SUBSTRUCTURES FOR HIGH CRITICAL CURRENT MELT TEXTURED YBa$_2$Cu$_3$O$_7$

F. Sandiumenge[*], X. Obradors and T. Puig
Institut de Ciència de Materials de Barcelona (CSIC)
Campus de la UAB, 08193 Bellaterra, Catalonia, Spain

J. Rabier
Université de Poitiers – CNRS, Laboratoire de Metallurgie Physique
Poitiers, Chasseneuil 86962, France

J. Plain[**]
Institut de Ciència de Materials de Barcelona (CSIC)
Campus de la UAB, 08193 Bellaterra, Catalonia, Spain;
and Université de Poitiers – CNRS, Laboratoire de Metallurgie Physique
Poitiers, Chasseneuil 86962, France

ABSTRACT

The introduction of dislocations acting as flux pinning centers in melt textured YBa$_2$Cu$_3$O$_7$, is a complex issue owing to the difficulty of inducing plastic deformation without degrading the microstructure in these brittle and metastable ceramics. This overview is devoted to processing routes having in common the use of a confinement gaseous medium and intermediate to low operating temperatures, as opposed to early conventional high temperature approaches. In this way, J$_c$ enhancements up to 180% at 5K have been obtained. An analysis of the flux pinning mechanisms is also presented, revealing that the newly created in-plane dislocation substructure is responsible for the total observed enhancement.

INTRODUCTION

Flux pinning by dislocations in plastically deformed superconductors [1,2] was the subject of renewed attention soon after the discovery of ceramic superconductors, in particular with the development of high critical current melt textured YBa$_2$Cu$_3$O$_7$ bulk ceramics. Pioneering work by Selvamanickam et al. indeed demonstrated significant J$_c$ enhancements in both, hot

---

[*] Corresponding author, e-mail: felip@icmab.es
[**] Present address: LEMA, Faculté des Sciences et Techniques, Parc de Grandmont, Université François Rabelais (Tours) France.

---

isostatically pressed [3] and uniaxially compressed [4] melt textured $YBa_2Cu_3O_7$-Y_2BaCuO_5 (1-2-3/2-1-1) composites. However, subsequent experiments in this direction could not systematically reproduce those initial optimistic results. It was soon realized that if, on the one hand, high temperatures were required to achieve sufficient plasticity in these brittle materials, on the other hand, such elevated temperatures also resulted in oxygen losses in $YBa_2Cu_3O_{7-\delta}$ and accordingly as deformed samples needed to be annealed in oxygen in order to recover their superconducting state. Careful studies of the microstructural evolution upon high temperature deformation under creep conditions and subsequent oxygenation indeed revealed a dramatic modification of the as-deformed microstructure [5], responsible for an observed concomitant degradation of the superconducting properties [6]. Since then, a large effort has been directed towards processing routes taking place in the orthorhombic phase. In this context, three kinds of approaches are distinguished: (a) Deformation in the high temperature regime (~900°C), (b) Deformation in the low temperature regime (300°C), and (c) Processing under high pO_2. In approach (a), two types of experiment have been performed: constant strain rate under oxidizing atmosphere [7], and deformation in a recessed anvil-type apparatus assembly in which the sample is in contact with annealed zirconia during the process at $2-5$ GPa and 700 - 1500°C and remains a 91K superconductor after deformation [8]. In this overview we focus on approaches (b) and (c), having as distinctive features the use of a confining gaseous medium and intermediate to low operating temperatures. Research on this newly developed strategies has allowed the largest J_c enhancements reported so far, up 180% at 5K.

ALTERNATIVE WAYS FOR THE GENERATION OF DISLOCATION SUBSTRUCTURES IN MELT TEXTURED 1-2-3/2-1-1 COMPOSITES

The objective of introducing dislocations in single domain melt textured 1-2-3 is to increase the density of flux pinning centers. For that purpose, the ideal dislocation substructure consists of a random distribution of dislocations, as opposed to subboundaries that constitute stable self organized arrangements of dislocations that, owing to their geometry, in most cases are likely to act as current limitation defects [9]. Unfortunately, the brittle behavior and metastability of 1-2-3 cuprates constitute serious impediments that limit the application of conventional plastic deformation routes. A further difficulty arises from the fact that those two problems cannot be solved independently. Indeed, at high temperatures (~900°C), where the plasticity may be enhanced [10], deformation takes place in the tetragonal phase [5] and therefore the samples need to be oxygenated to recover the superconducting orthorhombic phase. As a result of the metastability of 1-2-3, the as-deformed microstructure is severely affected during this stage, typically carried out at ~450°C under pO_2~1 bar, causing a degradation of the superconducting properties [6]. In order to understand microstructural

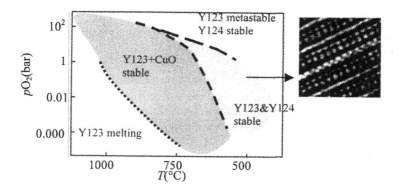

Fig. 1: Superposition of the stability fields of 1-2-3 and 1-2-4 in $pO_2 - T$ space using data from Refs. 11-13. HRTEM image shows a 1-2-4 like stacking fault viewed along the [100] direction (lattice period equals $d_{(001)}$).

consequences of phase metastability, Fig. 1 shows a superposition of the stability fields of 1-2-3 and its polytypoid 1-2-4, in $pO_2 - T$ space [11 - 13]. Three thermal regimes can be distinguished: (a) at high temperature, where a 1-2-3+CuO mixture is stable against 1-2-4, (b) at intermediate temperatures where 1-2-3 and 1-2-4 can coexist, and (c) at low temperatures, corresponding to typical oxygenation conditions (~450°C, 1 bar oxygen) where 1-2-4 is stable against a 1-2-3+CuO mixture. The 1-2-4 phase can be obtained by inserting an extra CuO_x chain layer in every unit cell [14]. However, since CuO appears as an inhomogeneously distributed impurity, presumably incorporated as a result of the decomposition of the 1-2-3 phase under oxidizing conditions [11] and presumably also during the solidification process, the 1-2-4 phase does not form as a bulk phase but in the form of non-ordered stacking faults as shown in the high resolution transmission electron microscopy image of Fig. 1. Such extrinsic faults nucleate preferentially at stress centers like 2-1-1/1-2-3 interfaces. The Burgers vector of the partial dislocation bounding 1-2-4 type faults is 1/6<301>. An important consequence of this scenario is that 1/6<301> faults appear as a response to a local phase transformation and can therefore expand over wide areas on the (001) plane, owing to a negative stacking fault energy (γ<0) [15].

Given the convergence, on the one hand, of strong difficulties to create perfect dislocations with, on the other hand, the facility to introduce 1/6<301> partials through the nucleation of stacking faults, we have envisaged a strategy that takes advantage of phase metastabilty to increase the dislocation density in melt textured 1-2-3. An obvious challenge emerging from the thermodynamic nature of 1-2-4 stacking faults introduced above, is the control of their size. Indeed, the desired effect of 1/6<301> partial dislocations may be counterbalanced by the concomitant increase of a weakly superconducting stacking fault surface area. Thus, the size of the newly introduced stacking faults must be as small as possible in order to maximize their perimeter to surface area ratio.

EXAMPLES OF TAILORED DISLOCATION SUBSTRUCTURES

In this section, we provide three examples illustrating the potential of using a gaseous confinement pressure at temperatures where the material retains its orthorhombic form. These examples cover a wide range conditions, useful for the increase of the 1/6<301> partial dislocation length while keeping the size of their associated stacking faults as small as possible.

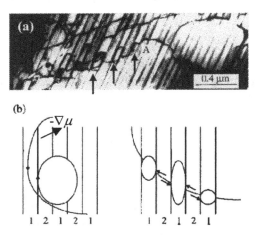

Fig. 2: TEM micrograph showing the selective dissociation of a perfect dislocation into stacking faults (arrowed in the image) in one twin variant (a). Schematic drawing of the basic mechanism (see text) (b).

Reorganization of stacking faults in one twin variant by Cold Isostatic Pressing (CIP)

By applying an argon pressure of 2 Kbar at 300 °C it is found that the dislocation density is increased preferentially around peritectic inclusions, where maximum values two to three orders of magnitude higher than in the virgin sample are obtained [16]. This demonstrates that the plastic anisotropy between the peritectic inclusions and the 1-2-3 matrix [17] is efficient for defect generation under CIP conditions. Besides such a local enhancement of dislocation density, the most striking feature found in CIP treated samples is a change in the morphology and distribution of stacking faults [16]. While in the virgin samples stacking faults typically expand across different twin variants, say 1 and 2, after CIP stacking faults expand selectively in one twin variant. Fig. 2a is a TEM micrograph of the (001) plane of a CIP processed sample. The dark area in the left hand side of the

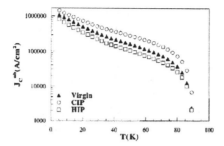

Fig. 3: Temperature dependence of the self-field critical current density for the virgin, CIP and HIP processed samples.

image is a 2-1-1 inclusion acting as nucleation site for the dislocations running along <100>. The banded structure corresponds to alternating twin domains 1 and 2, oriented parallel to the {110} planes. It can be observed that the lower dislocation (indicated by A) is decorated with stacking faults appearing at every other twin domain. This observation indicates that the stacking fault energy is not the same in the two twin variants [18]. Note that this distinction is only relevant to the orthorhombic twinned lattice and is associated to the different orientations of the displacement vector of the stacking fault relative to its neighboring matrix. The basic mechanism associated to the observed microstructural modification can be explained as follows (Fig. 2b). Consider first a stacking fault bounded by a partial dislocation with Burgers vector $b=1/6<031>$ expanding over several twin domains. Assuming that a deviatoric stress σ results from an anisotropy in the material, this yields to a Peach – Kolher force F on the partial dislocation. The total force acting on the dislocation in each twin variant is [16], $F+\gamma_i$ where γ is an apparent stacking fault energy [19] relevant to the intercalation of a CuO_x layer in the perfect matrix (the associated chemical potential of a CuO molecule is $\mu_i=\gamma_i S$, $i=1,2$, where S is the surface of a molecule) and subscripts 1 and 2 stand for the two twin variants. Assuming that the Burgers vector b_1 in variant 1 is $1/6[031]$ and that in variant 2 is $b_2=1/6[301]$, then $\gamma_1<\gamma_2$ since the former configuration corresponds to that found in the 1-2-4 structure [14]. As schematically shows in Fig. 2b, this difference yields to a driving force $-\nabla\mu$ for the diffusion of Cu-O species from variant 2 to variant 1, which exists in the absence of an applied stress. However this chemical force is only efficient when the total force applied on the dislocation is large enough to move the dislocation on the (001) plane. In the orthorhombic phase, this requires large applied stresses, which are obtained in the CIP process. Fig. 2b illustrates the case where the stacking fault interacts with a moving dislocation which is attractive with the partial loop, facilitating diffusion of Cu-O species through the core of the dislocation, leading to a

High Temperature Superconductors

"selective dissociation" [16] of the dislocation in variant 1 following the reaction $[010] \rightarrow \frac{1}{6}[031] + \frac{1}{6}[03\bar{1}]$. This leads to a configuration as that shown in dislocation labeled A in Fig. 2a.

The key point of this mechanism is that starting from a stacking fault spread out over several twin variants, it leads to a fragmentation of the stacking fault surface into several loops located in only one preferred twin variant. Thus, for a given stacking fault surface area in the crystal, this mechanisms results in an increase of the total partial dislocation length. Fig. 3 shows the inductive temperature dependence of J_c^{ab} for the virgin and CIP samples [16]. Results obtained after a hot isostatic pressing (HIP) process at 750°C followed by an oxygenation treatment at 450°C for 24h are included for comparison. A notorious increase of the critical currents is observed in the whole range of temperatures, which is maximum at 77K where an enhancement factor of 100% is achieved. It is also evident from this figure that a HIP process has a detrimental effect leading to reductions of J_c of about 60% at 77K.

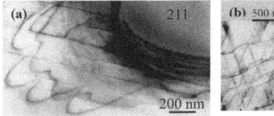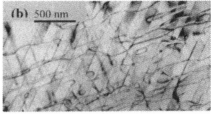

Fig. 4: TEM image of the microstructure on the (001) plane after 12 hours HOP processing at 400°C (a) and 600°C (b).

High Oxygen Pressure Processing: (HOP)

In HOP, the samples are submitted to $pO_2=100$ bar. The temperature, typically in the range 400 to 600°C, and process duration (1/2 hour to 12 hours) determine the kinetics and extend of the transformation [20 – 22]. Looking at Fig. 1 we find that under the experimental conditions used in HOP processing, the 1-2-3 phase is brought far from its stability field, within the stability field of 1-2-4. In contrast with CIP, where the 1-2-3 phase remains closer to its stability line and therefore no new staking faults are nucleated, under HOP conditions a large thermodynamic driving force is expected, which favors an enhanced nucleation and rapid growth of new stacking faults at preferential nucleation sites like 2-1-1 interfaces. Fig. 4a shows a TEM image of the typical microstructure on the (001) plane around the 2-1-1 particles obtained by HOP at 400°C for 12 hours. A characteristic feature is the development of small dendritic-like stacking faults, characteristic of rapid growth under strong non-equilibrium conditions. In

High Temperature Superconductors

this way, the dislocation density is preferentially enhanced around the 2-1-1 particles (the average partial dislocation density within a distance of 400 nm from the their interfaces may reach values of 2×10^{10} cm^{-2}, while the dislocation density in the bulk matrix remains almost unaltered [22]). Unlike the configurations generated under CIP where, as described above, the expansion of the stacking faults is constrained by the {110} twin walls, in the present case the stacking faults are found to expand across consecutive twin domains, their fast growing direction being the <100> in-plane screw component of their bounding partial dislocation [21]. By increasing the processing temperature up to 600°C, a dramatic increase of the partial dislocation density in the bulk matrix takes place, up to an average value of 2.5×10^{10} cm^{-2} [21,22] (Fig. 4b). In this case, after 12 hours, large stacking faults expand through the whole matrix.

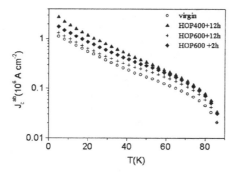

Fig. 5: Temperature dependence of J_c^{ab} for HOP samples processed under various conditions: at 600°C for 2 and 12 hours, and at 400°C for 12 hours. The curve corresponding to the virgin sample is included for comparison.

The temperature dependencies of J_c^{ab} for HOP samples processed at 400°C 12 hours, and 600°C for 2 and 12 hours, are presented in Fig. 5. The corresponding curve for the virgin sample is included for comparison. It is interesting to note that all curves lie above the virgin one for the whole temperature range. The enhancement is maximum for the 400°C+12h sample, in which the partial dislocation length has been substantially increased around the 2-1-1 particles while keeping the size of the stacking faults small. The fact that for the 600°C+12h sample, where the increase of dislocation density is slightly larger and, most important, is more homogeneously distributed in the matrix, the J_c enhancement is almost null, nicely illustrates the counterbalancing effect of the stacking fault surface area against the dislocation length. Indeed, as commented above, in this case TEM observations indicated that stacking faults overlap through the whole observable areas of the thin foils (thickness around 50 to 100 nm) [21,22]. In fact, the existence of large stacking faults is likely to result in enhanced thermal activation effects, as signaled by a downwards shift of the irreversibility line experienced by samples submitted 12 hours at 600°C [22]. Moreover, the expansion of fast growing stacking faults at 600°C can be controlled through the process duration, as illustrated by the curve obtained by reducing the processing time from 12 to 2 hours in the 600°C regime.

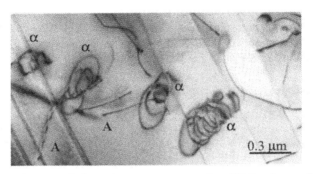

Fig. 6: TEM image of the microstructure on the (001) plane after uniaxial deformation in a gaseous confinement medium at 300°C revealing dislocation configurations involving perfect dislocations and 1/6<301> partial loops associated to them. A: **b**=[100], α: **b**=1/6[301].

Low temperature uniaxial deformation under a gaseous confinement pressure
    Uniaxial deformation under a gaseous (Ar) confining pressure of 0.3 GPa is carried out in a Paterson apparatus operated at 300°C and a strain rate of 2×10^{-5} s^{-1} [23]. The stress is oriented such as to favor dislocation glide on (001), the easy glide plane, as well as on (100) and (010). Microstructural observations of deformed samples reveal an increase of the density of perfect dislocations as well as an unexpected increase of the density of stacking faults in the bulk matrix. This can be seen in Fig. 6 where the dislocation substructure is imaged on the (001) plane with the **g**=200 diffraction vector. Diffraction contrast analysis reveals four 1/6<301> partial dislocation sources (labelled α) associated to perfect dislocations (**b**=<100>, labeled A) exhibiting long segments out of the easy glide plane (001). Strikingly, in each source the stacking faults appear grouped and stacked along the c-axis, suggesting that they are related by the same nucleation mechanism. The geometry of the configurations is consistent with a previously proposed theoretical model for stress driven formation of 1-2-4-type polytypoids [19]. The different stages of this partial dislocation Frank-Read source is schematized in Fig. 7 [23]. Consider a **b**=<100> dislocation gliding out of the primary glide plane (001) but having a short screw segment on that plane (a). This dislocation segment is then liable to dissociate on (001) following the reaction <100>→1/6 <301>+ 1/6 <30-1> provided that it is acted on by a

High Temperature Superconductors

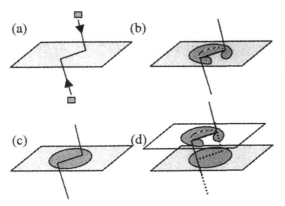

Fig. 7: Schematic drawings illustrating the mechanisms of the partial Frank-Read source. (a) CuO diffusion along an out of plane perfect dislocation having a segment on the (001) plane (CuO indicated by a shadowed square), (b) precipitation of CuO and dissociation of the perfect in plane dislocation segment, (c) closing of the loop by moving the partials around the pinning points and restoration of the initial perfect screw segment by recombination of the two partials, and finally, (d) cross-slip of the recombined perfect screw segment to an adjacent (001) plane and dissociation of this segment on this (001) plane. The process continues until no more CuO is available.

chemical force for CuO precipitation onto this perfect segment (b). Note that while the two partials are submitted to the same mechanical force $\mathbf{F}_m$ for gliding on (001), the chemical driving force $\mathbf{F}_c$ promoting the expansion of the stacking fault surface, acts in opposite direction on the two partials, $\gamma < 0$ [15], on the (001) plane. As a result, the chemical forces acting on the two partials are in opposite directions. This leads to a driving force $\mathbf{F}_c+\mathbf{F}_m$ on the leading partial and a driving force $\mathbf{F}_c-\mathbf{F}_m$ on the trailing one, so that the leading partial can expand leaving behind the trailing one. By moving around the pinning points, the leading partial generates a faulted loop and can ultimately recombine with the trailing partial for restoring the initial $\mathbf{b}=<100>$ dislocation segment (c). Note that the CuO feeding process of the faulted loop by the parent dislocations stops at this stage. This mechanism cannot operate once again by moving the leading partial in the faulted area. Rather the process can be perpetuated by cross-slip of the recombined $\mathbf{b}=<100>$ screw segment to the next (001) plane were the same mechanism can start again from stage (a) yielding a new faulted loop on that (001) plane (d). Therefore, a single out of plane dislocation segment can act as source for a large number of small stacking fault loops. This mechanism is non conservative and shuts off when no more copper oxide is available. The perfect dislocation segments located out of (001) have two functions in this mechanism: they act as pinning points for dislocation segments moving on (001) and they provide fast diffusion paths for CuO feeding the stacking fault. Furthermore those dislocations help to collect CuO by sweeping up the matrix during glide.

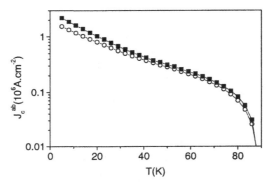

Fig. 8: Temperature dependence of the self-field critical current density for the samples prior to deformation (open circles) and after uniaxial deformation under a gaseous confinement pressure (black squares).

From a flux pinning perspective, a relevant aspect of this mechanism is that it generates a stack of small faulted loops rather than a large stacking fault expanding over wide areas on the (001) plane. This mechanism therefore allows the generation of stacking faults having a large perimeter to surface area ratio which is essential in order to optimize the pinning efficiency by stacking faults. The self field temperature dependence of the critical current density $J_c^{ab}(T)$ is shown in Fig. 8 for the virgin sample and the deformed one [24]. It can be clearly observed that $J_c^{ab}(T)$ is enhanced after the deformation particularly in the low temperature regime, reaching a value of 2.2×10^6 A/cm^2, i.e. an enhancement factor of 45% at 5K.

ANALYSIS OF FLUX PINNING MECHANISMS

The above three examples indeed reveal a clear correspondence between processing conditions promoting the increase of dislocation density and critical current enhancements. Deeper insights into the flux pinning mechanism responsible for the observed behavior can be extracted from a careful analysis of the $J_c^{ab}(T)$ dependencies. A detailed discussion including also the field dependencies can be found in Ref. 22. The basic idea for modeling the selective contribution of in-plane dislocations to the critical current density of melt processed 1-2-3/2-1-1 composites, when the flux line lattice is oriented along the c-axis, is the distinction of two major contributions, namely, a weak pinning mechanism (WP) described by $J_{WP} \exp(-T/T_0)$ [25] and a correlated disorder pinning mechanism (CD)

High Temperature Superconductors

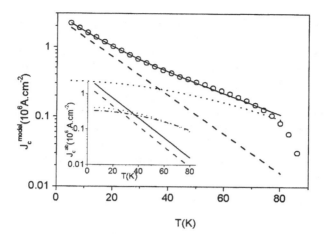

Fig. 9: Separation of the weak pinning (---) and correlated disorder (...) contributions to the critical current density in the deformed sample, as obtained by fitting eq. (1) to experimental data. The inset shows the weak pinning contribution in the deformed (solid line) and as-grown (dash line) samples, and the correlated disorder contribution in the deformed (dot line) and the as-grown (dash-dot line) samples.

described by $J_{CD}exp[(-3T/T^*)^2]$ [26], where J_{WP} and J_{CD} are factors proportional to the number of pinning centers and T_0 and T^* describe the pinning energy. The WP mechanism accounts for the pinning by point-like defects [27] and as shown recently, it also describes well the pinning by in-plane dislocations when the flux line lattice is oriented along the c-axis [22]. CD pinning is associated to linear or planar defects lying parallel to the magnetic field and in melt processed 1-2-3/2-1-1 composites provides an appropriate description of the contribution of the 2-1-1/1-2-3 interfaces [22,28]. Therefore, the self field temperature dependence of the critical current density can be described by the sum of those two contributions,

$$J_c^{ab}(T)= J_{WP}exp(-T/T_0)+ J_{CD}exp[(-3T/T^*)^2] \qquad (1).$$

As an example, Fig. 9 shows the fits of the experimental $J_c^{ab}(T)$ curves to eq. (1) as well as the two contributions separately, for the uniaxially deformed sample [24]. It can be observed that eq. (1) fits well the experimental curve up to T=80K, where a sharp drop of $J_c^{ab}(T)$ occurs. From the separate plots of the weak pinning and correlated disorder contributions, it becomes evident that WP dominates the overall $J_c^{ab}(T)$ behavior up to T=35K and that above this temperature the relative weight of the two contributions is inverted. An analysis of the temperature dependence of the weak and strong pinning contributions (see inset) clearly reveals that after processing, a selective enhancement of WP, associated to the newly generated in-plane partial dislocations, takes place, while the CD pinning contribution is kept constant within experimental accuracy, as expected taking into account that during the processing the total 2-1-1/1-2-3 interface area remains unaltered. A similar analysis for HOP samples yields identical results [22], strongly suggesting that regardless the processing route, flux pinning by in-plane partial dislocations nucleated within the bulk matrix constitute the main contribution to the enhancement of $J_c^{ab}(T)$, and that these in-plane partial dislocations in fact do behave as point-like pinning centers when the flux line lattice is oriented parallel to the c-axis.

CONCLUSIONS

This overview demonstrates that it is possible to take advantage of phase metastability effects for the generation of partial dislocation substructures efficient as flux pinning centers, through the controlled nucleation of 1-2-4 type stacking faults. The driving force for the nucleation and growth of such faults requires two well differentiated components: one thermodynamic force inducing the precipitation of CuO layers, and a mechanical force for the movement of their partial dislocations on the (001) plane. The examples presented here cover a range of experimental situations in which the relative weight of these two components is inverted. Thus, under HOP the main component is thermodynamic, whereas under CIP and low T uniaxial compression the driving force is mainly of mechanical nature. Among these latter ones, CIP represents an extreme situation of null thermodynamic component. Accordingly, no new stacking faults are nucleated. In this special case, the increase of partial dislocation length is driven by the minimization of the stacking fault energy in one twin variant, which ultimately results in a chemical force for the diffusion of Cu-O species, coupled with the dominant mechanical component. In all cases a significant increase of the critical current density is obtained. An analysis of the critical current behavior assuming a single vortex approximation, strongly suggests that, regardless the processing route, the newly generated in-plane partial dislocation substructures are responsible for the observed critical current enhancements.

ACKNOWLEDGEMENTS

This work has been supported by Supercurrents (EU – TMR Network ERBFMRXCT98 – 0189), CICYT (MAT99 – 0855) and Generalitat de Catalunya – CNRS (PICS 1015). Uniaxial deformation under gaseous confinement is performed at the Bayerisches Geoinstitut under the EU "IHP - Access to Research Infrastructures" Programme (Contract No. HPRI-1999-CT-00004 to D.C. Rubie).

High Temperature Superconductors

REFERENCES

[1] A. M. Campbell and J. E. Evetts, *Adv. Phys.* **21**, 199 (1972).

[2] D. Dew – Hughes and M. J. Witcomb, *Phil. Mag.*, **26**, 73 (1972).

[3] V. Selvamanickam, M. Mironova, S. Son and K. Salama, *Physica C*, **208**, 238 (1993).

[4] V. Selvamanickam, M. Mironova and K. Salama, *J. Mater. Res.* **8**, 249 (1993).

[5] N. Vilalta, F. Sandiumenge, J. Rabier, M. F. Denanot and X. Obradors, *Philos. Mag. A* **76**, 837-55 (1997).

[6] N. Vilalta, F. Sandiumenge, E. Rodríguez, B. Martínez, S. Piñol, X. Obradors and J. Rabier, *Philos. Mag. B* **75**, 431 (1997).

[7] M. Ullrich, A. Leenders, J. Krelauss, L.-O. Kautschor, H. C. Freyhardt, L. Schmidt, F. Sandiumenge and X. Obradors, *J. Mater. Sci. Eng. B*, **53**, 143 (1998).

[8] A. Prikhna, W. Gawalek, V. E. Moshchil, Ch. Wende, V. S. Melnikov, F. Sandiumenge, N. Vilalta, P. A. Nagorny, T. Habisreuther, S. N. Dub, and A. B. Surzenko, *Supercond. Sci. Technol.* **11**, 1123 (1998).

[9] F. Sandiumenge, N. Vilalta, J. Rabier and X. Obradors, *Phys. Rev. B* **64**, 184515 (2001).

[10] D. Rodgers, K. White, V. Selvamanickam, A. McGuire and K. Salama, *Supercond. Sci. Technol.* **5**, 640 (1992).

[11] R. K. Williams, K. B. Alexander, J. Brynestad, T. J. Henson, D. M. Kroeger, T. B. Lindemer, G. C. Marsh, J. O. Scarbrough and E. D. Specht, *J. Appl. Phys.* **70**, 906 (1991).

[12] D. E. Morris, A. G. Markelz, B. Fayn and J. H. Nickel, *Physica C*, **168**, 153 (1990).

[13] T. B. Lindemer, F. A. Washburn, C. S. MacDougall, R. Feenstra and O. B. Cavin, *Physica C* **178**, 93 (1991).

[14] H. W. Zandbergen, R. Gronsky, K. Wang and G. Thomas, *Nature*, **331**, 596 (1998).

[15] J. Rabier, P. D. Tall and M. F. Denanot, *Phil. Mag. A*, **67**, 1021 (1993).

[16] B. Martínez, F. Sandiumenge, T. Puig, X. Obradors, L. Richard and J. Rabier, *Appl. Phys. Lett.* **74**, 73 (1999).

[17] A. Goyal, W. C. Oliver, P. D. Funkenbusch, D. M. Kroeger and S. J. Burns, *Physica C* **183**, 221 (1991).

[18] J. Rabier, *Philos. Mag.* **73**, 753 (1996).

[19] J. Rabier, *Philos. Mag.* **75**, 285 (1997).

[20] T. Puig, J. Plain, F. Sandiumenge, X. Obradors, J. Rabier and J. A. Alonso, *Appl. Phys. Lett.* **75**, 1952-4 (1999).

[21] J. Plain, F. Sandiumenge, J. Rabier, M. F. Denanot and X. Obradors, *Philos. Mag. A*, **82**, 337-48 (2002).

[22] J. Plain, T. Puig, F. Sandiumenge, X. Obradors and J. Rabier, *Phys. Rev. B*, **65**, 104526 (2002).

[23] J. Rabier, F. Sandiumenge, J. Plain and I. Stretton, *Philos. Mag. Lett.*, in press.

[24] J. Plain, F. Sandiumenge, J. Rabier, A. Proult, I. Stretton, T. Puig and X. Obradors, manuscript submitted for publication.

[25] G. Blatter, M. V. Feigel'man, V. G. Geshkenbein, A. I. Larkin, V. M. Vinokur, *Rev. Mod. Phys.* **66**, 1125 (1994).

[26] D. R. Nelson and V. M. Vinokur, *Phys. Rev. B* **48**, 13060 (1993).

[27] L. Civale, A. D. Marwick, M. W. McElfresh, T. K. Worthington, A. P. Malozemoff, F. H. Holtzberg, J. R. Thompson and M. A. Kirk, *Phys. Rev. Lett.* **65**, 1164 (1990).

[28] B. Martínez, X. Obradors, A. Gou, V. Gomis, S. Piñol and J. Fontcuberta, *Phys. Rev. B* **53**, 2797 (1996).

FLUX PINNING and PROPERTIES OF SOLID-SOLUTION $(Y,Nd)_{1+x}Ba_{2-x}Cu_3O_{7-\delta}$ SUPERCONDUCTORS

T. J. Haugan, M. E. Fowler, J. C. Tolliver, P. N. Barnes
Air Force Research Laboratory, 2645 Fifth St. Ste. 13, Wright-Patterson AFB,
OH 45433-7919

W. Wong-Ng, L. P. Cook
National Institute of Standards and Technology, Materials Science and
Engineering Laboratory, Gaithersburg, MD 20899-8520

ABSTRACT
The effect of chemical composition variations on the flux pinning and physical properties of $(Y,Nd)_{1+x}Ba_{2-x}Cu_3O_{7-\delta}$ superconductors was studied in powders processed by solid-state reaction and equilibrated in air at 910 °C. An extended region of single-phase crystal structures was determined for compounds with Ba = 2.0 to 1.7 and Nd = 0 to 1.3. At 77 K, the composition $(Y_{0.6}Nd_{0.4})Ba_2Cu_3O_{7-\delta}$ had the highest J_c and flux pinning properties above 2 T, and showed over a magnitude of improvement for applied fields > 3 T compared to $YBa_2Cu_3O_{7-\delta}$. At 65 K, the $YBa_2Cu_3O_{7-\delta}$ composition had the highest $J_c(H)$ properties. With decreasing Ba content (< 2.0), the T_c and J_c were reduced for the processing conditions tested.

INTRODUCTION
Copper based (Rare-Earth)$Ba_2Cu_3O_{7-z}$ (RE123) superconductors are being considered for applications including thin film coated conductors and bulk devices because of their high superconducting transition temperatures (T_c) > 92 K, and high critical current density (J_c) at 77 K in useful magnetic fields. While these materials have many desirable attributes at 77 K, it is of interest to increase the J_c in applied magnetic fields even further by increasing the flux pinning properties of the superconductor. Many methods can be considered to introduce flux pinning defects into the superconductors, including irradiation and introduction of second-phase defects or precipitates. One method used

particularly for bulk applications is to introduce different rare-earth cations into the RE123 compound. Many variations of this theme have been studied, including substitution in (Y,R)123 with R = Ho,Dy, Gd, Eu, and Pr, and various other combinations of rare-earths such as (Gd,Sm,Eu)123 [1-15]. Possible mechanisms by which such substitutions increase the flux pinning include: (a) addition of second-phase defects by precipitation or composition changes, (b) formation of finely distributed lower T_c components from the mixed solubility of RE with each other and for Ba or other mechanisms, or (c) randomly distributed oxygen-deficient zones which also cause lower T_c components to form [7,8]. The finely distributed lower T_c components are suggested as a cause of the so-called 'fishtail' effect, where as the magnetic field is increased, the J_c increases as the lower T_c components transition to normal behavior before the J_c decreases at much higher applied magnetic fields [7,8]. The peak of J_c maximum in these materials typically occurs at ~ 2-3 T applied fields.

Studies of the system $(Y_{1-x}Nd_x)Ba_2Cu_3O_{7-\delta}$ have, to our knowledge, been limited thus far to melt-processed or single crystal materials [9-14], and powders [15]. Single crystals in this system demonstrated very high J_c and the 'fish-tail' effect for Nd content varying from 0.1 to 0.4 after varying oxygenation treatments [13].

In this paper, the effect of composition changes in $(Y,Nd)_{1+x}Ba_{2-x}Cu_3O_{7-\delta}$ is studied in solid-state powders annealed in air to achieve chemical equilibrium. The powders prepared in this work are expected to have different properties than (Y,Nd)123 melt-processed or crystals in previous studies, where non-equilibrium processes can affect the physical properties. In melt-processed materials it is almost impossible to eliminate the formation of second-phase defects such as RE211, which can affect pinning. With powder processing, it possible to completely eliminate second-phase defects, and investigate other causes of pinning more related to the intrinsic properties of the crystal structure.

For the studies herein, powders were annealed in air, which may have an advantage for flux pinning by increasing the substitution of Nd onto the Ba site, and causing lower T_c solid-solutions to form [8,15,16]. While further treatments of the powders in varying O_2 pressures is predicted to improve the superconductivity and possibly increase the fish-tail effect, such treatment may decrease the pinning effect by reducing substitution of Nd onto the Ba site. The effects of annealing in air are studied in this paper to examine the role of chemical substitution prior to oxygen treatments. This paper also provides J_c and flux pinning results for $(Y,Nd)_{1+x}Ba_{2-x}Cu_3O_{7-\delta}$ for a large range of compositions, including Ba < 2.0.

EXPERIMENTAL**

Superconducting powders were prepared by the solid-state reaction method, using starting reactants of Nd_2O_3, Y_2O_3, $BaCO_3$, CuO ($\geq$ 99.95% purity). The powders were dehydrated at 450 °C prior to weighing. The powders were mixed and ground with mortar and pestle, calcined by slow heating 650 °C to 850 °C at 25 °C/h, and subsequently annealed with intermediate grinding at 880 °C and 910 °C. Powders were annealed at 910 °C with intermediate grinding until phase equilibrium was reached (~3-4 annealings), as determined by X-ray diffraction (XRD). The powders were reacted in ~1 cm diameter pellets (0.5 – 1 g batches), formed by lightly pressing (~5-10x10^6 Pa) in molds. The powder with composition $YBa_2Cu_3O_{7-\delta}$ was purchased from Superconductive Components Inc. with nominal purity 99.9% [17]. X-ray diffraction was performed with a Rigaku diffractometer. A step size of 0.03 ° was used for the θ-2θ scans.

Superconducting properties of powders were measured with a SQUID magnetometer (Quantum Design**, MPMS/MPMS^2). Magnetization-applied field (M-H) hysteresis loops at different temperatures were made by heating samples to 100 K and zero-field cooling (ZFC) to the measurement temperature. The magnetic J_c was estimated using the extended Bean critical current model $J_c = 15(\Delta M)/R$ where ΔM is the magnetic hysteresis difference [emu/cm^3], and R is the radius of the superconducting volume roughly approximated as ~ 0.00005 cm for the finely reacted powders [18].

To characterize the superconducting properties of powders, field-cooled (FC) Meissner and zero-field-cooled (ZFC) measurements were performed from 5 to 100 K [18]. The SQUID magnet was reset to zero before any measurements. The superconducting volume percentages were calculated using $\chi(\%) = 4\pi\chi_v/(1-D*4\pi\chi_v)$, where $\chi_v = M/H_{appl}$ is the measured magnetic susceptibility [emu/cm^3•gauss], and D = 1/3 is the demagnetization factor assuming a spherical particle distribution [18]. The applied magnetic field was $H_{appl} = $ 10 Oe - H_{rem}, where H_{rem} is the remnant field of the magnet after resetting to zero, determined for each sample by measuring M(H) from 10 Oe to -5 Oe at 1 Oe intervals and plotting when M = 0 ($\pm$ 0.1 Oe accuracy). The transition temperature of the largest volume fraction of powder was determined by finding the temperature when $(d^2\chi/dT^2) \cong 0$ upon cooling from 100 K.

** "Certain commercial equipment, instruments, or materials are identified in this paper in order to specify the experimental procedure adequately. Such identification is not intended to imply recommendation or endorsement by the Air Force Research Laboratory or the National Institute of Standards and Technology, nor is it intended to imply that the materials or equipment identified are necessarily the best available for the purpose."

RESULTS

The range of compositions studied is shown in Figure 1. All of the compositions studied were nominal single-phase as determined by XRD, in agreement with previous results [15], except for a small region (Ba = 2.0 and Y = 0.3 to 0.4). For this small region, it was not possible to reach the single-phase composition within ~ 3 grinding and annealing cycles at 910 °C, and secondary phases including $BaCuO_{2+x}$ were observed to remain in the composite. Rather than achieving a new equilibrium phase assemblage, it is probable that reaction kinetics were too slow for these compositions, and higher temperatures might be required to form the single-phase assemblage.

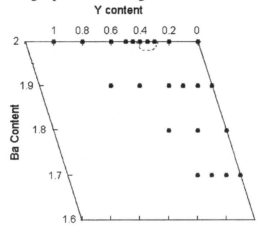

Figure 1. Compositions of $(Y,Nd)_{1+x}Ba_{2-x}Cu_3O_{7-\delta}$ studied. All compositions were nominal single-phase, excluding Ba = 2 and Y = 0.3 to 0.4.

The superconducting transitions of Ba = 2.0 compositions are shown in Figure 2. The transitions were broadened for Nd > 0.4, and the transition for the largest volume fraction of the powder was reduced to ~ 73 K for the Nd = 1.0 composition. A reduction and broadening of T_c for the Nd = 1.0 composition is usually observed without further processing in reduced oxygen partial pressures [15,16]. The transition of the bulk component for several different compositions is shown in Figure 3. With increasing Nd and reduced Ba content, the T_c of the bulk powder is decreased, in agreement with trends reported previously for $(Y,Nd)_{1+x}Ba_{2-x}Cu_3O_{7-\delta}$ and single-phase $Nd_{1+x}Ba_{2-x}Cu_3O_{7-\delta}$ [15,16]. Assuming previous optimization methods can be applied, it's possible the T_cs for the Nd-rich compositions can be increased and sharpened by further processing in reduced O_2 atmosphere [7,8,15,16].

Figure 4 plots the J_cs of the powders for different composition and temperature. For Ba = 2.0, the Nd = 0.4 composition had a higher J_c than the Nd

= 0.0 composition. The increase of J_c for Nd = 0.4 composition is similar to results achieved for single-crystals [13]. The trend of decreasing J_c for Nd > 0.4 has not been published elsewhere in the literature, to our knowledge. Also the decrease of J_c for Nd = 1.0 (Nd123) compared to Nd = 0.0 (Y123) is consistent with overall trends reported for thin films: Nd123 $J_c \sim 1$ MA/cm^2 compared to Y123 $J_c \sim 6$-9 MA/cm^2 at 77 K [19,20].

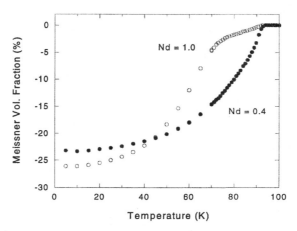

Figure 2. Meissner volume fraction (%) for Ba = 2.0 and different Nd content.

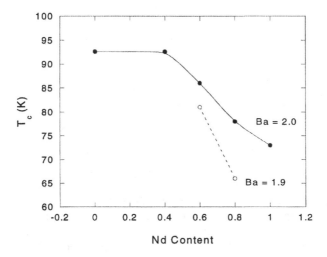

Figure 3. Transition temperature of the bulk superconducting volume fraction.

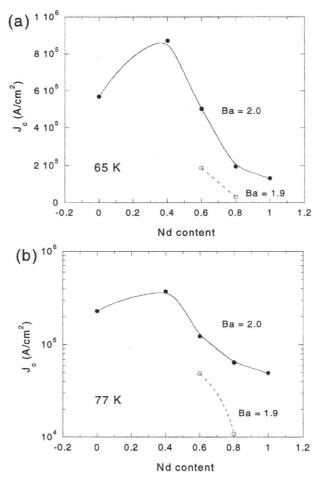

Figure 4. Critical current density estimated from M-H loops for varying Ba and Nd content.

The effect of applied magnetic field on the J_c for varying compositions is shown in Figure 5. At 77 K and for $H_{appl} > 2$ T, the Nd = 0.4 composition had markedly better pinning properties ($J_c(H)$) than other compositions tested. For $H_{appl} > 3$ T, the improvement is over one order magnitude higher compared to the Y123 compound (Nd = 0). The powders showed a very small fish-tail effect, with $J_{c\text{-max}}$ occurring at 0.06 to 0.2 T depending on composition. By comparison, the peak of maximum J_c in oxygenated melt-processed materials usually occurs around 2 to 3 T [7,8]. This suggests the compositions are homogeneous in nature without the oxygen-deficient regions thought to cause the fish-tail effect [7,8]. A homogenous distribution of oxygen is expected as a consequence of the small size of the powders, and the near-equilibrium conditions used for processing.

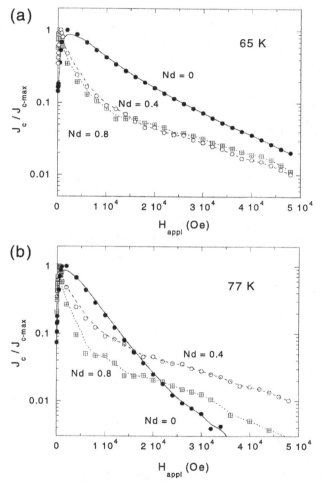

Figure 5. Normalized ($J_c/J_{c\text{-max}}$) for Ba = 2.0 and varying Nd content.

CONCLUSIONS

The flux pinning and physical properties of $(Y,Nd)_{1+x}Ba_{2-x}Cu_3O_{7-\delta}$ powders annealed in air were studied for varying compositions. At 77 K, the composition $(Y_{0.6}Nd_{0.4})Ba_2Cu_3O_{7-\delta}$ had the highest $J_c(H)$ values for $H_{appl} > 2$ T, with J_c over a magnitude greater compared to $YBa_2Cu_3O_{7-\delta}$ for $H_{appl} > 3$ T. At 65 K, the $YBa_2Cu_3O_{7-\delta}$ composition had the maximum $J_c(H)$ values. With decreasing Ba content (< 2.0), the T_c and J_c were reduced for the processing conditions tested, consistent with trends in the literature. While processing in air is generally considered to lower T_c transitions for Nd-rich 123 compounds, the results here show that considerable improvements of $J_c(H)$ can be made at least for partial substitution of Y with Nd.

ACKNOWLEDGEMENTS

The authors would like to thank T. Spry for assistance with sample preparation, and R. Drew for SQUID measurements.

REFERENCES

[1] Y. Feng, A. K. Pradhan, Y. Zhao, Y. Wu, N. Koshizuka, and L Zhou, "Influence of Ho substitution for Y on flux pinning in melt-processed YBCO superconductors," *Physica C* **357-360** 799-802 (2001).

[2] A. R. Devi, V. S. Bai, P. V. Patanjali, R. Pinto, N. H. Kumar, and S. M. Malik, "Enhanced critical current density due to flux pinning from lattice defects in pulsed laser ablated $Y_{1-x}Dy_xBa_2Cu_3O_{7-\delta}$ thin films," *Supercond. Sci. Technol.* **13** 935-939 (2000).

[3] H. H. Wen, Z. X. Zhao, R. L. Wang, H. C. Li, and B. Yin, "Evidence for lattice-mismatch-stress-field induced flux pinning in $(Gd_{1-x}Y_x)Ba_2Cu_3O_{7-\delta}$ thin films," *Physica C* **262** 81-88 (1996).

[4] Y. Li, G. K. Perkins, A. D. Caplin, G. Cao, Q. Ma, L. Wei and Z X Zhao, "Study of the pinning behaviour in yttrium-doped Eu-123 superconductors" *Supercond. Sci. Technol.* **13** 1029-1034 (2000).

[5] H. H. Wen, Z. X. Zhao, Y. G. Xiao, B. Yin, and J. W. Li, "Evidence for flux pinning induced by spatial fluctuation of transition temperatures in single domain $(Y_{1-x}Pr_x)Ba_2Cu_3O_{7-\delta}$ samples," *Physica C* **251** 371-378 (1995).

[6] E. S. Reddy, P. V. Patanjali, E. V. Sampathkumaran, R. Pinto, "Fabrication and superconducting properties of ternary $REBa_2Cu_3O_y$ thin films," *Physica C* **366** 123-128 (2002).

[7] M. R. Koblischka, M. Muralidhar, M. Murakami, "Flux pinning sites in melt-processed $(Nd_{0.33}Eu_{0.33}Gd_{0.33})Ba_2Cu_3O_y$ superconductors," *Physica C* **337** 31-38 (2000).

[8]M. Jirsa, M. R. Koblischka, T. Higuchi, M. Muralidhar, M. Murakami, "Comparison of different approaches to modeling the fishtail shape in RE-123 bulk superconductors," *Physica C* **338** 235-245 (2000).

[9]C. Varanasi, P. J. McGinn, H. A. Blackstead, and D. B. Pulling, "Nd Substitution in Y/Ba Sites in Melt Processed $YBa_2Cu_3O_{7-\delta}$ Through Nd_2O_3 Additions," *Journal of Electronic Materials,* **24** [12] 1949-1953 (1995).

[10]D. N. Matthews, J. W. Cochrane, G. J. Russell, "Melt-textured growth and characterization of a $(Nd/Y)Ba_2Cu_3O_{7-\delta}$ composite superconductor with very high critical current density," *Physica C* **249** 255-261 (1995).

[11]P. Schätzle, W. Bieger, U. Wiesner, P. Verges and G. Krabbes, "Melt Processing of (Nd,Y)BaCuO and (Sm,Y)BaCuO composites," *Supercond. Sci. and Technol.* **9** 869-874 (1996).

[12]A. S. Mahmoud and G. J. Russell, "Large crystals of the composite Y/Nd(123) containing various dopants grown by melt-processing in air," *Supercond. Sci. and Technol.* **11** 1036-1040 (1998).

[13]X. Yao, E. Goodilin, Y. Yamada, H. Sato and Y. Shiohara, "Crystal growth and superconductivity of $Y_{1-x}Nd_xBa_2Cu_3O_{7-\delta}$ solid solutions," *Applied Superconductivity* **6** [2-5] 175-183 (1998).

[14]D. K. Aswal, T. Mori, Y. Hayakawa, M. Kumagawa, "Growth of $Y_{1-z}Nd_zBa_2Cu_3O_x$ single crystals," *Journal of Crystal Growth* **208** 350-356 (2000).

[15]H. Wu, K. W. Dennis, M. J. Kramer, and R. W. Mccallum, "Solubility Limits of $LRE_{1+x}Ba_{2-x}Cu_3O_{7+\delta}$," *Applied Superconductivity* **6** [2-5] 87-107 (1998).

[16]R. W. McCallum, M. J. Kramer, K. W. Dennis, M. Park, H. Wu, and R. Hofer, "Understanding the Phase Relations and Cation Disorder in $LRE_{1+x}Ba_{2-x}Cu_3O_{7+\delta}$," *J. of Electr. Mater.* **24** [12] 1931-1935 (1995).

[17]Superconductive Components Inc., 1145 Chesapeake Ave., Columbus, OH 43212.

[18]*Magnetic Susceptibility of Superconductors and Other Spin Systems*, Plenum Press, New York, 1991.

[19]T. J. Haugan, P. N. Barnes, R. M. Nekkanti, I. Maartense, L. B. Brunke, and J. P. Murphy, "Pulsed Laser Deposition of $YBa_2Cu_3O_{7-\delta}$ Thin Films in High Oxygen Partial Pressures," submitted to MRS Conf. Proceedings, Nov. 2001.

[20]Los Alamos National Laboratory, as reported at Fall MRS 2001 meeting (for ND123).

STUDIES OF GRAIN BOUNDARIES IN MELT TEXTURED $YBa_2Cu_3O_x$

B. W. Veal, H. Claus, Lihua Chen and A. P. Paulikas

Materials Science Division, Argonne National Laboratory, Argonne Illinois 60439-4830

ABSTRACT

[001] tilt grain boundaries were studied in bi-crystal samples of melt textured $YBa_2Cu_3O_x$. Grain boundary critical current densities J_{CB} were obtained from SQUID magnetization measurements on ring samples that contain the grain boundary. The dependence of J_{CB} on oxygen stoichiometry and oxygen ordering were investigated and preliminary studies of grain boundary doping with selected cations, including Ca, Sr, and Bi were undertaken.

INTRODUCTION

It is well known that grain boundaries in the high Tc superconductor $YBa_2Cu_3O_x$ (YBCO) severely inhibit supercurrent flow, with losses being most severe at high grain boundary (GB) mismatch angles [1,2]. Consequently, considerable effort is being devoted to (1) develop methods for producing YBCO materials (both bulk and film) that are relatively free of GBs, especially high angle GBs, and to (2) develop methods to minimize the effect of GBs on supercurrent transport. Associated with these efforts is the ongoing scientific challenge; to determine the local structure and to understand associated electronic behavior, including those effects which influence supercurrent flow.

One approach for minimizing deleterious transport effects is to exploit GB dopants. One obvious and important GB dopant is oxygen (or oxygen vacancies) which necessarily must be properly adjusted for optimal transport performance. Another possibility is to decorate the GBs with suitable impurity atoms which might favorable affect electronic transport. Success with this approach has been reported with Ca doping of GBs [3,4].

In this study, we extensively exploit SQUID measurements on ring samples to measure the grain boundary J_C. Effects of oxygen stoichiometry are examined, as well as the role of selected cation dopants that are diffused into GBs. GBs are investigated in bulk YBCO.

High Temperature Superconductors

It is expected that a large variety of dopant atoms might diffuse into YBCO GBs under suitable processing conditions. Further, it seems likely that these impurity atoms will generally affect supercurrent transport across the GB. The demonstration of (beneficial) GB doping with Ca has generated an interest in understanding the mechanisms which influence supercurrent transport.. Hole doping [3] and strain effects in dislocation cores [5] have been proposed as controlling mechanisms. Systematic studies of dopants in YBCO GBs might provide new insights into GB doping mechanisms and could lead to new methods for minimizing the deleterious effect of GBs on supercurrent transport.

In addition to electronic effects (hole doping) and strain effects in dislocation cores, it may be that effects of vortex pinning will be important [2]. Such effects are likely to depend strongly on GB misorientation since the nature of GB vortices changes dramatically as misorientation increases. GB vortices at small angles approximate Abrikosov vortices, with relatively localized vortex cores, but convert to Josephson vortices, with ill-defined cores, at large misorientation. Thus, the effectiveness of particular defects as pinning centers is likely to vary dramatically with GB misorientation.

Some dopants might exert significant bonding effects, perhaps establishing a local oxygen coordination shell about the dopant atom. However, in order to controllably dope GBs using diffusion, it will be necessary that GB diffusion dominates over bulk diffusion or chemical attack, with attendant destruction of the YBCO structure. Presumably, strong bonding behavior will affect supercurrent transport, perhaps varying with bond strength and coordination shell.

EXPERIMENTAL

Sample Preparation

Melt-textured cylindrical YBCO monoliths were produced using a top-seeding method as described in ref 6. To obtain [001] tilt GBs, pressed powder cylindrical samples were melt-textured using two Sm123 seeds, with their c-axes aligned parallel to the axis of the cylinder. The crystals were rotated about their c-axes so that a-b planes had the desired misorientation. Simultaneous seeding, initiated at the two crystals, leads to a melt textured "bi-crystal" with the desired [001] tilt misorientation. Some of the textured samples were densified using a processing step in O_2 [6], while some were textured, in air, without the densification step.

J_{CB} Measurements on Ring Samples

Grain boundary critical current densities J_{CB} were obtained from SQUID magnetization measurements on ring samples that were core-drilled from the bi-crystal samples. Rings, with ~ 1x1 mm^2 cross section and 5 mm outer diameter, were cut so that the grain boundary plane cuts across the ring parallel to the ring axis. To acquire the SQUID measurements, a sample was cooled, in zero field, to

High Temperature Superconductors

a temperature well below T_C, then a magnetic field was applied parallel to the ring axis and the sample was warmed at constant field. Fig. 1 displays SQUID **dc**

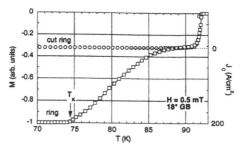

Fig. 1   Magnetic moment of a ring, containing an 18° grain boundary (see text). The data were taken on warming in a magnetic field of 0.5 mT after initially cooling in zero field to below 70 K. The right scale indicates the critical current density of the grain boundary. Also shown is the magnetic signal of the ring after it had been cut, preventing any circulating currents.

magnetization measurements taken from a ring sample that contains an 18° [001] tilt boundary.

At the lowest temperatures, where M is independent of temperature, flux is completely excluded from the ring bore. As the temperature is increased (Fig. 1), the critical current across the GB will fall, and the current that has been induced in the ring will eventually exceed the critical current across the boundary. At this point flux will begin to leak into the ring, decreasing the observed magnetization. This onset of flux penetration into the ring produces the abrupt decrease in M (T) that is clearly observed in Fig. 1 at the temperature labeled T_K (kink temperature). Considering the ring to be a single-turn solenoid, the induced current providing shielding of the ring bore can be estimated, to good approximation, from the simple expression I=D*H, where D is the outer diameter of the ring and H is the applied magnetic field [6,7]. Thus, I(A)=4*H (mT) for our ring geometry.

As the temperature is further increased above the kink temperature T_K, the shielding current flowing around the ring becomes limited by the declining critical current across the boundary and flux increasingly leaks through the GB into the bore of the ring. The nearly linear decrease in M vs. T above T_K reflects a nearly linear decrease of J_{CB} with temperature.

The plateau in M vs. T reached at about 30% of the full shielding signal indicates that the critical current in the junction has dropped to zero and the field inside the bore now almost equals the field outside. The remaining signal (i.e., at 90 K) is due to flux expulsion from the annulus, which has a transition temperature of about 92 K. The plateau in M(T) coincides with the diamagnetic signal from the ring after a slit has been cut into the ring to prevent any circulating shielding currents [6].

RESULTS AND DISCUSSION
J_{CB} Versus Misorientation Angle
Fig. 2 shows measurements of J_{CB} vs misorientation angle for [001] tilt grain

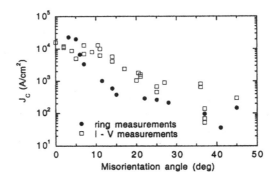

Fig. 2   Critical current at 77 K of grain boundaries versus misorientation angles (see text). Open squares are from resistivity measurements (see ref 2). The closed circles are from ring measurements like that in Fig. 1.

boundaries with misorientations spanning the range from 0 to 45°. Open symbols, taken from the literature, are J_{CB} obtained from conventional I-V measurements [2]. Solid symbols identify data acquired using the SQUID technique. The SQUID results are generally lower. This results, in part, from a much smaller voltage criterion for the SQUID measurements [6]. Also, some of the shift might result from a depression of J_{CB} by the applied measurement field; this appears to be most important at high misorientations [8].

Dependence of GB Transport on Magnetic Field
Numerous measurements have been reported for the zero field dependence of GB J_C on misorientation angle, especially for [001] tilt GBs [2]. However, it is also the case that J_{CB} is a very strong function of magnetic field, especially for large mismatch angles and temperatures relatively close to T_C; e.g., at 77 K [8]. We note that SQUID measurements on ring samples that contain a GB are particularly convenient for field dependent studies of grain boundary J_C. For these experiments, we cool the sample in an applied field of H-δ. Then, with the sample cooled to a temperature below Tc, the field is increased to H+δ. A

shielding current will be induced to screen the 2δ incremental field. At the kink temperature, $J_C = 400*(2\delta)$ for a ring sample with a 1 mm cross section, where J_C is expressed in A/cm^2 and δ is in mT.

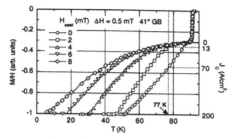

Fig. 3 Magnetic moment of a 41° grain boundary versus temperature in magnetic fields between zero and 8 mT. The data were taken on warming after initially cooling the ring in a field of H-δ to low temperature and then increasing the field to H+δ before warming, where 2δ = 0.5 mT. The right scale indicates the critical current density of the grain boundary.

Fig 3 shows M(T) for a GB with 41 degrees of misorientation when H+δ is varied from zero to 8 mT. The incremental field δ is 0.25 mT. The grain boundary J_C is shown on the right hand scale of Fig 3. Note that, at 77 K, the value of J_C falls by more than a factor of 5 as the applied field is increased from zero to 2 mT. Miller, et al. [8] observed comparably large field dependent changes in 77K measurements of transport J_C for misorientations up to 25°.

Dependence of GB Transport on Oxygen Stoichiometry

The level of oxygenation profoundly affects the superconducting and normal state properties of YBCO. However, the role of (excess or deficient) oxygen in the GBs has not been clearly established. In this study, we repeatedly process ring samples cut across [001] tilt GBs at appropriate temperatures and oxygen pressures so that the bulk oxygen stoichiometry is changed in a controlled fashion.

Fig. 4 shows T_K as a function of annealing time at 450°C, in O_2, for a 27 ° grain boundary. Measurements were made in a field of 0.5 mT. The sample was given repeated oxygenation exposures until T_K saturated at about 76 K. A treatment of 10 to 20 hrs appears to be adequate for essentially full oxygenation of the 1 mm^2 grain boundary. As Fig 4 attests, the oxygenation behavior is remarkably well behaved. No apparent damage to the GB is introduced by the repeated thermal cycling. The solid line in Fig 4 is a fit of the data to the exponential expression shown in Fig. 4 with $T_{K0} = 53$ K, $\Delta T = 23$ K and $\tau = 3.4$ h.

As expected, the oxygenation rate is strongly temperature dependent. Preliminary studies indicate that about 300 hrs are needed to fully oxygenate the

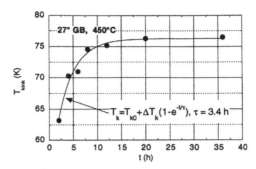

Fig. 4   Variation of the kink temperature (see Fig. 1) with oxygenation time at 450°C. The data are for a 27° grain boundary. The solid line is an exponential fit to the data.

GB at 350°C. It appears, furthermore, that J_{CB} continues to rise as the oxygenation level is increased, even as the bulk material becomes "overdoped" and T_C shows a small decline. This behavior is illustrated in Fig. 5 where measurements were taken after extended treatments at 450°C and at 350°C in O_2. J_{CB} increases as the oxygenation temperature is reduced, increasing the oxygen

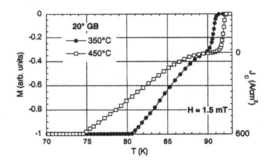

Fig. 5   Magnetic moment in 1.5 mT of a ring containing a 20° grain boundary versus temperature, after the ring was oxygenated first at 350°C then at 450°C. The right scale indicates the critical current density of the grain boundary.

stoichiometry, while the bulk T_C declines.

It is usually the case, at temperatures well below the melting temperature, that GB diffusion is significantly faster than bulk diffusion. Thus we expect the GB to become oxygenated more rapidly than the surrounding bulk. However, in order to make measurements such as those in Fig 4, it must be that oxygenation into the bulk is sufficiently fast so that an oxygenated shell of thickness exceeding the superconducting penetration depth is achieved. If bulk pinning is strong, the SQUID shielding measurement will show complete flux exclusion giving a full diamagnetic signal even though the central region of the ring might remain nonsuperconducting. In order to detect the full value of I_c across the GB, it must also be that the bulk material, in proximity to the GB, is also sufficiently oxygenated to carry the transmitted I_c to the ring surface. This internal bulk

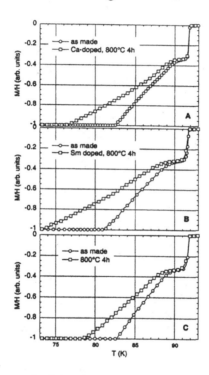

Fig. 6   Magnetic moment in 0.5 mT of three rings all containing a 21° grain boundary. Each panel shows one ring in the as made state (oxygenated at 450°C) and after heat treatment at 800°C for 4 hours followed by a 450°C oxygenation. The rings in panel **A** and **B** were immersed in $CaCO_3$ and Sm_2O_3 powders respectively during the 800°C treatment. The ring in panel **C** was given the same 800°C treatment in air.

oxygenation that occurs in close proximity to the GB is supplied through the fast diffusion GB path. The ring surface must also be sufficiently oxygenated to carry the full I_c. For the data of Fig 4, the bulk was sufficiently oxygenated to carry the full I_c transmitted across the GB; i.e., the full diamagnetic signal from the bulk was observed after each stage of oxygenation.

The data in Fig. 4 indicate that the bi-crystal samples can be repeatedly processed with thermal treatments up to about 500°C, apparently without damage or significant change to the GB. Thermal excursions between the process temperature and cryogenic temperatures, where measurements are taken, apparently do not damage the GBs. The GB structure (aside from effects of oxygenation) apparently remains remarkably stable. However, nonreversible behavior is often observed when process temperatures exceed 500-600°C.

Nonreversible Behavior

GB transport is apparently affected by the thermal history of the sample, especially if it has experienced extended time treatments in the temperature range of 600-900°C. Consequently, if a diffusion experiment is conducted in this temperature range, it is mandatory that reference samples be given identical heat treatments, but without exposure to the GB diffusant.

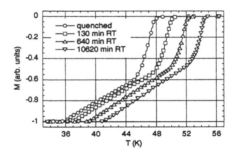

Fig. 7 Magnetic moment in 0.5 mT of a ring, containing an 11° grain boundary. The oxygen concentration was adjusted to 6.44 (see text). The ring was then equilibrated at 120°C and quenched into chilled ethanol followed by room temperature anneals for the time intervals as indicated.

A simple illustration is shown in Fig 6, where M(T) data are presented for three -crystal samples cut from the same dual seeded 21° GB. The samples were treated for 116 hrs at 450°C in flowing O_2 and M(T) data were recorded [open circles]. The samples were subsequently treated [open squares] for 4 hrs at 800°C in air while immersed (A) in a powder of $CaCO_3$, and (B) in a powder of Sm_2O_3. Sample (C) was simply reprocessed, as a 'placebo' standard for the same 4 hr air exposure at 800°C. All samples were given a final oxygenation treatment at 450°C for 95 hrs to return the bulk T_c to its optimal value.

Note that all three rings have quite similar GB J_c's at 77 K, both before and after the thermal processing. The behavior of the reference sample is nearly equivalent to the cation-treated samples, strongly suggesting that no significant concentration of impurity cations has diffused into the GBs under these processing conditions, even though J_c values across the GBs have changed quite dramatically.

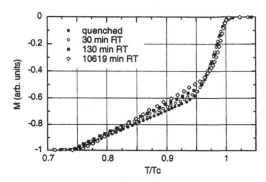

Fig. 8   Same measurements as in Fig. 7 but with the temperature scale normalized to T_C.

This behavior might result from altered GB pinning caused by the 800°C heat treatment. The reduced values of J_{CB} suggest that some pinning defects have been annealed out. Notably, a very similar alteration of the pinning behavior was encountered for each of the three samples. In any case, the thermally induced change does not require the presence of a GB diffused cation impurity. This illustrates the need to apply extreme care in interpreting the results of GB diffusion experiments. Also, it emphasizes that thermal history is likely to play a significant role in the GB transport.

Oxygen Ordering
    The state of order of oxygen atoms in the chain layer of oxygen deficient YBCO can have a profound influence on the superconducting and normal state properties of the bulk material [9]. Further, the state of order is strongly dependent on temperature. A disordered state will rapidly anneal, even at room temperature, with an associated rise in the superconducting T_c. The properties changes are consistent with, and have been attributed to, a change in the level of hole doping resulting from an increase in the concentration of 2-coordinated (monovalent) Cu atoms located in the chains [10].
    We have investigated the effect, on the grain boundary J_c, resulting from changing the state of order of oxygen vacancies residing in the chain layer of

YBCO. This experiment has an apparent analog in the study of GB doping with Ca. It is argued that Ca^{2+} doping increases the J_C as a consequence of substituting on Y^{3+} sites thereby increasing the hole doping at GBs [3]. Of course, doping associated with ordering occurs throughout the material, but may also affect GB properties [11].

M(T) measurements were made on a sample with stoichiometry x = 6.44, obtained by annealing for 190 hrs at 500°C in a flowing gas stream consisting of 0.17 % O_2 in N_2, followed by a quench to liquid nitrogen. To disorder the chain oxygens, the sample was then heated to 120°C and, after 10 min, was quenched into chilled ethanol. Then a series of M(T) measurements were recorded as the sample annealed at room temperature.

Fig 7 shows the M(T) data recorded after specified time intervals as room temperature annealing proceeded. The kink temperature, identifying a fixed value of J_C, increases with the rising T_C. Fig 8 shows the M(T) plotted vs the reduced temperature T/T_C. The fixed value of the kink temperature in the reduced temperature plot suggests that the GB J_C is primarily controlled by the value of T_C without significant effects that could be attributed to changes within the grain boundary. Thus, to first order, this doping mechanism does not appear to play an important role in influencing GB transport.

Cation Doping

Recent studies on thin film samples have demonstrated that Ca is an effective GB dopant to increase the supercurrent density across GBs in YBCO [3,4]. It is argued that Ca^{2+} ions substitute on Y^{2+} sites thereby increasing the hole concentration which serves to improve GB coupling increasing the GB J_C. Strain effects resulting from dopant ions in GB cores may also play an important role in influencing J_C [4,5].

The mechanisms responsible for improvement of GB transport resulting from Ca doping are not known. Consequently, it would appear to be worthwhile to investigate the effects caused by GB doping, for a variety of dopant ions, with variations in ion size, valence state and electro-negativity.

Using bi-crystal GBs produced with dual seeding techniques in melt textured bulk samples, we are searching for thermal treatment conditions that will permit controlled doping of the artificial GBs with selected cation dopants. If dopants can be placed within GB cores without reacting with the YBCO to form separate phases, the possibility exists that GB coupling can be improved so that supercurrent losses across GBs can be substantially reduced. Successful GB doping strategies could then be used for improving J_C in melt textured bulks and in coated conductors, both of which contain small angle GBs. Alternatively, with successful GB doping, it might be possible to utilize materials which contain larger mismatch angles than can presently be tolerated.

Ca doping: Since GB doping with Ca has been successfully demonstrated in film samples, we have attempted to achieve doping, and to demonstrate J_C enhancement, by diffusing Ca atoms into GBs prepared in MTG material. Results

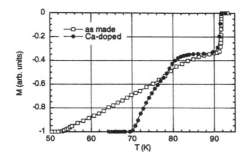

Fig. 9   Magnetic moment in 0.15 mT of a ring, containing a 38° grain boundary before and after Ca – doping at 970°C (see text). In both states the ring was fully oxygenated at 450°C.

of an apparently successful doping strategy are shown in Fig 9 for a 38° [001] tilt bi-crystal. The curve marked 'as-made' was measured before the ring sample was exposed to a Ca source. It was simply given an oxygenation treatment for 9 days at 450°C in flowing O_2. Following this measurement, the sample was immersed in $CaCO_3$ powder and was treated for 60 hrs at 970°C, followed by a 30 hr homogenization anneal in air, again at 970°C. The sample was then oxygenated for 9 days at 450°C in flowing O_2.

Fig 9 shows that, at temperatures below ~ 77 K, J_C is appreciably enhanced

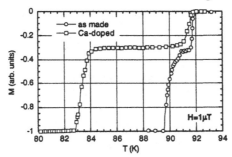

Fig. 10   Same as in Fig. 9 but in a field of 1 μT.

after the Ca treatment; at 70 K, J_C has approximately doubled. Fig. 10 shows M(T) measurements, taken at 1 μT, for both the original and Ca doped samples. At the kink temperature (both curves), the current density is about 0.4 A/cm^2. For the original sample, T_K appears at ~ 89.5 K and J_C falls to zero at the T_C of the bulk, near 91.5 K. For the doped sample, T_K has fallen to about 83 K. By ~ 84 K, the ring carries essentially no circulating shielding supercurrent. This suggests that the GB is somewhat excessively doped with Ca. Probably the bulk material in close proximity to the GB has become doped with Ca reducing its T_C to ~ 85 K. A slight broadening of the bulk transition is also apparent, possibly resulting from Ca doping, with some T_C depression, in the outer surface of the ring. A more carefully optimized heat treatment, utilizing Ca doping, would probably yield further improvement in the GB J_C, especially for operation at 77 K. It might also be, of course, that such beneficial effects will vary with misorientation angle.

Sr doping: Sr is of interest as a GB dopant because, like Ca, it is divalent, but it has a larger ionic radius which prevents substitution on the Y site. Sr can substitute on the Ba site, but since both Sr and Ba are divalent, no significant change in hole doping is expected. If GB decoration with Sr can be accomplished, and if there is a discernable effect on J_C, this might be taken as evidence that strain effects in GB cores provide a mechanism for controlling GB supercurrent transport.

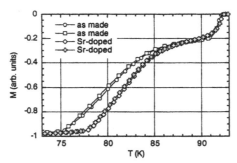

Fig. 11   Magnetic moment in 0.4 mT of four ceramic YBCO rings, two as made and two after Sr-doping (see text). All rings were fully oxygenated at 450°C before the measurement.

Our studies showed that the doping behavior was complex, generally showing minimal improvement in GB J_C. However, there did appear to be circumstances in which significant improvement could be realized. Low quality GBs; e.g., GBs with unusually low values of J_C, were typically improved, sometimes dramatically, with Sr treatment. Even open circuit rings could generally be restored to typical $J_C(\theta)$ values after Sr treatment. Occasionally, however, improvements were observed in these low quality GBs, even after a heat treatment with no Sr source available.

Dense polycrystalline ceramic ring samples [6] showed a small but apparently

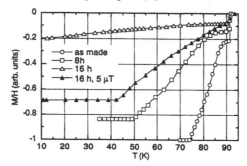

Fig. 12   Magnetic moment in 0.5 mT of a ring containing a 26° grain boundary, before and after Bi – doping at 800°C (see text). Before each measurement the ring was fully oxygenated at 450°C. For the 16 h treatment the critical current of the boundary is too small to show a kink. For this state, data were taken also at 5μT.

robust improvement after Sr treatment. Fig 11 shows measurements on four ring samples, two of which received Sr treatment (60 hrs at 970°C while immersed in $SrCO_3$ powder) two did not (all oxygenations at 450°C). At 77 K, an improvement of ~ 30% in J_C is observed. Note that, for the ceramic ring samples, a well defined and highly reproducible kink temperature is observed. This defines the weakest link for flux entry into the ring core. The J_C for transport across this complex path of linked GBs traversing the sample is remarkably similar for the different rings.

Bi doping: Fig. 12 shows M(T) for a ring containing a 26° GB before (open circles) and after treatments at 800 C in Bi_2O_3 powder. The ring was given two successive 8 hr treatments, with magnetization measurements acquired after each treatment. At 800°C, Bi_2O_3 slowly attacks the YBCO, generating an abrupt decomposition front as attack proceeds. The decomposed material is nonsuperconducting. After the first treatment (open squares), the diamagnetic signal from the bulk of the ring (i.e., the signal at 90 K), is about 60% of the original signal. This signal is approximately proportional to the undecomposed volume.

Below the kink temperatures, the circulating currents are nominally the same for the treated and untreated rings (the currents are induced by the 0.5 mT applied field). However, the grain boundary J_C is dramatically smaller after the Bi treatments. We see that, at 75 K, the grain boundary J_C has fallen, after the first Bi treatment, to about 25 % of the pretreatment value (compared to the 60% bulk diamagnetic signal). After the second (16 hr) Bi treatment, the bulk signal (at 90 K) has fallen to about 25 % of the original signal, while the grain boundary J_C (e.g., at 75 K) has fallen to about 2.5 % of its original value. The solid triangles show M measured at 5μT.

Thus, the current circulating around the ring falls much more rapidly than the bulk signal, indicating that the rate of Bi diffusion into the GB exceeds the progression rate of the attack front at the bulk. Clearly, Bi doping of the GB severely depresses supercurrent transport. Because of the strong tendency for chemical attack, it seems plausible that an isolated Bi ion residing in the GB might cause severe distortions of the local atomic structure and the electronic charge distribution. The net result is reduced GB coupling.

SUMMARY

Grain boundary supercurrent transport was studied in melt textured bi-crystal samples containing [001] tilt grain boundaries. J_{CB} data were obtained from SQUID magnetization measurements on ring samples containing a GB. The study included investigation of J_{CB} vs misorientation angle, the field dependence of J_{CB} in small externally applied fields, dependence of J_{CB} on oxygen stoichiometry and oxygen ordering. Preliminary studies of grain boundary doping with selected cations, including Ca, Sr, and Bi were undertaken. Additional systematic studies with related dopants, coupled with atomistic calculations, might provide a clearer picture of the role of dopant cations in grain boundaries and their influence on superconducting coupling.

ACKNOWLEDGEMENTS

This work is partially supported by the U.S. Department of Energy, Basic Energy Sciences-Materials Sciences-Metals, Ceramic and Engineering Sciences and Energy Efficiency and Renewable Energy, Superconductivity Program for Electric Systems, under contract W-31-109-ENG-38.

REFERENCES

[1] D. Dimos, P. Chaudhari and J. Mannhart, "Orientational Dependence of Grain-Boundary Critical Currents in YBa$_2$Cu$_3$O$_{7-x}$ bi-Crystals," *Phys. Rev. B* **41**, 4038-4041 (1990).

[2]K. E. Gray, M. B. Field and D. J. Miller, "Explanation of low critical currents in flat, bulk versus meandering, thin-film [001] tilt bi-crystal grain boundaries in $YBa_2Cu_3O_7$", Phys. Rev. B**58**, 9543 (1998), and refs. therein.

[3]G. Hammerl, A. Schmehl, R.R. Schulz, B. Goetz, H. Bielefeldt, C.W.Schneider, H. Hilgenkamp and J. Mannhart, "Enhanced Supercurrent Density in Polycrystalline $YBa_2Cu_3O_{7-x}$ at 77 K from Calcium Doping of Grain Boundaries," *Nature* **407**, 162-164 (2000).

[4]K. Guth, H. U. Krebs, H. C. Freyhardt and Ch. Jooss, "Modification of transport properties in low-angle grain boundaries via calcium doping of thin films", Phys. Rev. B**64**, 140508 (2001), and refs. therein.

[5]A. Gurevich and E. A. Pashitskii, "Current transport through low-angle grain boundaries in high-temperature superconductors", *Phys. Rev.* B**57**, 13878-13893 (1998).

[6]H. Claus, U. Welp, H. Zheng, L. Chen, A. P. Paulikas, B. W. Veal, K. E. Gray, G. W. Crabtree, "Critical Current across Grain Boundaries in Melt-Textured YBCO Rings," *Phys. Rev. B* **64**, 144507-1-9 (2001).

[7]H. Zheng, H. Claus, L. Chen, A.P. Paulikas, B.W. Veal, B. Olsson, A. Koshelev, J. Hull, and G.W. Crabtree, "Transport currents measured in ring samples: a test of superconducting weld," *Physica* C **350**, 17-23 (2001).

[8]D. J. Miller, V. R. Todt, M. St. Louis-Weber, X. F. Zhang, D. G. Steel, M. B. Field and K. E. Gray, "Microstructure and transport behavior of grain boundaries in $YBa_2Cu_3O_y$: a comparison between thin films and bulk bi-crystals", Mat. Sci. and Eng. B**53**, 125 (1998).

[9]B. W. Veal, A. P. Paulikas, H. You, H. Shi, Y. Fang, and J. W. Downey, "Observation of Temperature-Dependent Site Disorder in $YBa_2Cu_3O_{7-\delta}$ Below 150°C", Phys. Rev. B **42**, 6305 (1990).

[10]B. W. Veal and A. P. Paulikas, "Dependence of Hole Concentration on Oxygen Vacancy Order in $YBa_2Cu_3O_{7-\delta}$: A Chemical Valence Model", Physica C **184**, 321 (1991).

[11]B. H. Moeckly, D. K. Lathrop, and R. A. Buhrman, "Electromigration study of oxygen disorder and grain-boundary effects in $YBa_2Cu_3O_{7-\delta}$ thin films", Phys. Rev. B **47**, 400 (1993).

HIGH-Tc BULK-SUPERCONDUCTOR-BASED MEMBRANE-MAGNETIC SEPARATION FOR WATER PURIFICATION

Norihide Saho, Takashi Mizumori and Noriyo Nishijima
Mech. Eng. Res. Lab., Hitachi, Ltd.,
502 Kandatsu-machi, Tsuchiura, Ibaraki 300-0013, Japan

Masato Murakami and Masaru Tomita
Superconductivity Research Laboratory, ISTEC,
1-16-25 Shibaura, Minato-ku, Tokyo 105-0023, Japan

ABSTRACT

We have developed a new, continuous water- purification system that uses bulk high-Tc superconductors to rapidly remove contaminants from water supplies. Its new membrane-magnetic separator consists of (1) a preapplication treatment unit in which magnetic flocs made up of suspended solids and seeded with ferromagnetic particles are coagulated in a flocculation vessel, (2) a membrane separator that purifies water by using a rotating net to gather the flocs, and (3) a magnetic separator that recovers the magnetic flocs on the net. The maximum magnetic field of the magnetized bulk superconductors is 3.2 T. This system simultaneously removed 98% of the contaminants from sewage of our works and recovered sludge at a concentration of 40,000 mg/l.

INTRODUCTION

The side effects of urbanization are deteriorating water quality in lakes and rivers, in which flows of domestic waste water and the excessive propagation of phytoplankton have resulted in advanced eutrophication. Conventional magnetic separation technologies that use permanent magnets are unable to process broad streams of waste water because of their small narrow-range magnetic field, so they can only process small flows. Hopes have thus been placed in purification technologies that remove the organic ma0tter and phosphorous that nourish the plankton and nutrients that cause the propagation. The development[1-4] is underway of continuous purification systems that use low- Tc superconducting solenoid magnets to achieve magnetic separation. The advantages of using high-temperature superconducting (HTS) bulk magnets as a mean of generating magnetic fields like those of permanent magnets[5] are their broader range and much stronger magnetic field than permanent magnets. Accordingly, we have developed a prototype membrane-magnetic separation system combining membrane separation and magnetic separation and confirmed its high performance experimentally.

PROTOTYPE MEMBRANE-MAGNETIC SEPARATION SYSTEM
Structure of Membrane and Magnetic Separator

The structure of the new membrane-magnetic separation system combines membrane separation and magnetic separation. As shown in Figure 1, it divides purification into a series of three components: a preapplication treatment unit that gathers the targeted contaminants in the waste water into magnetic flocs; a membrane separator that filters the magnetic flocs thus formed with a membrane in order to obtain purified water; and a magnetic separator that magnetically collects the magnetic flocs deposited on the surface of the membrane, washes the surface of the membrane for reuse, and recovers the magnetic flocs as highly concentrated sludge.

In the preapplication-treatment component, first, ferromagnetic magnetic particles (Fe_3O_4), a flocculant ($Fe_2(SO_4)_3$-nH_2O), and a polymer are introduced into the waste water taken in. They are then stirred and mixed in to the waste water to generate magnetic flocs (containing contaminant particulates and ferromagnetic particles) in order to magnetize nonmagnetic contaminants, such as fine organic matter. Three to four minutes is an adequate time for stirring and mixing.

Since most of the particulates in the generated magnetic flocs have a diameter of several hundred microns, we considered a single membrane that would trap and filter the magnetic flocs sufficient if it had a micropore diameter of several tens of microns. The membrane unit with a frame and a wire net is shown in Figure 2. A stainless-steel wire net with a pore diameter of 43 μ m is used as a membrane, and the width of an aperture inside of frame is 200 mm. Twelve membrane units forming the rotating micropore membrane, are located on the outer circumference of a rotating shell with an outer diameter of 400 mm.

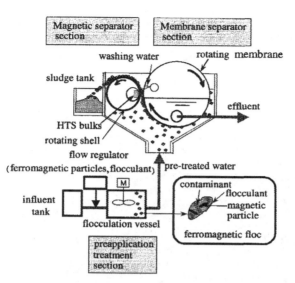

Fig. 1. Structure of high-Tc bulk superconductor-based membrane-magnetic separator for water purification The preapplication treatment waste water including magnetic flocs is passed to the membrane-magnetic separation section for purification and the concentrated sludge on the rotating recovery shell is recovered in the magnetic separation section.

To provide continuous purification, the configuration adopted is that of a rotating membrane fitted to the outermost circumference of a rotating drum, and the pre-treated water is passed from the outside to the inside of the membrane. The magnetic flocs are trapped and accumulated on the surface of the rotating membrane, the waste water is purified and flows to the inside of the drum, and the purified water is released to the outside of the system. The magnetic flocs thus accumulated on the membrane in the water migrate from the rotating membrane towards the HTS bulk magnet positioned near the surface of the pre-treated water. The magnetic flocs are separated from the membrane by the high magnetic field and the shower-like flow of wash water from inside the membrane near the surface, so the membrane goes through continuous purification and is continuously readied for reuse.

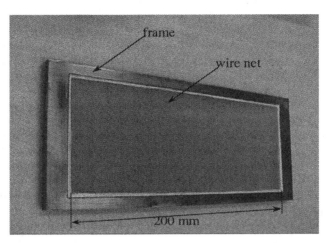

Fig. 2. One membrane unit with a frame and a wire net. A stainless steel wire net with a pore diameter of 43 μ m is used as the membrane, and the width of the aperture inside of the stainless steel frame is 200 mm.

In the magnetic separator, the HTS bulk magnet, magnetized in advance and cooled by a cooler inside a vacuum adiabatic chamber, is fixe inside a nonmagnetic rotating cylinder. The separated magnetic flocs adrift in the magnetic field near the surface migrate swiftly to the magnet, drawn by the strong attraction of the HTS bulk magnet. The migrating magnetic flocs adhere to the surface of the cylinder and are then transferred to the atmosphere above the surface of the water by the rotation of the cylinder. At this point, the surplus water in the magnetic flocs falls downwards due to gravitation, and the concentration of the magnetic flocs rises, resulting in highly concentrated sludge. The sludge is continuously stripped from the surface of the cylinder by a claw and drops of its own weight into a sludge reservoir. The surface of the cylinder is continuously readied for reuse by the claw. Through this series of operations the waste water is continuously purified, and the byproduct is highly concentrated sludge.

Since magnetic flocs may be magnetically drawn to the surface of the cylinder at high speed by the strong magnetism of the HTS bulk magnet, a large volume of magnetic floc can be separated per time unit. Therefore, if the number of revolutions is increased, the new purification system can purify large volumes of pre-treated water even with a small rotating membrane; therefore, a small purification system could purify large volumes of water.

Configuration of the HTS Bulk Magnet System

Figure 3 is a sketch of the 33-mm-square, 20-mm-thick $YBa_2Cu_3O_7$ impregnated high-temperature superconducting bulk superconductor by epoxy resins[6] used in the system. We used eleven such bulks to build a 387-mm-long trial HTS bulk magnet system. Figure 4 shows the configuration of the magnet system and its magnetization method. A small, single-stage Gifford-McMahon helium cryocooler cools the inside of the adiabatic vacuum chamber of the HTS bulk to a temperature of approximately 35 K. For connection to the magnetizing unit, the HTS bulks are embedded in the tip of a copper, thermal-conductive bar. The other end of the thermal-conductive bar is joined to the cold station of the cooler by a flange via an indium sheet. The low-temperature unit is wrapped many times with laminated heat-insulating material to prevent the penetration of radiation heat.

Split solenoid superconducting magnets are used to magnetize the HTS bulks. The magnetic field in the tunnel between the split magnets is 70 mm in diameter and approximately 100 mm long with a maximum field of 5.0 T. As the configuration of the magnetizing system in Figure 3 shows, the HTS bulk section of the HTS bulk magnet system is inserted into the tunnel between the split superconducting magnets before excitation. After cooling the bulks to a temperature of approximately 100 K, just above its critical temperature Tc, the split superconducting magnets are excited so they emit a prescribed magnetic field. Since the bulk does not reach a state of superconduction, the magnetic field penetrates the bulk unassisted. When the bulk is then cooled further, the temperature falls below its Tc and the internal magnetic flux gradually begins to

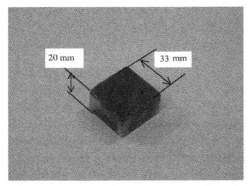

Fig. 3. YBa$_2$Cu$_3$O$_7$ bulk superconductor impregnated with epoxy resin (20 mm thick). Eleven such bulk magnets are used to build a 387-mm-long trial HTS bulk magnet.

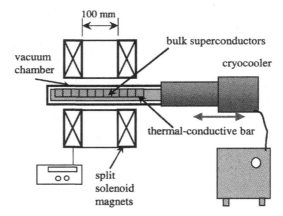

Fig. 4. Magnetization system of bulk superconductors. The magnetic field in the tunnel between the split magnets is 70 mm in diameter and approximately 100 mm long.

be trapped. The split superconducting magnets are demagnetized at a few degrees above the lowest temperature, and the temperature of the bulk then dropped to the lowest temperature. These operations complete the zero-field cooling magnetization process. Finally, the HTS bulk magnet system is extracted from the solenoid magnets (with the cooler still running) and mounted it in the sludge-recovery cylinder of the prototype membrane-magnetic separation system.

EXPERIMENTAL RESULTS
Magnetization of HTS Bulk Magnets

The magnetization characteristics of the six HTS bulk magnets (approximately 200 mm long, magnetized in a 5.0-T magnetic contained field in a section 100 mm long) are shown in Figure 5(a). The diagram shows the distribution of a magnetic field on the wall of a vacuum adiabatic chamber. The trapped maximum magnetic field was from 3.0 to 3.2 T. These values were about 70% of the excitation magnetic field. Because the width of the membrane unit is 200 mm, the peak value of each magnetic field is 1.5 to 3.2 T. The measured cooling characteristics of the cooling system are listed in Table 1. The cool-down time of the bulk magnets was about four hours from near room temperature. Under steady-state conditions, the temperature of the bulk magnets was 34 K and temperature difference between the bulk magnets was under 0.2 K. This small temperature difference shows that both magnets were cooled uniformly. The electrical consumption of the GM cryocooler is 2.8 kW.

Table 1. Performance of bulk-superconductor cooling system

bulks temperature	34 K
cool-down time	4 hours
electric power consumption	2.8 kW

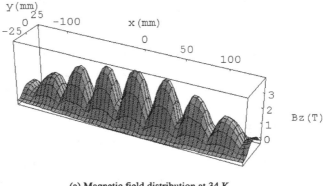

(a) Magnetic field distribution at 34 K

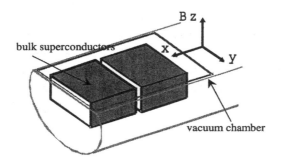

(b) 3D axis for magnetic field
measurement

Fig. 5. Measured magnetic field distribution of bulk superconductors   on the
surface of the vacuum chamber by zero-field cooling. The split solenoid
superconducting magnets can generate 5.0 T and was used for magnetization and
the trapped maximum magnetic field was 3.2 T.

Treatment of Sewage

The influent used in our experiment with a prototype mode of the new system was sewage from our works. In a preapplication treatment section, ferromagnetic particles with a concentration of 30 mg/L, flocculant with a concentration of 30 mg/L, and polymer with a concentration of 2 mg/L were added to the influent. The preapplication treatment detention time was about four minutes. After the preapplication treatment of the sewage, i.e., generating magnetic flocs, the water was passed to the membrane-magnetic separation system and purification trials were performed.

We operated this prototype system continuously, automatically rotating and backwashing the membranes, demonstrating its excellent ability for water purification and sludge recovery at a flow rate of 100 m³/day. Table 2. lists the

Table 2. Purrification test results for sewage

	Influent	Effluent	Removal efficiency (%)
SS (mg / L)	90	<1	>98
BOD (mg / L)	65	<1	>98
COD (mg / L)	61	1.7	97
Fe (mg / L)	2.3	0.18	92
total - phosphorus (mg / L)	2.7	0.07	97
turbidity (-)	44	<1	-
chromaticity (-)	85	2.4	-
concentration of solids in recovered sludge (mg / L)	-	40,000	-

removal efficiency of suspended solids (SS), biochemical oxygen demand (BOD), chemical oxygen demand (COD), and total phosphorus (T-P) when filtering water from a sewage. More than 98% of the SS in the sewage were removed at a flow rate of 100 m³/day. The concentration of BOD in the influent was 65 mg/L, and decreased to less than 1 mg/L in the effluent, indicating a removal efficiency in excess of 98%. The removal efficiencies of COD and T-P were both 97%. The turbidity results in Table 2. show that the effluent became transparent water. Concentration of Fe in Table 2. shows that since the concentration of the iron of effluent is smaller than influent, flocculant hardly leaks to the effluent.

The concentration of the recovered sludge on the rotating recovery shell with the outer diameter of 150 mm was over 40,000 mg/L. The time required to recover high-concentration-sludge while simultaneously obtaining purified water from the sewage was only four to five minutes. We therefore conclude that the system is capable of high-speed purification at approximately 20 times faster than conventional flocculant deposition methods (which require several hours).

SUMMARY

We developed a continuous membrane-magnetic separator using a HTS bulk magnet system for generating a magnetic field. Its main features are summarized as follows:

• A simple structure for trapping magnetic flocs with a single micropore membrane, and magnetic separation and recovery of magnetic flocs accumulated on the membrane with HTS bulk magnets

• Continuous purification and recovery of sludge by a rotating micropore membrane and a sludge-recovery cylinder

• High-concentration sludge recovery by dewatering; that is, the migration of sludge to the upper part of the surface of the water after magnetic separation.

Water purification trials using a prototype model of the separator found that more than 98% of the contaminants in sewage could be removed and simultaneously sludge could be continuously recovered at a concentration of 40,000 mg/L at a flow rate of 100 m³/day. Treatment time was only four to five minutes, and it was we found that the new system is capable of purification and sludge recovery at higher speeds than conventional flocculant deposition methods.

We developed this technology with the aid of an Industrial Technology R&D Implementation Technology Development Grant from the New Energy and Industrial Technology Development Organization (NEDO) of Japan.

REFERENCES
[1]N. Saho, H. Isogami, and T. Takagi, "Development of a Superconcucting Magnetic Separator with An Integral Refrigerator for Blue-Green Algae," *Journal of the Society of Electrical Engineers,* Japan, vol. 8, no. 874, pp. 31-33, 1996

[2]H. Isogami, N. Saho, and M. Morita, "A Superconcucting Magnetic Separator with An Integral Refrigerator for Blue-Green Algae, "*Proc. of the 16th Int. Cryo. Eng. Conf.,* pp. 1125-1128, 1996

[3]N. Saho, H. Isogami, T. Takagi, M.Morita, Y. Yamaoka, and H. Takayama, "Development of Continuous Superconducting-Magnet Filtration System,"*The Fourth International Conference on Environmental Management of Enclosed Coastal Sea.,* vol. 3, pp. 1399-1410, 1999

[4]N. Saho, H. Isogami, T. Takagi, and M. Morita, "Continuous Superconducting-Magnet Filtration System,"*IEEE Transactions on Applied Superconductivity,* vol. 9, no. 2, pp. 398-401, 1999

[5]J. H .P. Watoson, "High Temperature Superconducting   Permanently Magnetized Discs And Rings Prospects For Use In Magnetic Separation, " *Minerals Engineering*, vol. 12, no. 3, pp. 281-290, 1999

[6]M. Tomita and M. Murakami: Improvement of the mechanical properties of Bulk superconductors with resin impregnation. *Supercond. Sci. Technol,* Vol.13 , pp.722 -724, 2000.722 -724.

BULK SUPERCONDUCTING FUNCTION ELEMENTS FOR ELECTRIC MOTORS AND LEVITATION

T. Habisreuther, D. Litzkendorf,
R. Müller, M. Zeisberger,
S. Kracunovska, O. Surzhenko,
J. Bierlich, and W. Gawalek
Institut fuer Physikalische Hochtechnologie,
Winzerlaer Str. 10, POB 100239,
07702 Jena, Germany

T.A. Prikhna
Institute for Superhard Materials
2 Avtozavoskaya St.
Kiev, 04074, Ukraine

ABSTRACT

Bulk melt-textured High Temperature Superconductors (HTS) can be applied in electric motors, magnetic bearings or other power applications. We prepare melt-textured YBCO in a batch process. Field mapping and levitation force measurements are performed as quality control. Differences between the evidence of these methods can be seen. Trapped fields higher than 1 T at 77 K were achieved in standard material, e.g. cuboids with an edge size of 3.5 cm. We prepare function elements from this material by assembling plates for electric motors, rings for magnetic bearings, or other elements defined by the application. Plates for electric motors up to 200 kW were constructed. Superconducting joining was successfully performed on a superconducting ring.

INTRODUCTION

Melt textured Yttrium-Barium-Copper-Oxide ($YBa_2Cu_3O_{7-x}$, YBCO) is the most promising High Temperature Superconducting (HTS) material for bulk applications in the range 60 K - 77 K. It is suitable for many magnetic applications, e. g. for electric motors [1,2] and generators, HTSC permanent magnets [3], levitation systems [4], and magnetic bearings [5-7].

The base for all applications is the availability of high quality material. In the last few years we developed and optimised a reproducible batch processing [8] technique using seed crystals. Monolithic material, cylinders with a diameter up to 5 cm and cuboids with an edge size of 5 cm, can be produced reproducibly. Integral magnetic characterisation techniques [9], levitation force measurements and field mapping, were adapted to the increasing size of the monoliths, to the increased production capacity, and to the function of the material in the special application.

For further optimisation of the material detailed investigations were performed. Polarisation microscopy [10] and scanning electron microscopy was used to detect the influence of the preparation on the micro-structure of the material. VSM measurements on samples in the range of some cubic millimeters detect the magnetic properties. These measurements give the information about superconducting properties, for example the critical temperature T_c, the critical current density j_c, the irreversiblitiy field H_{irr}, and the mechanisms of flux pinning.

The applications mentioned above define more complex shapes than the monoliths we prepare as a standard. Joining of several YBCO monoliths to build rings, plates or even more complex shapes is possible. Up to now the monoliths are bonded together [8]. Superconducting joining methods, that were shown already [11,12], will lead to improved function elements of melt-textured YBCO.

In this paper we will report about the latest results on the batch processing. Also we will demonstrate some difficulties in the interpretation of the integral magnetic characterisation techniques and show our first results on superconducting joining.

BATCH PROCESSING

As standard starting material we use commercially $YBa_2Cu_3O_{7-x}$ - powder and add 0.25 mol% Y_2O_3 and 1 wt.% CeO_2. The precursor powder are thoroughly mixed. Cuboids with an edge size of 40 - 45 mm and a height of 17 mm are uniaxially pressed. The YBCO plates were processed in a modified melt-textured growth process (MTG) [13], where the squares are placed in a six side heated and six side controlled box furnace with a quasi isothermal temperature distribution.

High Temperature Superconductors

Figure 1: Batch of melt-textured YBCO photographed directly after the melt-texturing process.

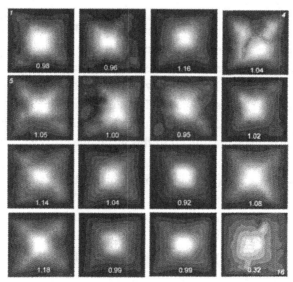

Figure 2: Remanent flux distribution and maximum trapped field of the batch shown in Fig. 1 measured in a distance of 0.8 mm at 77 K.

Before processing a melt-textured $SmBa_2Cu_3O_{7-x}$ seed is placed in the middle on top of the sample [14]. Low cooling rates of 0.2 - 1 K/h have to be applied to guarantee a single grain growth without spontaneous nucleation. Up to 16 YBCO cuboids can be fabricated in one furnace run. Shrinkage of about 15% is typical. Figure 1 is a photograph of a batch taken directly after the texturing. (Sample 1 is in the top left, sample 4 in the top right, and sample 16 in the bottom right corner). Cross structures on top of the monoliths indicate a single grain growth. Finally, by a separate procedure the material is oxygenated at a temperature of 450-600 °C in flowing O_2 for 150 hours.

BATCH CHARACTERISATION

To control the quality levitation force measurements and field mapping are performed on each sample. Both methods are non-destructive. Levitation force measurements need less time in comparison to the field mapping. The levitation quality of the monolith is given by one force value. This value is dependent on the experimental set-up, especially the shape of permanent magnet and the distance between magnet and superconductor. Usually the field mapping quality is given by the maximum trapped field and the shape of the remanent flux distribution. Therefore it contains more information of the material. The remanent flux distribution depends also on the experimental set-up. Here the distance between Hall probe and the monolith is very important. Also the size of the Hall probe, the magnetising technique the speed of the measurement and the temperature control influence the measurement [15]. Ideally the remanent flux distribution is a pyramid for rectangular shaped material. Detailed measurements in combination with numerical calculations can detect the distribution of the current in the monolith [16]. Due to the increased effort this technique [16] can not be applied to each monolith.

Figure 2 shows the results of field mapping on the batch shown in Fig. 1. The samples were magnetised in a field of 2 T at 77K. The distance between Hall probe and monolith was 0.8mm. Extrapolated to the surface the values would be about 15% higher [16]. Most of the monoliths are single domain. This means that the current is flowing through the whole volume. Sample 4 is not single domain, even the monolith shows the "cross"-structure. Cracks in the material prohibit a homogeneous current flow inside the grain. Sample 6 is also not pure single domain. Additional nucleation was initiated by impurities (insulation material from the furnace). Sample 16 seems to be single domain, but the trapped field is bad. Such an effect can happen if the oxygenation of the monolith is incomplete.

Figure 3 shows a comparison between the maximum trapped field and the standard zero field cooled levitation measurement, where a SmCo permanent magnet with 25 mm diameter was used. The magnetic induction on the surface is 0.4 T. In general the data are comparable. But there are some differences. The "best" sample according to levitation forces is sample 4, but this monolith is not single domain. In contrast single domain sample 3 has a quite bad levitation value. The difference of the measurements is caused by the following effect: High levitation forces are achieved if the superconductor beneath the edge of the

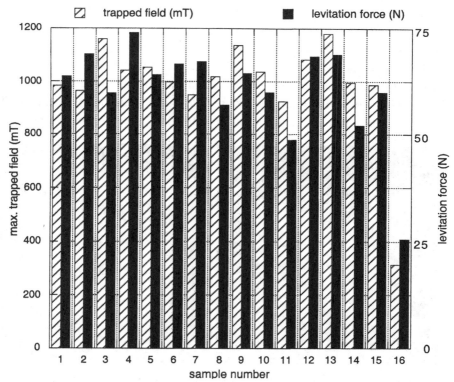

Figure 3: Comparison between maximum trapped flux and zero-field cooled levitation forces, both at 77K, for the batch of Fig. 1.

permanent magnet is very good. For sample 4 the three domains match with the edge of the permanent magnet. Therefore the gradient of the magnetic field between the superconductor and permanent magnet is high. This leads to high forces. The edge of the good domain of sample 3 seems to be less attached to the edge of the magnet. This leads to the effect that flux from the permanent magnet is pushed around the monolith. So the gradient is smaller and the force is lower. The critical current densities of the monoliths are comparable, because both samples trap about 1 T.

The comparison was made to demonstrate difficulties in the quality control.
In levitation applications, for example fly-wheels, high forces are requested. Here the magnets or the magnet systems differ from the one we selected for the quality control. Combinations of iron and permanent magnets are proposed. The advantage of these systems (samplers) are smoothing of magnetic inhomogeneities and creating of steep magnetic field gradients. Figure 4 and 5 show the comparison of field mapping and levitation force measurements with a sampler unit with three poles and a maximum induction of about 1 T on the middle pole. Sample A is single domain, sample B is multi-domain material. Even the material is different, the levitation forces (zero field cooled) are nearly identical. The maximum force that can be achieved is defined by the permanent magnets. It is

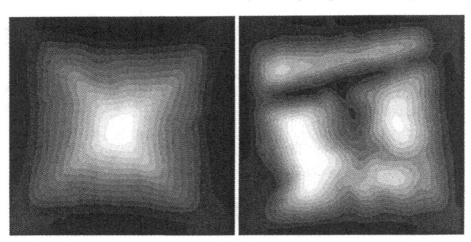

Figure 4: Remanent flux distribution for a single domain (left) and a multi-domain monolith (right).

High Temperature Superconductors

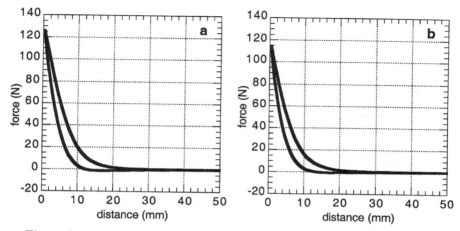

Figure 5: Levitation forces for the samples shown in Fig. 4, a - single domain, b - multi domain.

given by measuring repulsive forces between identical magnets. This experiment represents a magnet above a superconductor with infinite size and very high critical current density. Because the superconductors are smaller the force is reduced (about 10%, depending on the set-up) even if the current density would be infinite. The quality of melt-textured material even if it is not single domain was good enough to achieve about 80% of this limit. Single domain material at 77K reaches about 85-90% of the maximum force.

From this point of view material with domains adapted to the poles of permanent magnets can be used for the levitation application. The influence of non single domain material on losses in bearings is unknown up to now. It will be investigated next.

FUNCTION ELEMENTS

For levitation applications the superconducting elements have to be attached to the magnetic poles of the system. So the requirements on the shape of the monoliths are low, cuboids can be used. In levitated trains [4] melt-textured YBCO is placed in the vehicle. The rail is magnetic. The vehicle levitates above the rails and can move linear.

For rotation applications, especially bearings in fly-wheels, the low friction is given by homogeneous ring-shaped magnetic fields. The superconductors have to be placed in the stray field of the magnet system. The superconducting element has to be ring-shaped or the whole area beneath the magnet should be covered by superconducting material. So the application demands rectangular or hexagonal monoliths [6] to cover an area or trapezoidal monoliths, if a ring has to be formed.

Several designs for electric motors using bulk HTS elements were developed in the last years. Rings were used in Hysteresis motors [1]. The second design was to build reluctance motors [1]. Here an increased magnetic anisotropy in the rotor due to combinations of HTS and magnetic steel allowed to achieve double of the force density in the motors air gap in comparison to conventional electric reluctance motors [1].

At present two large engines are under construction. At the MAI Moscow a motors with a designed output power of 100 kW will be tested. For this machine plates from melt-textured YBCO were build. The elements for 1/4 of the motor are shown in Fig. 6. The testing of 1/4 of the motor is going on. This machine contains a liquid nitrogen cryostat for the rotor to reduce the cooling costs. The horizontal test bench is shown in Fig. 7. At the OSWALD GmbH a 200 kW motor was designed. In Fig. 8 some plates for this machine are shown. The machine will

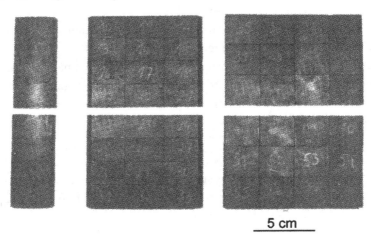

5 cm

Figure 6: Single domain melt-textured YBCO cut to plates for 1/4 of the 100 kW rotor.

Figure 7: Test bench for 1/4 of the 100 kW HTS motor. The rotor works within a cryostat at 77K.

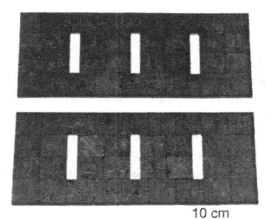

10 cm

Figure 8: Single domain melt-textured YBCO cut to plates for the rotor of the 200 kW motor.

run in liquid nitrogen. This cooling method does not increase the losses of the motor due to friction. Also the superconducting material in the rotor is heated not more than about 1K.

For all these machines single domain monoliths were prepared and cut to exact plates. In the rotor they will be fixed mechanically. No superconducting connection between the monoliths exists. If the whole plates would be superconducting, motor parameters (efficiency, output power, and reactive power) will be optimised. Therefore the monolith size has to be increased or superconducting joining has to be developed.

Another rotor design from the University of Stuttgart is shown in Fig. 9. This is an example for a more complex geometry. The monoliths c-axis are arranged in radial direction. After melt-texturing several steps of mechanical processing (cutting, grinding, bonding, ...) were applied.

Figure 9: YBCO elements with complex geometry for a rotor.

SUPERCONDUCTING JOINING

As mentioned above superconducting joining is one possibility to improve the size of melt-textured YBCO monoliths and thus their quality. Several successful approaches were demonstrated already [11,12]. We started our experiments using Thullium-Barium-Copper-Oxyde (TmBCO) powder as solder. A temperature between the peritectic decomposition temperatures of TmBCO (about 950°C) and YBCO (about 1030°C) has to be applied. In a cooling step the decomposed TmBCO re-crystalises epitaxial on the YBCO and so superconducting joining is created.

The soldering procedure is demonstrated on a ring with a outer diameter of 8 mm and a height of 3 mm. Magnetic measurements on rings are very significant to show whether the soldering was successful [11]. The size of the ring was chosen to perform VSM and field mapping (0.8 mm distance) on the ring. The results are shown in Fig. 10 and Fig. 11. From the VSM measurements a critical current density of about 20000 A/cm^2 was calculated. The quality after soldering

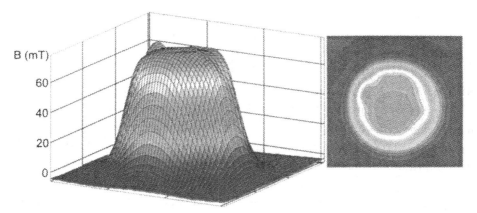

Figure 10: Remanent flux distribution of a ring (Ø=8mm, h=3mm, T=77K) after soldering. The undisturbed profile indicates soldered areas with high j_c.

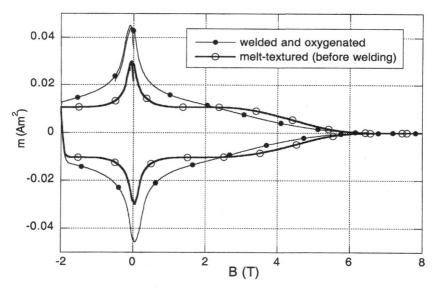

Figure 11: Magnetisation curves of the ring before and after welding (T = 77 K)

seems to be a little better than after melt-texturing. This effect can be related to the oxygenation. After soldering the material has to be oxygenated again. This can lead to changes in the magnetisation loops and thus to changes in dependence of j_c on the magnetic field. Further development has to be done to join large individual grown monoliths.

SUMMARY

We showed that melt-textured YBCO of high quality was prepared reproducibly. Careful interpretation of standard characterisation measurements has to be performed. Field mapping contains more information about the material than Levitation forces. For the application of melt-textured YBCO the characterisation method has to be adapted to the function of the HTS material in the special application. Strong motors with melt-textured YBCO elements will be tested. From these test we will get the information which material parameters have to be improved further. Superconducting joining was demonstrated on a small sample. Here further development will improve the function of melt-textured material in the application.

ACKNOWLEDGEMENT

This work is supported by the German BMBF under Grant No. 13N6854A3 and by the German BMWi under Grant No. 0327305J. The authors wish to thank P. Dittmann, M. Wötzel, S. Manzel, H. Steinmetz, Ch. Schmidt, G. Bruchlos, M. Arnz, and J. Dellith for technical assistance and their colleagues from the OSWALD Elektromotoren GmbH, MAI Moscow and Uni Stuttgart for designing and testing the electric motors.

REFERENCES

[1]   L. K. Kovalev, K. V. Ilushin, V. T Penkin, K. L. Kovalev, S. M.-A. Koneev, V. N. Poltavets, A. E. Larionoff, K. A. Modestov, S. A. Larionov, W. Gawalek, T. Habisreuther, B. Oswald, K.-J. Best, and T. Strasser, "Hysteresis and Reluctance Electric Machines with Bulk HTS Elements. Recent Results and Future Development", *Inst. Phys. Conf. Ser.* **No 167** 1037-1042 (2000)

[2]   B. Oswald, M. Krone, M. Söll, T. Strasser, J. Oswald, K.-J. Best, W. Gawalek, and L. Kovalev, "Superconducting Reluctance Motors with YBCO Bulk Material", *IEEE Transactions of Appl. Superconductivity*, **Vol. 9, No. 2** 1201 - 1204 (1999)

[3]   St. Gruss, G. Fuchs, G. Krabbes, P. Verges, P. Schaetzle, K.-H. Mueller, J. Fink, and L. Schultz, "Trapped fields beyond 14 Tesla in bulk $YBa_2Cu_3O_{7-x}$", *IEEE Transactions of Appl. Superconductivity*, Vol. **11**, No. 1 3720-3723 (2001)

[4]   T. Habisreuther, M. Zeisberger, D. Litzkendorf, O. Surzhenko, and W. Gawalek, "Developing Melt-Textured YBCO", *JOM* **Vol 52** No6 26-28 (2000)

[5]   Z. Xia, Q. Y. Chen, K. B. Ma, C. K. McMichael, M. Lamb, R. S. Cooley, P. C. Flowler, and W. K. Chu, "design of superconducting magnetic bearings with high levitating force for flywheel energy storage systems", *IEEE Transactions of Appl. Superconductivity*, Vol. **5** 622-625 (1995)

[6]   T. M. Mulcahy, J. R. Hull. K. L. Uherka, R. G. Abboud, and J. J. Juna, " Test Results of 2-kWh Flywheel Using Passive PM and HTS Bearings", *IEEE Transactions of Appl. Superconductivity*, Vol. **11** No. 1 1729 - 1732 (2001)

[7]   R. Koch, R. Wagner, M. Sander, H. J. Gutt, "Development and Testing of a 300Wh/10kW Flywheel Energy Storage System", *Inst. Phys. Conf. Ser.* No **167** 1055 - 1058 (2000)

[8]   D. Litzkendorf, T. Habisreuther, M. Zeisberger, O. Surzhenko, and W. Gawalek, "Preparation of melt-textured YBCO with optimized shape for cryomagnetic applications", *Inst. Phys. Conf. Ser.* No **167** 91 - 94 (2000)

[9]   T. Strasser, *Magnetische Charakterisierung von schmelztexturierten $YBa_2Cu_3O_{7-x}$ Hochtemperatursupraleitern*, PAPIERFLIEGER, Clausthal-Zellerfeld (1999)

[10]   P. Diko, "Growth-related microstructure of melt-grown REBa2Cu3Oy bulk superconductors", *Supercond. Sci. & Technol.* **13** 1202 (2000)

[11   B. W. Veal, H. Zheng, H. Claus, L. Chen, A. P. Paulikas, A. Koshelev, G. W. Crabtree, "Critical Currents and Weak Links in Melt-Textured R123", *Proceedings of the 2000 International Workshop on Superconductivity*, June 19-22, Matsue-Shi, Shimane, Japan 211 - 214 (2000)

[12]   H. Walter, Ch. Jooss, F. Sandiumenge, B. Bringmann, M.P. Delamare, A. Leenders, H.C. Freyhardt, "Large Intergranular Critical Currents in Joined YBCO Monoliths" *Europhysics Letters* **55** 100 (2001).

[13]   K. Salama, S. Sathyamurthy, "Melt texturing of YBCO for high current applications", *Appl. Supercond.*, Vol **4**, No 10-11 547 - 561 (1996)

[14] S. Schauroth, "Samarium-Barium-Kupfer-Oxid-Impfkristallherstellung zur Schmelztexturierung von massiven $YBa_2Cu_3O_{7-x}$ - Hochtemperatursupraleitern", thesis, FH Jena (2001)

[15] T. Habisreuther, O. Surzhenko, M. Zeisberger, D. Litzkendorf, S. Kracunovska, R. Müller, and W. Gawalek "Experimental influences on the significance of field mapping", *3rd International Workshop on Processing and Applications of Superconducting (RE)BCO Large Grain Materials*, July 11 - 13, Seattle, Washington, USA, 2001

[16] M. Zeisberger, T. Habisreuther, D. Litzkendorf, R. Mueller, O. Suezhenko, and W. Gawalek, "Experimental and theoretical investigations on the field trapping capabilities of HTSC-bulk material", *paper presented at the EUCAS2001*, Aug 26-31, Kopenhagen, Dennmark (2001), proceedings in press.

PROCESSING AND PROPERTIES OF Gd-Ba-Cu-O BULK SUPERCONDUCTOR WITH HIGH TRAPPED MAGNETIC FIELD

S. Nariki, N. Sakai and M. Murakami
Superconductivity Research Laboratory, ISTEC
1-16-25 Shibaura, Minato-ku, Tokyo
105-0023 Japan

ABSTRACT

We report on the performance of large-grain Gd-Ba-Cu-O bulk superconductors. Employing fine Gd-211 powder as a starting material remarkably reduced the size of 211 particles dispersed in the Gd-123 matrix, leading to a dramatic enhancement of the critical current density and field trapping ability. The trapped magnetic field of single-grain Gd-Ba-Cu-O/Ag bulk 50 mm in diameter reached 2.6 T at 77 K. Furthermore, the trapped field was 3.5 T at 77 K when it was measured between two large bulk samples in order to minimize the demagnetizing effect.

INTRODUCTION

Large grain RE-Ba-Cu-O (RE: rare earth elements) bulk superconductors can trap large magnetic fields and thus can function as a strong quasi-permanent magnet.[1-9] For the enhancement of trapped magnetic field of bulk superconductors, it is important to increase the critical current density (J_c) and/or the size of single-domain. LRE-Ba-Cu-O (LRE: light rare earth: Nd, Sm, Eu and Gd) superconductors fabricated in a reduced oxygen atmosphere exhibit high J_c values in a high field region compared to Y-Ba-Cu-O superconductors, since the magnetic field dependence of J_c exhibits a secondary peak effect at intermediate fields. As a result, the trapped magnetic fields of large grain Sm-Ba-Cu-O and Nd-Ba-Cu-O bulks exceeded those of Y-Ba-Cu-O superconductors with similar sizes.[3, 5, 6, 10-13]

Earlier works on Gd-Ba-Cu-O materials showed that the trapped fields were only

0.5 – 0.7 T at 77 K for the sample 32 mm in diameter[14,15] and 0.9 T for the sample of 45 mm in diamter.[16] These values were comparable to those of Y-Ba-Cu-O.

The J_c values of RE-Ba-Cu-O superconductors are strongly dependent on the microstructure. The introduction of excess RE_2BaCuO_5 (RE-211) contributes to the enhancement of J_c in a low field region. Finely dispersed RE-211 particles in the superconducting $REBa_2Cu_3O_y$ (RE-123) phase act as a strong pinning center. In the earlier Gd-Ba-Cu-O bulk samples, however, the particle size of Gd-211 was about 2 μm, which is larger than that of Y-211 particles (about 1 μm) in Y-Ba-Cu-O samples.

Thus we employed fine Gd-211 starting powder and found that the J_c values could dramatically be enhanced. Microstructural observation revealed that the size of Gd-211 inclusions was reduced to submicron level.[17,18] With such a treatment, very high trapped-fields were achieved in large-grain Gd-Ba-Cu-O bulk superconductors.[19-22]

In this paper, we report the performance of recently developed large-grain Gd-Ba-Cu-O superconductors with finely distributed Gd-211 inclusions.

REFINEMENT OF Gd-211 PARTICLES IN Gd-Ba-Cu-O BULK MATERIALS

In this section, we show the effect of the size of Gd-211 starting powders on the microstructure for small Gd-Ba-Cu-O bulk samples. As listed in Table I, three kinds of Gd-211 starting powders with different particle sizes were prepared from commercial Gd_2O_3, BaO_2 and CuO powders. The powders A and B were prepared by the calcination of mixed powders at 1323 K for 4 h and 1173 K for 8 h, respectively. The respective average particle sizes of these powders were 2.7 and 1.0 μm as determined by BET specific area measurements. The powder C was prepared by ball-milling the powder B using ZrO_2–Y_2O_3 balls in acetone for 2 h. The average size of powder C was reduced to 0.1 μm with this treatment. These Gd-211 powders were added to commercial Gd-123 powders in a molar ratio of Gd-123 : Gd-211 = 5 : 2. 0.5 wt.% Pt was also added to the mixtures. The mixed powders were uni-axially pressed into pellets 10 mm in diameter and 12 mm in thickness. An MgO (100) seed was placed at the center top of the pellet, which was then melt-processed in 99%Ar - 1%O_2 mixture gas. The pellets were partially melted at 1353 K for 30 min, slowly cooled from 1273 K to 1233 K at a rate of 1 K/h, and finally cooled to room temperature at a rate of 100 K/h.

Figure 1 shows SEM (scanning electron microscope) photographs of the polished surface of the melt-textured samples. One can see that the Gd-211 particles in the Gd-123 matrix can be refined with decreasing the particle size of Gd-211 starting powders. In particular, the average diameter of Gd-211 inclusions in the sample made from powder C was reduced to 0.3 μm as shown in Fig. 1 (c).

The observation of Gd-211 particles at the partial melting stage will be informative for understanding the difference in the final microstructure. Thus the samples were quenched to room temperature after melted at 1373 K for 30 min in air.

Table I.   Gd-211 starting powders

Powder	Preparation condition	Average particle size (μm)
A	Calcination at 1323 K for 4 h	2.7
B	Calcination at 1173 K for 8 h	1.0
C	Ball-milling of powder B	0.1

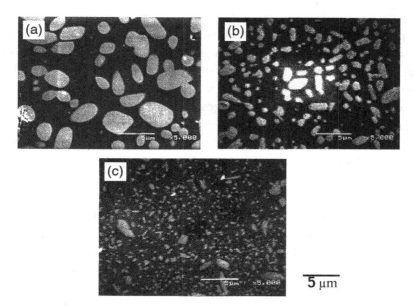

Figure 1   SEM photographs for the polished surfaces of Gd-Ba-Cu-O bulk samples fabricated using various Gd-211 starting powders: (a) powder A; (b) powder B; and (c) powder C listed in Table I.

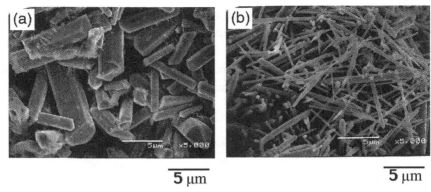

5 μm   5 μm

Figure 2   SEM photographs for the fracture surfaces of melt-quenched samples, which were prepared using (a) powder A and (b) powder C in Table I.

Figure 2 shows SEM photographs of the melt-quenched samples prepared from the Gd-211 starting powders A and C.   The morphology and the size of Gd-211 particles at the partial melting stage depended on the size of initially-added Gd-211 particles. Rod-like Gd-211 particles are observed in the sample made from powder A as shown in Fig. 2 (a).   In contrast a large number of fine Gd-211 needles were formed in the sample made from powder C, as shown in Fig. 2 (b).   From these results, we can propose that an addition of fine Gd-211 powders increased the number of nucleation sites for Gd-211 during the peritectic decomposition of Gd-123 phase.   This will lead to very fine dispersion of Gd-211 inclusions in the melt-textured bulk sample as shown in Fig. 1 (c).

Superconducting properties were characterized with the specimens of 1.5 x 1.5 x 1.0 mm^3 cut from the melt-grown blocks.   After the specimens were annealed in oxygen atmosphere at 673 K for 100 h, the measurements of critical temperature (T_c) and J_c were performed with a SQUID magnetometer.   DC-susceptibility measurements showed that onset T_c of all the samples was about 92 K with sharp transition.   Figure 3 shows the J_c-B curves at 77 K. J_c was estimated based on the extended Bean model[23] from the magnetization loop.   It is worth nothing that a dramatic increase in J_c was achieved with the size reduction of Gd-211.   The J_c of the sample prepared using powder C exhibited an extremely high value of 1.3 x 10^9 A/m^2 at 77 K in self-field.   Figure 4 displays the J_c values at 0.05 T as a function of

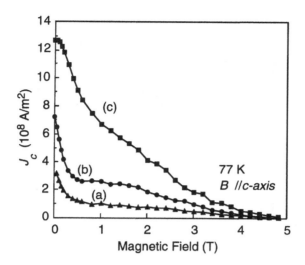

Figure 3 J_c-B curves at 77 K for the Gd-Ba-Cu-O bulk samples fabricated using (a) powder A; (b) powder B; and (c) powder C in Table I.

V_f/d, where V_f is the volume fraction of Gd-211 phase and d is their mean size. V_f/d reveals the effective interface area of Gd-211/Gd-123 per unit volume.   A linear relation between J_c and V_f/d suggests that the dominant pinning in a low field region is Gd-123/Gd-211 interfacial pinning.[24]

FABRICATION OF LARGE SINGLE GRAIN Gd-Ba-Cu-O/Ag BULKS AND THEIR TRAPPED MAGNETIC FIELDS

The defects such as cracks and high-angle grain boundaries deteriorate the field-trapping ability.   It is known that an addition of Ag is effective in inhibiting the crack formation.[3, 5, 6]   We also confirmed that the macro-cracks in Gd-Ba-Cu-O samples decreased with Ag addition, thus the mechanical and field trapping properties were greatly improved.[25, 26]   In this section, we report on the fabrication and trapped field of large-grain Gd-Ba-Cu-O with employing Ag and ultra-fine Gd-211 starting powders.

The fabrication procedure of large Gd-Ba-Cu-O sample is as follows.   The Gd-211 starting powders with the average size of 0.2 and 0.1 μm were prepared by ball-milling the calcined Gd-211 powder.   The powders of Gd-123 and Gd-211 were mixed in a molar ratio of Gd-123: Gd-211 = 2: 1.   0.7 wt.% Pt and 10 - 15 wt.%

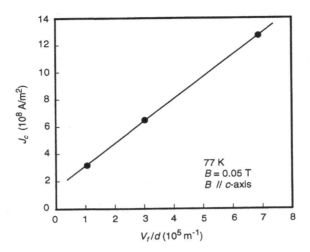

Figure 4   Plot of the J_c at 0.05 T versus V_f/d of Gd-211 particles in Gd-Ba-Cu-O samples.

Ag_2O were also added to the mixture.   These mixed powders were first uni-axially pressed into pellets 30, 40 and 60 mm in diameter, and subjected to cold-isostatic pressing (CIP) under a pressure of 200 MPa.   Melt processing was performed in an atmosphere of $1\%O_2$ in Ar.   The precursor was heated to 1373 - 1388 K and held for 30 min, cooled to 1298 K in 1 h, and then a Nd123 (001) seed crystal was placed on the top of the partially-molten pellet.   After seeding, the pellet was cooled to 1253 K in 1 h, slowly cooled at a rate of 0.2 - 0.5 K/h to 1228 K, and finally cooled at a rate of 20 - 50 K/h to room temperature. The melt-textured samples were annealed at 673 - 723 K for 200 - 900 h in flowing pure oxygen gas.

Figure 5 shows photographs of the top surface of as-grown Gd-Ba-Cu-O samples, which were made from a precursor pellet 40 mm in diameter.   The final diameter of melt-grown sample was reduced to 32 mm.   Unfortunately, as shown in Fig. 5 (a), the grain growth from the seed was inhibited by nucleation from the edge, when the size of Gd-211 starting powder was 0.1 μm.   On the other hand, the sample fabricated using the Gd-211 powders 0.2 μm in size could be grown into a single grain (Fig. 5 (b)).   From the SEM observation for the sample of Fig. 5 (b), the average diameter of Gd-211 inclusions is determined to be 0.5 μm.

High Temperature Superconductors

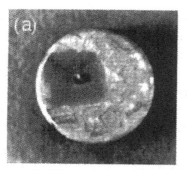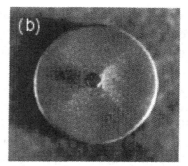

Figure 5   Photographs of the Gd-Ba-Cu-O/Ag bulks 32 mm in diameter. These bulk samples were fabricated using Gd-211 starting powders with the particle sizes of (a) 0.1 μm and (b) 0.2 μm.

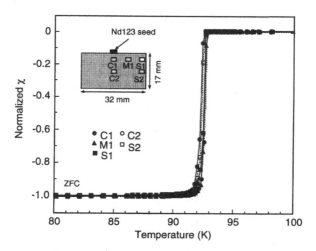

Figure 6   Temperature dependence of DC-susceptibility for the specimens cut from various locations in a single grain bulk.   The location of each specimen was sketched in this figure.   The bulk sample was fabricated from fine Gd-211 starting powders with the particle sizes of 0.2 μm.   The magnetization values were normalized with the value at 10 K.

It is known that superconducting properties of large grain bulk samples vary with relative positions within a bulk. Hence we measured T_c and J_c-B properties for small specimens cut from various locations in the bulk of Fig. 5 (b) to study spatial distribution. Figure 6 shows the temperature dependence of the DC susceptibility in the zero-field-cooled (ZFC) process in the presence of a magnetic field of 1 mT. The location of the specimens is displayed in the inset. The onset-T_c was around 92.5 K for each specimen. Figure 7 shows the J_c-B curves at 77 K for the specimens. The J_c values in a low field region were very high and reached 7.5 – 8.5 x 10^8 A/m^2 in a top side area of the bulk (for example, M1 and S1). It is also interesting to note that the present bulk has high J_c values at the applied field of 0.5 - 3.0 T. Figure 8 shows the normalized pinning force $f = F_p/F_{p,max}$ for specimen C1 and S1, where $F_p = J_c$ x B, as a function of $b = B/B_{irr}$, where B_{irr} denotes the irreversibility field. A good scaling was obtained in the temperature range of 77 - 84 K. The position of peak is $b_0 = 0.39$ for C1 and 0.37 for S1. These values were larger than the peak position reported for Y-Ba-Cu-O ($b_0 = 0.33$),[27] in which the

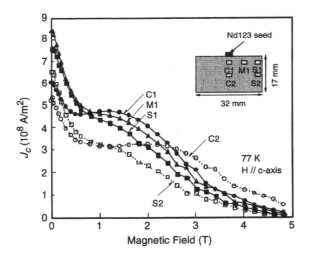

Figure 7 J_c-B curves at 77 K for the specimens cut from various locations in a single grain bulk, which was fabricated using fine Gd-211 starting powders with the particle sizes of 0.2 µm.   The notations are identical to those in Fig. 6.

High Temperature Superconductors

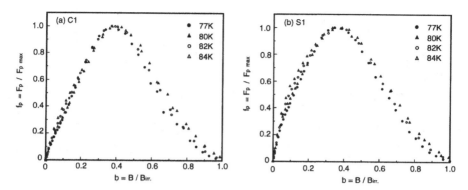

Figure 8 Scaling of the volume pinning forces, f_p versus $b=B/B_{irr}$, where B_{irr} is the irreversibility field, for the specimens C1 and S1 cut from a single-grain bulk.

dominant pinning is Y-211 interfacial pinning. Especially, the peak position for C1 is almost equal to the value reported for Nd-Ba-Cu-O bulk sample ($b_0 = 0.40$),[27] which suggests the contribution of field-induced pinning by Gd/Ba substitution region with depressed T_c.

The trapped field of the bulk sample was measured with an automated Hall probe measurement system.[28] The sample was placed at the center of bore in a 10 T superconducting magnet such that the c-axis of the sample was aligned parallel to the field. The sample was field-cooled to liquid nitrogen temperature (77 K) in the presence of 7 T. After removing the external field, the profile of trapped magnetic flux density was measured by scanning Hall probe sensors. The total gap between the polished top surfaces of the samples and the active area of the Hall sensor was adjusted to be 1.2 mm, which included the thickness of a mold of the sensor, 0.7 mm. Figure 9 shows the trapped field distributions for the Gd-Ba-Cu-O bulk samples with various diameters. The trapped field was enhanced with the enlargement of grain size. As presented in Figs. 9 (a) and (b), the maximum values of trapped field were 1.4 T for the sample 25 mm ø and 2.05 T for the sample 32 mm ø. Furthermore, we succeeded in fabricating a large single grain bulk 50 mm in diameter and 20 mm in thickness with 15 wt.% Ag_2O addition. The trapped field reached a high value of 2.4 T at 1.2 mm above the sample surface as shown in Fig. 9 (c), and it was 2.6 T when the Hall sensor was directly placed on the bulk surface. This trapped field far exceeds those of conventional RE-123 type bulk superconductors. Since the

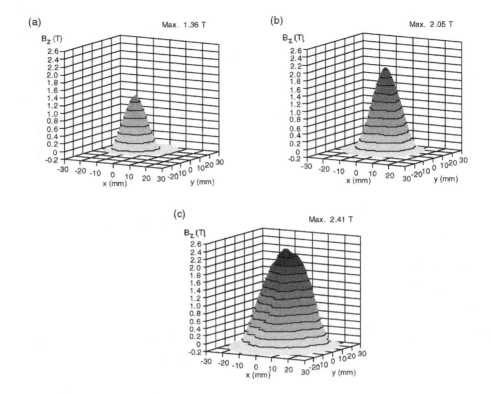

Figure 9   Trapped field distributions of Gd-Ba-Cu-O /Ag bulk samples at 77 K. The sample dimensions are: (a) 24 mm in diameter and 12 mm in thickness; (b) 32 mm in diameter and 14 mm in thickness; and (c) 50 mm in diameter and 20 mm in thickness.

surface field of a single sample is affected by the demagnetizing effect,[29] we also measured the trapped field between two blocks 50 mm in diameter.   As shown Fig. 10, the trapped flux density of 4.1 T was recorded immediately after reducing the external field to zero.   This value was reduced to be 3.5 T after 60 min and then almost stayed constant, which shows that the bulk Gd-Ba-Cu-O has a potential to generate a field of 3.5 T at 77 K.

High Temperature Superconductors

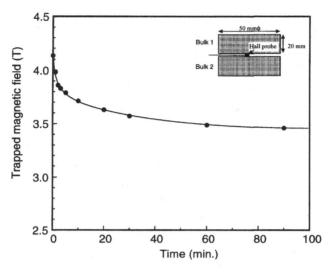

Figure 10   Trapped magnetic field recorded between two Gd-Ba-Cu-O bulks 50 mm in diameter as a function of time after reducing the external field to zero.

CONCLUSIONS

We studied superconducting and field-trapping properties of Gd-Ba-Cu-O bulk superconductors fabricated in a controlled reduced oxygen atmosphere.   The size of Gd-211 particles dispersed in the bulk was reduced to submicron level with the employment of ultra-fine Gd-211 starting powders, leading to an increase in the J_c values.   The field-trapping performance of Ag-added large Gd-Ba-Cu-O sample could be enhanced by using ultra-fine Gd-211 powders of 0.2 μm diameter.   The single grain bulk Gd-Ba-Cu-O 50 mm in diameter and 20 mm in thickness exhibited a trapped field of 2.6 T at 77 K, when the Hall sensor was in contact with the surface. In addition, the trapped field between two bulks reached 3.5 T at 77 K.   These results imply that Gd-Ba-Cu-O bulk superconductors have a significant potential for various engineering applications as very strong permanent magnets.

ACKNOWLEDGEMENTS

This work is supported by the New Energy and Industrial Technology Development Organization (NEDO) as Collaborative Research and Development of Fundamental Technologies for Superconductivity Applications.

REFERENCES

[1] M. Morita, M. Sawamura, S. Takebayashi, K. Kimura, H. Teshima, M. Tanaka, K. Miyamoto and M. Hashimoto, "Processing and Properties of OMG Materials," *Physica C*, **235-240**, 209-12 (1994).

[2] M. Morita, K. Nagashima, S. Takebayashi, M. Murakami and M. Sawamura, "Trapped Field of YBa$_2$Cu$_3$O$_7$ QMG Bulk Magnets," *Materials Science and Engineering*, **B53**,159-63 (1998).

[3] H.Ikuta, A. Mase, Y. Yanagi, M. Yoshikawa, Y. Itoh, T. Oka and U. Mizutani, "Melt-Processed Sm-Ba-Cu-O Superconductors Trapping Strong Magnetic Field," *Superconductor Science and Technolology*, **11**, 1345-47 (1998).

[4] S.Gruss, G. Fuchs, G. Krabbes, P. Schätzle, J. Fink, K. H. Müller and L. Schultz, "High Trapped Fields in Melt-Textured YBa$_2$Cu$_3$O$_{7-d}$," *IEEE Transactions on Applied Superconductivity*, **9**, 2070-73 (1999).

[5] H.Ikuta, A. Mase, U. Mizutani, Y. Yanagi, M. Yoshikawa, Y. Itoh and T. Oka, "Very High Trapped Field in Melt-Processed Sm-Ba-Cu-O," *IEEE Transactions on Applied Superconductivity*, **9**, 2219-22 (1999).

[6] H. Ikuta, T. Hosokawa, M. Yoshikawa and U. Mizutani, "Synthesis and Superconducting Properties of c-Axis Aligned, Single-Domain NdBCO/Ag, Melt-Processed Superconductors," *Superconductor Science and Technology*, **13**, 1559-68 (2000).

[7] G. Krabbes, G. Fuchs, P. Schätzle, S.Gruss, J. W. Park, F. Hardinghaus, G. Stöver, R. Hayn, S.-L. Drechsler and T. Fahr, "Zn Doping of YBa$_2$Cu$_3$O$_7$ in Melt-Textured Materials: Peak Effect and High Trapped Fields," *Physica C*, **330**, 181-90 (2000).

[8] S. Nariki, N. Sakai and M. Murakami, "Enhancement of Critical Current Density of Dy-Ba-Cu-O Bulk Superconductor by the Refinement of Dy211 Particles," *Physica C*, **357-360**, 814-16 (2001).

[9] S. Nariki and M. Murakami, "Peak Effect in Field Trapping Properties of Large Grain Dy-Ba-Cu-O/Ag Bulk Superconductors," *Physica C*, in press.

[10] M. Murakami, N. Sakai, T. Higuchi and S. I.. Yoo, "Melt-Processed Light Rare Earth Element-Ba-Cu-O," *Superconductor Science and Technology*, **9**, 1015-32 (1996).

[11] M. Matsui, T. Miyamoto, S. Nariki, N. Sakai and M. Murakami, "Fabrication of Large-Grain Nd-Ba-Cu-O Superconductor," *Physica C*, **357-360**, 694-96 (2001).

[12] W. Lo and K. Salama, "Growth and Properties of NdBCO-Ag Large Grains for

Device Applications," *Physica C*, **341-348**, 2293-96 (2000).

[13]M. Muralidhar, S. Nariki, M. Jirsa, Y. Wu and M. Murakami, "Strong Pinning in Ternary (Nd-Sm-Gd)Ba$_2$Cu$_3$O$_y$ Superconductors", *Applied Physics Letters*, **80**, 1016-18 (2002).

[14]H. Hinai, S. Nariki, S. J. Seo, N. Sakai, M. Murakami and M. Otsuka, "Melt Processing and Superconducting Properties of Bulk Gd123 Superconductors," *Journal of Cryogenic Society of Japan*, **34**, 584-86 (1999).

[15]H. Hinai, S. Nariki, S. J. Seo, N. Sakai, M. Murakami and M. Otsuka, "Superconducting Properties of Gd123 Superconductor Fabricated in Air," *Superconductor Science and Technology*, **13**, 676-78 (2000).

[16]S. Haseyama, S. Kohayashi, J. Ishiai, S. Nagaya and S. Yoshizawa, "Preparation and Properties of Melt-Processed Large RE (Eu, Gd)-System Suprconductor Bulk in Air", p.p. 701-704 in *Advances in Superconductivity XI*, Edited by K. Koshizuka and S. Tajima, Springer, Tokyo, 1999.

[17]S. Nariki, S. J. Seo, N. Sakai and M. Murakami, "Influence of the Size of Gd211 Starting Powder on the Critical Current Density of Gd-Ba-Cu-O Bulk Superconductor," *Superconductor Science and Technology*, **13**, 778-84 (2000).

[18]S. Nariki, N. Sakai and M. Murakami, "Preparation and properties of OCMG-Processed Gd-Ba-Cu-O Bulk Superconductors with Very Fine Gd211 Particles," *Physica C*, **357-360**, 811-13 (2001).

[19]S. Nariki, N. Sakai and M. Murakami, "Fabrication of Large Melt-textured Gd-Ba-Cu-O Superconductor with Ag Addition," *Physica C*, **341-348**, 2409-12 (2000).

[20]S. Nariki, N. Sakai and M. Murakami, "Processing of GdBa$_2$Cu$_3$O$_{7-y}$ Bulk Superconductor and Its Trapped Magnetic Field," *Physica C*, **357-360**, 629-34 (2001).

[21]S. Nariki, N. Sakai and M. Murakami, "Processing of High-Performance Gd-Ba-Cu-O Bulk Superconductor with Ag Addition," *Superconductor Science and Technology*, **15** (2002) in press.

[22]S. Nariki, N. Sakai, M. Murakami, "Development of Gd-Ba-Cu-O Bulk Magnets with Very High Trapped Magnetic Field ," *Physica C*, in press.

[23]E. M. Gyorgy, R. B. van Dover, K. A. Jackson, L. F. Schneemeyer and J. V. Waszczak, "Anisotropic Critical Currents in Ba$_2$YCu$_3$O$_7$ Analyzed Using an Extended Bean Model", *Applied Physics Letters*, **55**, 283-85 (1989).

[24]M. Murakami, K. Yamaguchi, H. Fujimoto, N. Nakamura, T. Taguchi, N. Koshizuka and S. Tanaka, "Flux Pinning by Non-superconducting Inclusions in

Melt-Processed YBaCuO Superconductors ," *Cryogenics* , **32**, 930-35 (1992).

[25]S. Nariki, N. Sakai, M. Tomita and M. Murakami, "Mechanical Properties of Melt-Textured Gd-Ba-Cu-O Bulk with Silver Addition,"*Physica C*, in press.

[26]S. Nariki, N. Sakai, M. Matsui, M. Murakami, "Effect of Silver Addition on the Field Trapping Properties of Gd-Ba-Cu-O Bulk Superconductors,"*Physica C*, in press.

[27]M. R. Koblischka and M. Murakami, "Pinning Mechanisms in Bulk High-T_c Superconductors,"*Superconductor Science and Technology*, **13**, 738-44 (2000).

[28]K. Nagashima, T. Higuchi, J. Sok, S. I. Yoo, H. Fujimoto and M. Murakami, "The Trapped Field of YBCO Bulk Superconducting Magnets," *Cryogenics*, **37**, 577-81 (1997).

[29]H. Fukai, M. Tomita, M. Murakami and T. Nagatomo, "Nymerical Simulation of Trapped Magnetic Field for Bulk Superconductor", *Physica C*, **357-360**, 774-76 (2001).

SYNTHESIS AND SINTERING OF MgB_2 UNDER HIGH PRESSURE

T. A. Prikhna, Ya. M. Savchuk,
N. V. Sergienko, V.E. Moshchil,
S. N. Dub, and P. A. Nagorny
Institute for Superhard Materials
2 Avtozavodskaya St.
Kiev, 04074, Ukraine

W. Gawalek, T. Habisreuther,
M. Wendt, A. B. Surzhenko,
D. Litzkendorf, Ch. Schmidt and
J. Dellith
Institut für Physikalische
Hochtechnologie
Winzerlaer Str. 10
Jena, D-07743, Germany

V. S. Melnikov
Institute of Geochemistry,
Mineralogy and Ore-Formation
34 Palladin Pr.
Kiev, 03142, Ukraine

ABSTRACT

High-pressure (HP) synthesis and sintering are promising methods for manufacturing of the bulk MgB_2 superconductive material. The available high-pressure apparatuses with 100 cm^3 working volume can allow us to use the bulk MgB_2 for practical applications such as electromotors, fly-wheels, bearings, etc. We have found that the Ta presence during HP synthesis (especially) or sintering process (in the form of a foil that covered the sample and as an addition of Ta powder of about 2-10 wt.% to the starting mixture of B and Mg or to MgB_2 powder) increases the critical current density (j_c) in the magnetic fields up to 10 T and the fields of irreversibility (H_{irr}) of MgB_2-based bulk materials. We observed the strong evidences that Ta absorbs hydrogen and nitrogen during synthesis and sintering to form Ta_2H, TaH and $TaN_{0.1}$ and prevents or reduces the formation of MgH_2 (both with orthorhombic and tetragonal structures). Ta also essentially reduces the amount of impurity nitrogen in black Mg-B (most likely, MgB_2) crystals of the matrix phase. The presence of Ta during synthesis allowed us to obtain a bulk materials with the following critical current densities j_c in 1 T field: 570 kA/cm^2 at 10 K, 350 kA/cm^2 at 20 K and 40 kA/cm^2 at 30 K. Besides, the materials demonstrated high mechanical characteristics (microhardness, fracture toughness, Young modulus).

INTODUCTION

Magnesium diboride can be synthesized under ambient, elevated or high pressures. But the characteristics that are important for practical applications of a material as a superconductor, such as critical current density, irreversible magnetic field, etc. are very sensitive to the material density, impurity content and structural defects. As it was reported by A. Serquis et al.[1], the weak connectivity between domains and the presence of impurities in the grain boundaries in the MgB_2 could be the reason for the limited j_c at high fields, although not as strong as in high-Tc superconductors. They are hot isostatically pressed (HIPed) at 200 MPa MgB_2 samples demonstrated the improved field dependence of j_c through better connectivity of the grains, the generation of dislocations, and the destruction of MgO at MgB_2 boundaries, which were redistributed in the form of fine particles inside the MgB_2 matrix. These defects on authors' opinion can act as effective flux pinning centers. The HIPed samples had also a higher irreversibility field.

C.U. Jung et al.[2] after the investigation of a very pure single crystal of MgB_2 have suggested that the strong bulk pinning reported for polycrystalline material might be due to entirely extrinsic pinning sites, such as grain boundaries and crystallographic defects. This is consistent with the absence of core pinning, even at T~0.5xT_c, for bulk samples.

As a result of atomic resolution study of synthesized MgB_2 (experimental Z-contrast images and EEL spectra) R.F. Klie et al.[3] found ~20-100 nm sized precipitates that were formed by the ordered substitution of oxygen atoms onto boron lattice sites and that the basic bulk MgB_2 crystal structure and orientation were preserved. The periodicity of the oxygen ordering was dictated by the oxygen concentration in the precipitates and primarily occured in the (010) plane. The presence of these precipitates correlated well with an improved critical current density and superconducting transition behavior, implying that they act as pinning centers.

V.V. Flambaum et al.[4] investigating the hydrogenation of MgB_2 powder have found out that incorporation of hydrogen into MgB_2 structure can influence the superconductive properties (transition temperature) of the compound. They observed the increase in the transition temperature (T_c) of the MgB_2H_x powder above that of the original powder, at least by $\sim 0.5K$ for the $MgB_2H_{0.03}$ material. Authors suggest that these results confirm the assumption of a T_c dependence on high frequency modes for an electron-phonon mediated superconductor. But whether hydrogenation can substantially increase the T_c and the material, especially in the solid polycrystalline form, can be stable it's not studied yet.

The aim of the present investigations was to find the high-pressure-temperature conditions of synthesis (from Mg and B) and sintering (from MgB_2 powder) materials with high critical current density j_c and irreversible field H_{irr} and to study the correlations between the materials structure and properties. We continued also to study the positive effect of Ta presence during synthesis on superconductive properties of MgB_2 that was previously reported by us[5].

We succeeded in the synthesis of MgB_2-based materials with j_c and H_{irr} higher than those reported by Kijoon H. P. Kim et al.[6] and A. Serquis et al.[1] In Fig.1 we compared our data on the high pressure synthesized MgB_2 with the data on high-pressure sintered MgB_2 by Kijoon H. P. Kim et al.[6].

EXPERIMENTAL

In the experiments on synthesis, Mg and B have been taken in the stoichiometric ratio of MgB_2. To study the influence of Ta, the Ta powder has been added to the stoichiometric mixture of Mg and B powders in the amount of 2 and 10 wt. %. Then we mixed and milled the components in a high-speed activator with steel balls for 1-3 min. The obtained powder was compacted into tablets. In our experiments, we used two types of initial amorphous boron: type "A" – 95-97 % purity and type "B" – commercial boron produced about 25 years ago that during storing was partly transformed into H_3BO_3. For the experiments on sintering the commercial MgB_2 powder of Alfa Aesar company have been used.

The high-pressure conditions have been created inside a high-pressure apparatuses (HPA) of the recessed-anvil and cubic (six punches) types, described elsewhere.[7] Both types of apparatuses are usually used for diamond synthesis. They were slightly modernized for our purposes in order to measure temperature by thermocouples and to prevent the contact of the sample with the graphite heater. The working volume of the biggest cube-type HPA is of about 100 cm^3 (sample can be up to 60 mm in diameter). In

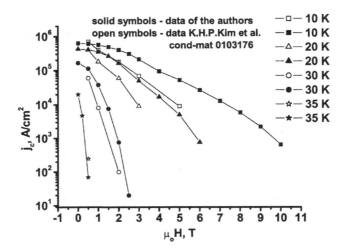

Fig.1. Data on critical current density j_c vs. magnetic field μH_o for the bulk MgB$_2$ samples:

solid symbols - high pressure synthesized at 2GPa, 800 – 900 °C, 1h from Mg and B (of type "A") by the authors;

open symbols - high pressure sintered from MgB$_2$ powder at 3 GPa, 950°C by Kijoon H. P. Kim et al. [6]

our experiments, the sample was in contact with a compacted powder of hexagonal BN or enveloped in a Ta-foil and then placed inside the compacted BN or ZrO$_2$.

The structure of materials was studied using SEM and energy dispersive X-ray analysis, polarizing microscopy, X-ray structural and phase analysis. The j_c was estimated from magnetization hysteresis loops obtained on an Oxford Instruments 3001 vibrating sample magnetometer (VSM) using Bean's model[8]. Hardness was measured on a Matsuzawa Mod. MXT-70 microhardness tester by a Vickers indenter. Nanohardness and Young modulus were investigated using Nano Indenter-II, MTS Systems Corporation, Oak Ridge, TN, USA. The fracture toughness was estimated from the length of the radial cracks emanating from the corners of an indent.

High Temperature Superconductors

RESULTS AND DISCUSSION

Figure 2 shows VSM measurements of sintered MgB_2 powder. Figure 3 shows data of materials synthesized from magnesium and different types of boron. Figure 4 illustrates the structure of sintered (a) and synthesized (b) MgB_2 samples under a polarizing microscope; the structure of MgB_2 synthesized from boron type "A" in contact with BN (c) and with 10 wt.% of Ta addition (d) under a SEM.

Analysis of the obtained data allows us to conclude that the presence of Ta increases the critical current density (j_c) and irreversible field (H_{irr}) of the high pressure synthesized or sintered MgB_2-based material. Ta can be present during synthesis in the form of a foil that covers the sample or as addition to the starting mixture of B and Mg or to MgB_2 powder.

As X-ray, SEM and VSM studies show the superconductive properties (j_c, H_{irr}) of the materials are strongly influenced by the impurity content (of oxygen, hydrogen, nitrogen etc.) in the initial boron or magnesium diboride.

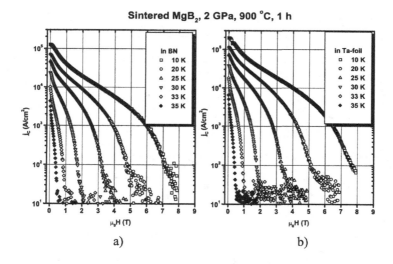

Fig.2 Critical current density j_c vs. magnetic field μ_oH for the high pressure sintered MgB_2 samples at 2 GPa, 900 °C, 1 h in BN (a) and in Ta-foil (b).

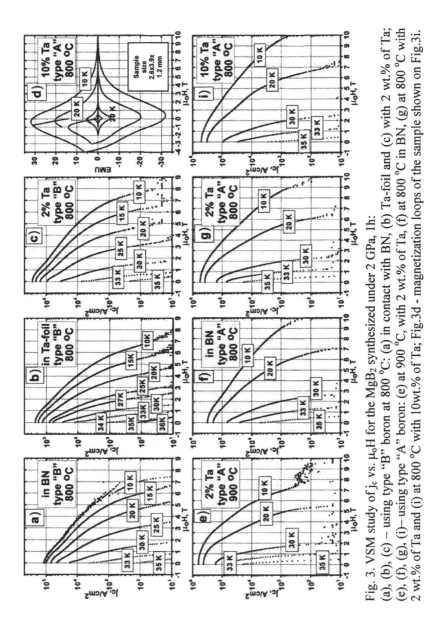

Fig. 3. VSM study of j_c vs. μ_0H for the MgB$_2$ synthesized under 2 GPa, 1h:
(a), (b), (c) – using type "B" boron at 800 °C: (a) in contact with BN, (b) Ta-foil and (c) with 2 wt.% of Ta;
(e), (f), (g), (i)– using type "A" boron: (e) at 900 °C, with 2 wt.% of Ta, (f) at 800 °C in BN, (g) at 800 °C with
2 wt.% of Ta and (i) at 800 °C with 10wt.% of Ta; Fig.3d - magnetization loops of the sample shown on Fig.3i.

High Temperature Superconductors

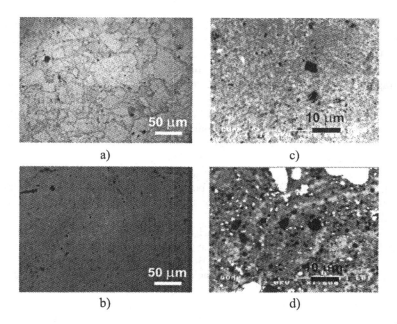

a)   c)

b)   d)

Fig.4. Photos of the structures of MgB_2 obtained at 2 GPa, 900 °C, 1 h by sintering of MgB_2 powder (a) and by synthesis from Mg and B (type "A") with 2 wt.% of Ta (b) taken under a polarizing microscope;

SEM pictures (composition image) that show the different concentrations of black Mg-B (evidently, MgB_2) grains in the samples synthesized from magnesium and type "A" boron at 2 GPa, 1h: (c) at 950 °C in contact with BN (without Ta presence) and (d) at 800 °C with 10 wt.% of Ta addition.

High-pressure synthesized and sintered samples have a multiphase nanostructure. As SEM study shows the matrix phase of the samples consists mainly of Mg, B, O. The black grains or single crystal inclusions (from the micron or even less to dozen microns in size) of Mg-B phase (MgB_2) are distributed in the matrix. The distribution of black MgB_2 grains in the matrix in high-pressure synthesized material is more homogeneous than that in the sintered one (Fig.4 a, b).

We observed the strong evidence that Ta absorbs gases during the manufacturing process:

a) Ta_2H and TaH were detected after the synthesis and $TaN_{0.1}$ was detected after the sintering in the materials, while no Ta-Mg, Ta-B or Ta-Mg-B compounds have been found;

b) we haven't observed the MgH_2 presence either with orthorhombic or tetragonal structure in the samples synthesized from boron "A" (without impurity of H_3BO_3) into which Ta (2 - 10 wt.%) was added. While in the samples synthesized without Ta addition the MgH_2 (both structures) were found. When boron "B" (commercial, with a high amount of impurity H_3BO_3) have been used as starting material, the addition of about 20 wt.% of Ta only allowed us to avoid the MgH_2 formation;

c) black inclusions (or single crystals) of MgB_2 phase in the samples synthesized from "A" amorphous boron with Ta addition contain no impurity nitrogen and less impurity oxygen than those in the samples synthesized without Ta addition.

The samples with higher j_c and H_{irr} have the higher density of Mg-B (MgB_2) inclusions in their matrix, i.e. the higher amount of black grains (see, for example, Figs. 4 c, d). Besides, in the samples with better superconductive properties, these black Mg-B inclusions contain a higher amount of boron than that in the samples with worse superconductive characteristics while the amount of magnesium is about the same. The samples with better superconductive properties also contained a higher amount of unreacted magnesium in the matrix phase.

Positive influences of Ta are much more pronounced in the synthesis process than in the sintering one. The presence of Ta extends the temperature region of synthesis of a material with high j_c and H_{irr}.

The results of investigations of mechanical properties are given in Table 1. The black inclusions (MgB_2 single crystals) have hardness higher than that of sapphire.

CONCLUSIONS

The positive effect of Ta presence during synthesis of MgB_2 on critical current density and field of irreversibility (evidences of which were shown by us earlier) has been confirmed. Ta does not react with Mg or B during the manufacturing process but absorbs gases: hydrogen, nitrogen, etc. By adding Ta and using amorphous boron free from H_3BO_3 impurity, the MgB_2-based material with high critical current density at temperatures 10 - 30 K in magnetic fields up to 2-10 T have been obtained. We observed a lot of differences in the structure of materials with different superconductive

Table 1. Results of micro- (Vickers indent) and nanohardness (Berkovich indent) investigations of the sintered MgB_2 samples

Characteristics	Matrix phase of the samples	Single crystal MgB_2	Sapphire, Al_2O_3
Indentation load: 60 mN			
Nanohardnes, H_B, GPa	17.4±1.1	35.6±0.9	31.1±2.0
Young modulus, E, GPa	213±18	385±14	416±22
Indentation load: 4.96 N			
Vickers microhardness, Hv, GPa	17.1±1.11	-	-
Indentation load: 147.2 N			
Fracture toughness, K_{1c}, $MN \cdot m^{-3/2}$	7.6±2.0	-	-
Vickers microhardness, Hv, GPa	10.12±0.2	-	-

properties and it seems that for the higher values of j_c and H_{irr}, the higher amount of black MgB_2 grains in the matrix phase (1) as well as the lower content of nitrogen and oxygen but higher boron in them (2), more uniform distribution of these black grains in the matrix (3), presence of some amount of unreacted magnesium (4) can be responsible. May be the important role played milling and mixing of the initial powders by the high-speed activator (5) and our worst results in high –pressure sintering are due to rather big sizes of grains of the starting MgB_2 powder. Because, as the polarizing microscope shows, the density of black grains in the sintered material is higher in the places of former boundaries between the initial particles, thus black grains seems to "repeat" the structure of the initial powder (Figs. 4 a, b).

The hardness of the black single crystalline MgB_2 grains turned out to be higher than that of sapphire single crystals.

The attained level of superconductive and mechanical properties of the high-pressure synthesized and sintered MgB_2 and the possibility to produce large bulk MgB_2 products make this material very promising for practical applications.

ACKNOWLEDGEMENTS

We are grateful to Prof. S.Abell (University of Birmingham) for the support of the present investigations.

REFERENCES

[1] A. Serquis, X.Y. Liao, Y.T. Zhu, J.Y. Coulter, J.Y. Huang, J O. Willis, D.E. Peterson, F.M. Mueller, N.O. Moreno J.D. Thompson, S.S. Indrakanti and V.F. Nesterenko, "The influence of microstructures and crystalline defects on the superconductivity of MgB_2", *cond-mat*/0201486.

[2] C.U. Jung, J.Y. Kim, P. Chowdhury, Kijoon H.P. Kim and Sung-Ik Lee, D.S. Koh, N. Tamura, W.A. Caldwell and J.R. Patel, "Microstructure and pinning properties of hexagonal-disc shap ed single crystalline MgB_2", *cond-mat*/0203123.

[3] R.F. Klie, J.C. Idrobo, N. D. Browning, A.C. Serquis, Y.T. Zhu, X.Z. Liao and F.M. Mueller, "Observation of coherent oxide precipitates in polycrystalline MgB_2", *cond-mat*/0203292.

[4] V.V. Flambaum, G.A. Stewart, G.J. Russell, J. Horvat and S.X. Dou, "Superconducting transition temperature of $MgB_2H_{0.03}$ is higher than that of MgB2", *cond-mat*/0112301.

[5] T.A. Prikhna, W. Gawalek, A. B. Surzhenko, N.V. Sergienko, V.E. Moshchil, T. Habisreuther, V. S. Melnikov, S.N. Dub, P.A. Nagorny, M. Wendt, Ya.M. Savchuk, D. Litzkendorf, J. Dellith, S. Kracunovska, Ch. Schmidt, "The high-pressure synthesis of MgB_2", *cond-mat*/0109216

[6] Kijoon H. P. Kim, W. N. Kang, Mun-Seog Kim, C.U. Jung, Hyeong-Jin Kim, Eun-Mi Choi, Min-Seok, Park & Sung-Ik Lee, "Origin of the high DC transport critical current density for the MgB_2 superconductor", *cond-mat*/0103176.

[7] T. Prikhna, W. Gawalek, V. Moshchil, S. Dub, T. Habisreuther, V. Melnikov, F. Sandiumenge, V. Kovylayev, A. Surzhenko, P. Nagorny, P. Schaetzle , A. Borimsky, "Improvement of properties of Y- and Nd-base melt textured high temperature superconductors by high pressure-high temperature treatment", p.p.153-158 in *Functional Materials,* Edited by K.Grassie, E. Teuckhoff, G.Wegner, J.Hausselt, H.Hanselka, EUROMAT99-V.13, 2000.

[8] C. B. Bean, "Magnetization of high-field superconductors," *Rev. Mod. Phys.*, **36** 31-36 (1964).

ADHESIVE COATED HTS WIRE AND OTHER INNOVATIVE MATERIALS

Anatoly Rokhvarger and Lubov Chigirinsky, TROG Corporation, Brooklyn, NY

ABSTRACT

Prepared colloid suspension intermixture in toluene solvent of YBCO or another HTS ceramics ultra fine powder, metal dope (mostly silver) and multi-purpose silicone rubber additive. This homogeneous slurry is very suitable for an application of all known ceramics forming methods. One method is pulverizing or dip adhesive coating deposition of HTS compound on silver, metal alloy or quartz glass screen or strand/thread substrates. The second one is ink printing/laser drawing of electronic circuits. After firing shrinkage, a thickness of the precipitated adhesive compound layer can be 3--12 microns as desired. Ceramics filled silicone polymer allows electrical or magnetic grain orientation of HTS ceramic particles. Silicone emulsion thin films prevent initial ceramic degradation while polymerized ceramic-silicone semi-products can be long-stored before firing. After firing and oxygenation, produced HTS coated wire, screen, and electronic circuits have the required values of electrical current carrying capacities and increased durability and reliability in air and liquid nitrogen ambiences. Two US patents have been received and several more patents are pending.

INTRODUCTION

A few institutions and affiliated scientific groups have applied a wet method of the Ceramic Engineering for HTS ceramic material preparation and substrate tape coating by superconductor ceramic suspension [1-7]. All authors pointed out their motivation: wet ceramic techniques should result in less expensive manufacturing than use of their dry counterparts. They produced one-sided coated silver tape substrate with restricted electrical current carrying capacity, reliability and durability. However, these groups just discredited "ceramic approach" since they did not pay enough attention to used additives and therefore they could not solve adhesion and HTS ceramic degradation problems. Additionally, no group could receive patents for what they had developed.

Most "superconductor" specialists are involved in very complicated and therefore professionally prestigious PIT and IBAD "dry" methods and their modifications. Being focused on these "metallurgical" or "physical" paradigms, these specialists are skeptical about a possibility of the application of a well-known ceramic engineering approach for effective fabrication of superconductor materials, such as HTS ceramic wire. Additionally, they believe that the "ceramic method" proposed by us is "too obvious and simple" to be attractive for long-term R&D and capital investments by the US Government and the private sector. Below is a summary of technical problems that are exhibited and/or were listed as substantiating the doubts of some physicists and engineers.

NEW HTS-CSP TECHNOLOGY
We formulated and solved (eliminated) the following technological problems:
1. The available HTS ceramic powder consists of rigid particles and therefore such powder is not formable for continuous and flexible HTS wire, tape, coil, and shaped bulk products, such as continuous rods, beams or rails. The ceramic articles are brittle and therefore cannot be cut, polished or drilled.
2. Original HTS ceramics superconductivity degrades both during the production process and service period in both a conventional air environment and liquid nitrogen coolant.
3. Original HTS ceramics lose oxygen from its crystal lattice during the ceramic firing process that results in an unavoidable loss of superconductor properties.
4. Existing methods of material preparation and forming of the bulk-shaped HTS article from dry powder yields a significant scrap percentage due to unavoidable irregular distribution of ceramic particle sizes and shapes/forms. Such irregularities of the raw poly-crystal ceramic powder cause cracks and non-homogeneity of the sintered ceramic body shrinkage, which interrupt electrical current flux. If we add some powder dope, the non-homogeneity of the multi-component mixture will be increased.
5. Existing methods of forming ceramic articles from dry powder can not now yield commercially acceptable product and a cost acceptable to the market. That material cannot be produced in a conveyor and fully automatic production system resulting in scrap-free and quality assured continuous HTS wire and tape, as well as cable, rods, beams, and rails.
6. Colloid chemistry phenomena of the ceramic particles impact adhesion to other substrates while appropriate adhesion additives and technique were not being developed.
7. Employment of condensing and drying of the slurry suspension to provide either injection-molding/extrusion or dry pressing forming of HTS bulk shaped products need an appropriate binder additive.
8. Special grain orientation and alignment techniques need to be developed to impact ceramic crystals within the ceramic body that should result in at least 50x higher superconductivity of the HTS products.

High Temperature Superconductors

9. It is necessary develop a unique ceramic firing mode, which results in high density and full integrity of the HTS ceramic compound body since cracks, pores and voids interrupt electrical current carrying.

10. It is necessary to develop a new ceramic formulation and an appropriate thermal sintering mode that will result in crystal growing and size unification following decreased grain boundary areas.

11. It is necessary to develop the new ceramic formulation and related thermal sintering mode that will result in ceramic body density at least 97-98% of the theoretical ceramic compound density; with no cracks, voids and pores.

12. While a dope should significantly increase the electrical current density of the HTS composite, optimal type/composition and percentage of the dope are still under discussion.

13. In applying dip coating or other wet forming techniques, previously used dispersants and binders in other alcohol solvents produce HTS ceramics suspensions that do not result in proper adhesion to the metal, alloy, ceramics, and quartz glass substrates.

14. Presently used PIT/OPIT and IBAD/RaBiTS methods of HTS tape production are associated either with precise jewelry/metallurgy or vapor deposition physics techniques. Meanwhile both techniques are very expensive in comparison with conventional techniques of copper wire production.

15. The commonly used silver sheaths of the HTS-PIT methods or multi-layer rare-earth substrates / templates of IBAD methods are very expensive. At any anticipated volume ratio HTS-ceramics: silver-sheath/template-substrate, the total raw material cost is too high and these HTS products never can be usable in broad commercial practice.

16. A list of promising inorganic materials with superconductor and specific physical-chemical properties is increasing, which makes all existing HTS technologies temporary.

17. To be competitive with conventional electrical copper wire, HTS wire has to have similar reliability and durability, has to be combined/assembled from non-restricted length and number of wound/twisted round strands.

18. To be competitive with copper wire, HTS electrical wire has to have superconductor transition temperature $T_C > 77K$, magnetic susceptibility $T > 0.1$ Tesla, electrical current carrying capacity $J > 20kA/cm^2$, and cost-performance ratio C/P < \$55/kA-m for wire cross-section one cm^2.

19. High electrical current carrying density causes specific problem for contacts of HTS leads/wires in a cable or distribution net system.

20. Most known and currently used wire insulation plastic materials became brittle and were destroyed at liquid nitrogen temperature.

21. To advance the state for the art in electronics and provide/bring specific superconductor effects and technical advances for existing and new products, devices and services, customized techniques have to be devised with the

High Temperature Superconductors

same or similar workability, as used in existing electronic and specific article production technologies.

22. There are not developed yet:

- The multi-step firing and oxygenating modes of the HTS ceramic materials
- Advanced technology for the conveyor fabrication of HTS round wire
- Advanced technology for the conveyor ink-printed 2D HTS electronics
- Advanced technology for laser drawing HTS 2D and 3D electronics
- Technology for spray coating HTS large radar/radiation screen/shield surfaces
- Advanced fabrication technology for dry pressed HTS bulk-shaped products
- Injection molding/extrusion forming HTS bulk-shaped products

Our newly invented *Ceramic Silicone Processing (CSP)* method [8, 9] solves all of the problems mentioned above and overcomes all disadvantages of the prior arts. Our CSP method uses the well-known wet ceramic processing technique. Employing both Chemistry and Physics phenomena, we demonstrate the technical workability and technical and cost advantages of the paradigm-shifting and revolutionary cost-effective Ceramic-Silicone-Processing (CSP) method. This method includes a highly workable material formulation suitable for versatile conveyor production of a new class of durable HTS materials.

First to be produced is flexible round HTS-CSP wire on the base of readily available superconductor ceramic powder, such as $YBa_2Cu_3O_{7-x}$ (YBCO) ceramics. This ceramic is now the premier choice of high temperature superconductor materials since YBCO has ten times higher electrical current carrying capacity and several times higher magnetic susceptibility than other HTS ceramics. Meanwhile, as stated in [8, 9], our CSP method is equally applicable for all ceramic and intermetallic solid powders, including those that may be coming for room temperature HTS (which would be developed at some point). MgB_2 or $CaCuO_2$ ceramics have significant disadvantages when compared to YBCO.

Figure 1. HTS-CSP sample of the flexible HTS strand. A part of the YBCO compound-coating layer of ~10μm thickness on the metal substrate was intentionally removed to demonstrate a strand structure.

We applied a ceramic engineering approach utilizing off-the-shelf superconductor ceramics raw materials and we used the wet method of raw material preparation and associated broad spectrum of workable and inexpensive forming methods. An application of the wet ceramic method usually includes a choice of the appropriate dispersant, binder and particular additives to control

High Temperature Superconductors

material formulation, including its technological behavior and required consumer properties. Our invented HTS-CSP material formulation includes silicone rubber HO-[-Si(CH₃)₂O-]-H as an empirically discovered key multi-functional component, which provides perfect adhesion of the initial HTS-CSP compound content suspension on metal and other substrate surfaces and plastic properties for condensed suspension. Additionally, this additive significantly influences both the processing and quality of HTS-CSP wire and other HTS-CSP materials.

Below are listed synergetic impacts of the silicone adhesive additive [10].

1. An inhibitor of the degradation of HTS ceramics at material preparation and forming of the HTS compound or article, such as wire

2. A dispersant in obtaining homogenization of the HTS compound suspension in toluene solvent

3. As an adhesive component in material suspension for adhesive precipitation of coating suspension particles on metal, alloy, ceramics, quartz glass or glassy carbon substrate structures applying a set of techniques for 2D and 3D coating. These structures can have any size and form, such as a continuous strand, fabrics and clothes, screen/shield, tape, and coil. Adhesive coating methods include dip coating, slip casting, brushing, ink printing, drawing, and surface spraying

4. A plastic binder for other forming methods including extrusion of the condensed suspension mass

5. A polymerization hardening material, which keeps semi-product form, including coating layer form, for further thermal treatment and temporary storage

6. A stimulator of the ceramic grain alignment in the form of spirals and a media for magnetic orientation of ceramic crystals

7. As a magnetic susceptive media filled by metal powder dope and HTS ceramics, which provides magnetic two-three dimensional crystal orientation within the green ceramic body.

8. A sintering aid since at temperatures higher than 850^0C silicon atoms react with oxide ceramic components of YBCO producing liquid eutectics of silicate glasses, which significantly decreases the sintering temperature and increases ceramic compound density and integrity through sticking grain boundaries due to incongruently point-melted ceramic body

9. A protective component in the form of silicate glass films which enhance durability and reliability of the HTS composite material in conventional air conditions and during direct contact with liquid nitrogen coolant, since it produced super-thin silicate glass films that coat the crystals of HTS ceramics

10. Silicate glass nano-size films increase ductility, flexibility and strain tolerance of the HTS composite material, which makes possible polishing, cutting and drilling of bulk-shaped HTS materials since the super-thin silicate glass films it produced are very thin and therefore flexible.

11. Adhesion self-controlled thickness of the sintered coating layer of the HTS/CSP strand is about 10-12 microns, which allows easy bending of such coated strands and winding them in a multi-strand wire/cable of any thickness. Meanwhile such coating layer on a substrate strand with diameter 0.06mm provides a substrate/superconductor volume ratio of 1:1. This makes HTS-CSP wire very effective from electrical engineering and economics points of views.

PROCESS AND QUALITY CONTROL

Developing process and quality control systems for HTS-CSP materials, we employed those colloid chemistry and superconductor effects/phenomena that can be controlled by particular engineering leverages [10]:

1. Using profilometer and optical microscopes we estimate quality and thickness of the adhesion coating layer

2. We provide magnetic impact for crystal/grain orientation within a green polymer compound body to obtain shorter pathways for the electrical current flow along each current lead. Practically in the lab, we attached to both wire or circuit ends powerful rare-earth magnets

3. At 450^0C +/- 100^0C we deliver an oxygen flow through the lower end of the tube furnace. Insertion-back of oxygen atoms results in rebuilding the orthorhombic superconductor crystal structures of the CSP compound. Only such crystals have superconductor properties

4. We apply a special firing mode including incongruent melting to increase grain boundary contacts and material density that make possible the effect of the Josephson junction phenomenon

5. We provide doping of the HTS compound to make possible proximity effect phenomenon to increase electrical carrying capacity

6. At high sintering/firing temperatures silicone residuals interact with HTS ceramic oxides producing liquid eutectics (incongruent melting points) These eutectics harden as silicon glass thin films of the angstrom-size thickness that adheres to grain boundaries and significantly increase material integrity and density being "penetrable" for Josephson junction. These silicon glass films also increase material durability and flexibility/ductility. Dopes and silicon glass impurities provide pinning centers to increase electrical carrying capacity.

TECHNICAL AND COST ADVANTAGES OF HTS-CSP TECHNOLOGY

For all electrical engineering applications the desirable/working level of heat/energy dissipation (W) of any electrical current lead is about 1-3Watt/cm^2 while, taking in account continuous circulation of LN coolant, such working level of W for HTS generators, motors, transformers and cables may be in 2-5x higher. International and US DOE have attempted to determine costs of electrical conduction wire in correspondence with its consumer properties in dollars per one meter of wire applicable for working electrical current density is kA/cm^2. A ratio *(C/P)* is the traditional and important consumer and market characteristic of electrical wire and other current leads. A milestone is C/P value of market dominated conventional copper wire, which has at room temperature electrical current carrying capacity J = 200A/cm^2 and C/P=\$55-\$60/kA-m. For comparison, for PIT/OPIT C/P = \$200–\$300/kA-m applying wholesale price of silver and achieved J [11, 12]. Meanwhile, the U.S. DOE goals for HTS tape are C/P=\$55/kA-m by 2004 producing just one km of HTS tape prototype.

Industrial conveyor production of HTS-CSP wire will demand and consume several hundred tons of YBCO powders. This and special reconstruction will allow "Superconductive Components, Inc., Columbus, OH significantly decrease a price of the supplying YBCO ceramic powder. Significant expenses cause electrical energy and oxygen consumption. Marketing/sale, administrative, labor, finishing and insulation work expenses would increase production cost of the HTS-CSP wire on ~15%. For HTS-CSP wire/tape we are achieving J = 30–40kA/cm^2 at negligible heat dissipation in liquid nitrogen coolant media just about 1.0Watt/cm^3. Using retail prices of substrate and raw materials, the total production cost of 0.5mm in diameter multi-filament HTS-CSP wire assembled from nickel or nickel alloy strand substrates is about \$1.3/m and correspondingly C/P = \$15/kA-m. Consequently, production cost of HTS-CSP tape on quartz filament fabric substrate would be \$0.7/m and correspondingly C/P = \$8/kA-m.

Electrical current carrying capacity J = 40kA/cm^2 of HTS-CSP wire is higher than J = 0.2kA/cm^2 of the copper wire in 200 times. This decreases a size (diameter for cable current lead) and weight of the HTS-CSP wiring cable, transformer, motor, and generator rotor in 10–15 times and costs in 5-10 times. It means the multiple business advantage/benefit of usage of HTS-CSP wire in comparison with copper wire can be estimated as >200x. Meanwhile used for HTS wire liquid nitrogen coolant is environmentally friendly and cheaper than coolant oil uses taking in account oil disposal and soil cleanup problems.

For SMES and microelectronics the C/P characteristic is not applicable while a required level of energy/heat dissipation W should be up to 1×10^{-3}W/cm^3. At such W HTS-CSP electrical current leads/strands transfer electrical current density J~10-15kA/cm^2, which is achieved now the critical value of J for HTS-CSP wire/strands. Application of such strands can miniaturize in 5-7 times sizes and/or thickness of electronic circuit leads. While we do not consider special superconductor effects in electronics, only HTS-CSP current leads can increase the electronic package densities at least 5 times [13-15].

OTHER ADVANCEMENTS OF HTS-CSP MATERIALS

During a period of nine months we repeated measurements of transported electrical current for the same HTS strand samples with the same measurement results while these samples were not protected or insulated during their exposure to the open air and dozens of immersions in liquid nitrogen media. Additionally, two CSP tablets have been measured with the same results during two years while we have been developing and testing HTS wire technology. CSP disks also kept their superconductor properties after at least 700 cyclic submerge into liquid nitrogen. It was compared with pure YBCO tablets prepared under the same conditions. They lost their superconductor properties after 100-140 cyclic submerges. These all exhibit the durability and reliability of the CSP materials.

The specific impact strength values of the CSP tablets are in the range of $0.5 \text{kg} \cdot \text{cm/cm}^2 - 2 \text{kg} \cdot \text{cm/cm}^2$. Before firing, polymerized CSP beam, disk or another bulk-shaped resin-like material can be easy machined (cut or drilled). Fired CSP material can be polished as is provided for other known ceramics. If a thickness of the sintered coating layer of the HTS-CSP on a flexible substrate is fewer 20 microns, such strand, tape, fabrics/cloth or foil can be bent/arched.

Applying our newly invented silicone compound, which is a unique polymer material since it can withstand 77K, such elementary HTS-CSP wire will be insulated/polymer coated on the same conveyor line similarly to any round copper wire used everywhere in Electrical Engineering and Electronics.

COMPARISON WITH TWO OTHER TECHNIQUES

The following table provides comparison of the major engineering and economic characteristics of TROG's Ceramics-Silicone Processing (CSP-method) with two other developing methods in the field of HTS wire. The first one, which is associated with the first generation of HTS wire, is Powder in Tube or Oxide Ceramics Powder Loaded In Silver Tube (PIT/OPIT-method). The second generation HTS wire is associated with Ion Beam Assisted Deposition (IBAD-method) and its modifications, such as the RABiTS method.

As one can see in Table I, all twenty-one-engineering characteristics of the CSP-method are the best.

CONCLUSION

Invented and proved in lab we know have a paradigm-shifting chemical ceramic technology named ceramics-silicone-processing – CSP. CSP allows production of adhesion coated wire, electronic circuits and systems and other HTS materials that meet all consumer requirements and can be manufactured at a performance cost which is very cost competitive with conventional copper wire.

HTS-CSP continuous and flexible round wire and other 2D and 3D materials are highly beneficial for all end products of the Electrical, Transportation, Electronics, Medical, Defence, Mechanical, Wireless Internet and Telecommunication Industries utilizing electrical and magnetic energy. Additionally CSP technology opens the door for beneficial applications of unique superconductor phenomena for many new techniques and services of 21st century.

Table I. Comparison of the PIT/OPIT, IBAD/RABiTS and CSP Technologies

#	Characteristics	PIT/OPIT	IBAD/RABiTS	CSP
1	Preferable geometrical form of the wire	Tape/Strip	Tape/Strip	Continuous round strand
2	Substrate materials and their geometrical form	Silver tube	Multi-layer composite tape	Metal alloy, quartz glass, and ceramics of any shape
3	Wire forming method	Rolling/drawing in silver tube	Atom-by-atom deposition on a tape template	Dip adhesive coating
4	Thickness of the HTS current lead layer, μm	5 – 20	0.01 – 1.5	> 2 (10 – 12 for dip coated strand)
5	Comparative carrying capacity for an equal wire cross section	1 (tape)	(0.3 for one layer) 1 (tape)	5 (round strand)
6	Volume ratio of the substrate to current lead	2 : 1	(50 -- 100) : 1	1 : 1
7	Ratio of production costs to raw material costs	(2 – 3) : 1	30 : 1	0.1 : 1
8	Comparison of capital/equipment costs	2 : 1	30 : 1	1
9	Critical current density of the sample, kA/cm^2	30 – 50	3 ,000 (for 0.3μm film)	20 – 40
10	Usable level of carrying capacity, kA/cm^2	2 -- 5	unknown	up to 20 – 40
11	Wire contacts and isolation sheath techniques	A	D	A
12	Comparison of AC current losses of wire	2	3	1
13	Scrap percentage during wire production cycle	30 – 50	30 – 50	~ 0
14	Estimation of wire flexibility	A	B	A
15	Cost/performance of wire (C/P), $/kA-m	200 -- 300	Not applicable	7 -- 14
16	Wire workability/customer suitability	B	D	A
17	Circuits for electronic systems and packages	Impossible	Impossible	Ink printing 2D and 3D circuits
18	Production of the bulk shaped materials	Impossible	Impossible	Extrusion or press forming all shapes
19	Workability for any oxide/non-oxide ceramics or intermetallic powders	A	B	A
20	Estimation of material ductility	Not applicable	Not applicable	A
21	Material durability in liquid nitrogen and air	A	D	A

High Temperature Superconductors

REFERENCES

1. Masur, L., et al., July 1999, "Long Length Manufacturing of BSCCO-2233 Wire for Motor and Cable Applications", *International Cryogenic Materials Conference, Montreal, Canada, www.amsuper.com,* 7pages

2. Buhl, D., et al., 1994, "Processing, Properties and Microstructure of Melt-Processed Bi-2212 Thick Films", *Physica,* C 235-240, pp. 3399-3400

3. Marken, K.R., et al., 1997, "Progress in BSCCO-2212/Silver Composite Tape Conductors", *IEEE Trans. Appl. Superconductivity,* vol. 7, pp. 2211-2214

4. Hasegawa, T. et al., June 1997, "Fabrication and Properties of $Bi_2Sr_2CaCu_2O_y$ Multilayer Superconducting Tapes and Coils", *IEEE Trans. Appl. Superconductivity,* vol. 7, No. 2, pp.1703-1706.

5. Ilyushechkin, A.Y., et al., 1999, "Continuous Production of Bi-2212 Thick Film on Silver Tape", *IEEE Trans. Appl. Superconductivity,* vol. 9, 1912-1915

6. Walker, M., et al., 1997, "Performance of Coils Wound from Long Length of Surface-Coated, Reacted, BSCCO- Conductor", *IEEE Trans. Appl. Superconductivity,* vol. 7, No. 2, pp. 889-892

7. Dai, W., et al., 1995, "Fabrication of High T_C Coils from BSCCO 2212 Powder in Tube and Dip Coated Tape", *IEEE Trans. Appl. Superconductivity,* vol. 5, No.2, pp. 516-519

8. Topchiashvili, M.I. and Rokhvarger, A.E., Date of Patent: Jan. 4, 2000, "Method of Conveyor Production Of High Temperature Superconductor (HTS) Wire, Coil, and Other Bulk-Shaped Products Using Compositions of HTS ceramics, Silver, and Silicone", *The U.S. Patent No. 6,010,983,* 32 claims, 3 drawings

9. Topchiashvili, M.I. and Rokhvarger, A.E., Date of Patent: May 29, 2001, "High Temperature Superconductor Composite Material", *The US Patent No. 6,239,079,* 18 claims, 8 drawings

10. Rokhvarger, A., Chigirinsky, L., and Topchiashvili, M., 2001, "Inexpensive Technology for Continuous HTS Round Wire", *The American Ceramic Society Bulletin,* Vol. 80, No. 12, pages 37-42

11. Malozemoff, A. P., et al., 1999, "HTS Wire at Commercial Performance Levels" *IEEE Trans. on Applied Superconductivity,* vol.9, #2, pp.2469-2473

12. Lindsay, D., Winter 2001, "Southwire High Temperature Superconducting Power Delivery System", *Superconductor & Cryoelectronics,* pp. 27-34

13. *Heat Management Chapter, The Electronics Handbook,* 1996, J. C. Whitaker, ed., CRC Press and IEEE Press, pp. 1133-1151

14. Ciszek, M., et al., 1995, "Energy Dissipation in High Temperature Ceramic Superconductors", *Applied Superconductivity,* vol. 3, issue 7-10, pp. 509-520

15. Paracchini, C. and Romano, L., 1996, "The Dissipation of a Superconducting BSCCO Film on the I-T Plane" *Physica C, Superconductiv.,* vol.262, pp. 207-214.

MELTING EQUILIBRIA OF THE BaF_2-CuO_x SYSTEM

W. Wong-Ng, L.P. Cook, and J. Suh
Ceramics Division
National Institute of Standards and Technology
Gaithersburg, MD 20899

ABSTRACT

Since the BaF_2-CuO_x system is a limiting binary of the overall quaternary reciprocal Ba,Y,Cu//F,O phase diagram, it provides an important frame of reference for studying the more complicated multi-component system. The phase diagrams of the BaF_2-CuO_x system were determined under 0.1 MPa O_2 (BaF_2-CuO) and under 0.1 MPa Ar (BaF_2-Cu_2O). The phase diagrams of these two systems are similar. No intermediate binary compound was found in either system. There is experimental evidence that supports a possible small two-liquid immiscibility field in the BaF_2-CuO system. A series of schematic ternary diagrams showing the oxidation/reduction aspects of the melting equilibria with increasing temperature is presented.

INTRODUCTION

The *ex-situ* BaF_2 process is currently one of the most promising methods of producing long-length coated conductors [1-10]. It is a 2-step process that involves the low-temperature deposition of precursor layers using either high rate e-beam deposition of Y, BaF_2 and Cu onto a substrate, or open-air solution techniques, followed by a post-annealing at high temperature in the presence of water vapor under reduced oxygen pressure. Several investigators have reported that the growth of the $Ba_2YCu_3O_{6+x}$ (Y-213) films may be assisted by the presence of a low-temperature liquid. Micrographs of intermediate *ex-situ* films deposited on $SrTiO_3$ substrates at 735°C consistently show an amorphous layer between the product Y-213 and the untransformed precursor [10]. The amorphous layer has a nominal composition of Ba:Y:Cu of 2:1:1.5. It still remains unanswered as to whether the amorphous layer was due to the presence of liquid. A liquid phase, if present, could be important for enhancing formation of Y-213 through increased chemical mobility, and also for promoting textured growth.

To determine whether the Y-213 film composition, and also the composition of the amorphous layer reported by the Brookhaven Laboratory [10] yield a low-temperature melt, it is necessary to study the phase equilibria of appropriate regions in the complex multicomponent Ba-Y-Cu-F-O system (Ba,Y,Cu//O,F reciprocal system) in the presence of water. Our previous work [11] illustrated use of a triangular prism to represent this reciprocal system (Fig. 1). The oxides are represented at the base, and fluorides at the top. On the basis of Gibbs energy minimization computations, this prism can be considered as divided into three tetrahedra. The BaF_2-BaO-Y_2O_3-CuO_x system forms the lower tetrahedron (within which the final defluorination of the precursor film takes place). The other two tetrahedra are BaF_2-YF_3-CuF_x-CuO_x and BaF_2-YF_3-Y_2O_3-CuO_x. We have previously reported the presence of low-temperature melts [11] - however, they occur in abundance only in the relatively fluorine-rich regions. In order to determine the origin of liquids in the tetrahedral systems, an understanding of the phase equilibria of the subsystems is important.

The BaF_2-CuO_x system (including BaF_2-CuO and BaF_2-Cu_2O) forms an edge of the overall quaternary reciprocal oxide/fluoride system, and therefore diagrams of this system can be used as part of the frame of reference for studying the more complicated reciprocal prism. The goal of this study is to generate preliminary phase diagrams of the system BaF_2-CuO_x. As oxidation-reduction of CuO_x plays an important role in phase equilibria of the reciprocal system, the experiments were conducted under both argon and oxygen to outline the range of effects due to varying oxygen pressure between the practical limits likely to be encountered in processing.

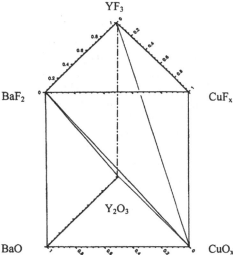

Fig. 1. A triangular prism used to represent the reciprocal system (Ba,Y,Cu//O,F)

High Temperature Superconductors

EXPERIMENTAL[1]

To investigate the system BaF_2-CuO_x, twenty samples each were prepared from BaF_2 and CuO, and from BaF_2 and Cu_2O, by weighing out the required amounts followed by extensive homogenization in a mortar with pestle. Starting reagents were of 99.99% purity (mass basis). Powdered BaF_2 was obtained by grinding fragments from a single crystal boule of optical quality material.

To verify subsolidus assemblages, annealed samples were characterized by powder X-ray diffraction using Cu Kα radiation on a Philips diffractometer with Siemens control and data acquisition software. Samples were studied by simultaneous differential thermal analysis (DTA) and thermogravimetric analysis (TGA) using a Mettler TA1 thermoanalyzer with Anatech digital control and data acquisition software and hardware. Samples were contained in MgO crucibles. During the experiments, temperature was ramped at 10 °C/min. Temperatures were measured with Pt/Pt10%Rh thermocouples calibrated against the melting points of NaCl, Au and Pd. Temperatures reported in this study have an estimated combined standard uncertainty of ± 5 °C (one standard deviation).

RESULTS AND DISCUSSION

Selected samples were annealed at 950 °C for 8 hr. The x-ray patterns of samples of a 50/50 mixture of BaF_2 and Cu_2O annealed in 0.1 MPa Ar, and of BaF_2 with CuO annealed in 0.1 MPa O_2 are shown in Fig. 2a and Fig. 2b, respectively. No intermediate phases in the subsolidus systems were found, indicating that BaF_2 is compatible with Cu_2O and with CuO.

A phase diagram of the BaF_2 – "CuO" system based on the results of the DTA/TGA experiments is shown in Fig. 3. It should be noted that as soon as temperatures above the eutectic are reached, the system loses oxygen, as evidenced by simultaneous TGA, and is therefore no longer binary. It is for this reason that "CuO" is used as the end member of the phase diagram. DTA data indicate an event just below 1100 °C which occurs without interruption across the diagram, and is consistent with a simple eutectic without intermediate compounds. An obvious feature of the diagram is the rapid rise of the BaF_2 liquidus as temperature increases above the eutectic, with a leveling off at higher temperatures. The DTA data in the region of CuO: BaF_2 = 45:55 (Fig. 4a) shows a classic indication of eutectic melting (the relatively sharp peak), followed by a gradually rising DTA signal through the two phase (liquid + crystal) region, finishing with penetration of the liquidus (the broad low curve with a truncated tail). At higher BaF_2 concentrations (CuO :BaF_2 = 5 : 95; Fig. 4b), the eutectic peak is diminished, and the higher temperature peak has become sharper. This is likely the result of the flattening of the liquidus, so that the effect on the DTA

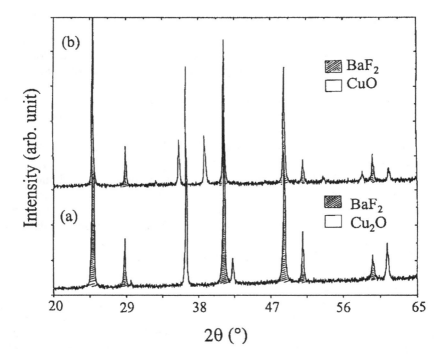

Fig. 2. X-ray diffraction patterns of a 50/50 mixture of BaF_2 and Cu_2O in Ar at 950 °C for 8 h (2a), and BaF_2 and CuO at 950 °C in oxygen for 8 h (2b).

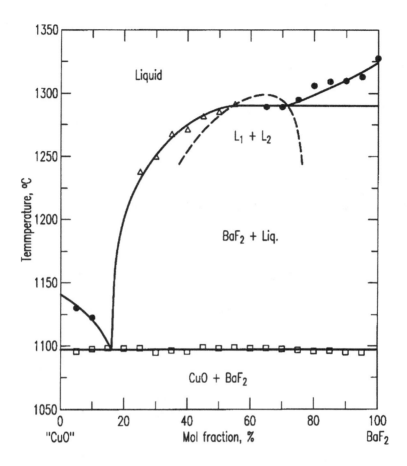

Fig. 3. A preliminary phase diagram of the "CuO"- BaF₂
system in 0.1 MPa O₂, based on the results of DTA/TGA
experiments. A small liquid immiscibility field is
proposed.

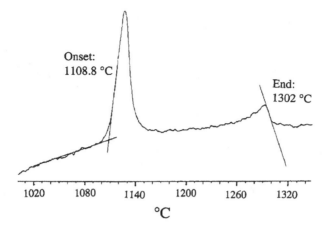

Fig. 4a. DTA data for the composition $CuO:BaF_2 = 45:55$

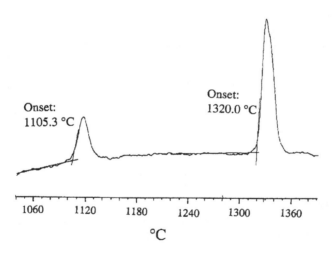

Fig. 4b. DTA data for the composition $CuO:BaF_2 = 5:95$

High Temperature Superconductors

signal is similar to that of a constant temperature reaction, leading to a sharper peak for penetration of this part of the liquidus.

There appears to be evidence that supports the existence of a small two-liquid field in the flat region of the liquidus. Figs 5a to 5c show optical micrographs of samples from three different parts of the BaF$_2$–"CuO" diagram, looking down into the bottom of the MgO crucibles, after the DTA experiments. All experiments were ramped to 1400 °C at 10 °C/min in flowing O$_2$ and then furnace-cooled. They all show smooth meniscuses, suggesting complete melting. The DTA experiments of the samples near the eutectic region (CuO: BaF$_2$ = 85 : 15; Fig. 5a) and from the BaF$_2$-rich part of the diagram (CuO: BaF$_2$ = 20 : 80; Fig. 5c) appear to be relatively homogeneous. However, the optical micrograph of the sample near the proposed two-liquid region (CuO: BaF$_2$ = 55 : 45; Fig. 5b), suggests segregation into two liquid phases, one with a lighter color and the other with a darker color. If a miscibility gap is present in the system, it will have important implications for low-melting liquids in the quaternary reciprocal system. Miscibility gaps generally expand with decreasing temperature, resulting in the possibility of metastable formation of liquids at low temperatures, at compositions which might not otherwise be possible.

The phase diagram of the reduced Cu$_2$O-BaF$_2$ system is similar to the CuO-BaF$_2$ case (Fig. 6), except that the eutectic temperature is about 100 °C higher. The S-shaped flattening of the liquidus is suggestive of incipient immiscibility. Although we have not yet found compelling evidence of the presence of two liquids in the Cu$_2$O-BaF$_2$ system, an immiscibility field may be a short distance away in p_{O2} – temperature – composition space.

To better portray the effect of copper oxidation/reduction, a series of schematic ternary diagrams showing the important aspects of the phase equilibria with increasing temperatures is given in Fig. 7(a) to Fig. 7(h). These diagrams are plotted using Cu$_2$O, O, and BaF$_2$ as end members. In Fig. 7(a), the Cu$_2$O-rich side of the diagram consists of the ternary Cu$_2$O-CuO-BaF$_2$ subsystem. By performing DTA/TGA while ramping the oxygen pressure at a value below 0.1 MPa, the eutectic melting temperature of this subsystem was determined to be at 1008 °C, which is substantially lower than that of the binary CuO-BaF$_2$ and Cu$_2$O-BaF$_2$ systems. Following the ternary eutectic melting, the liquid region expands, until it reaches the bottom Cu$_2$O-O edge (Fig. 7(b)), which corresponds to the Cu$_2$O-CuO eutectic at 1091 °C. As the temperature is increased slightly to 1097 °C (Fig. 7(c)), the BaF$_2$-CuO join is replaced by a liquid-gaseous oxygen tie line, which corresponds to the eutectic in the CuO-BaF$_2$ system. At a higher temperature of 1140 °C (Fig. 7(d)), solid CuO decomposes to liquid and O$_2$ gas. The liquid continues to expand further. At 1211 °C (Fig. 7(e)), the liquid reaches the Cu$_2$O-BaF$_2$ system, which corresponds to the eutectic of the Cu$_2$O-BaF$_2$ system. The next event, which takes place at 1229 °C (Fig. 7(f)), corresponds to

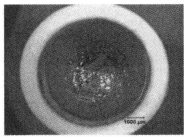

Fig. 5a. Optical micrograph of the melted sample near the eutectic region of the diagram $(CuO:BaF_2 = 85: 15)$, looking into the MgO crucible after the DTA experiment. Meniscus is smooth (reflects light) yet uniformly dark.

Fig. 5b. Optical micrograph of the melted sample near the proposed two-liquid region of the diagram $(CuO:BaF_2 = 55: 45)$, looking into the MgO crucible after the DTA experiment. Meniscus is smooth with dark region in center surrounded by light region at edges

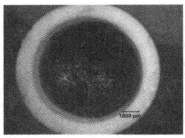

Fig. 5c. Optical micrograph of the melted sample near the BaF_2-rich part of the diagram $(CuO:BaF_2 = 20: 80)$, looking into the MgO crucible after the DTA experiment. Meniscus is smooth and uniformly dark.

High Temperature Superconductors

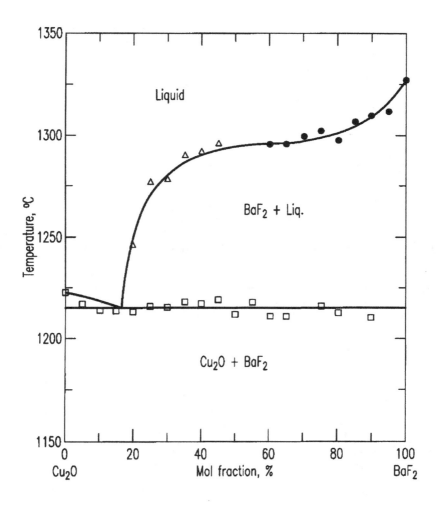

Fig 6. A preliminary phase diagram of the Cu_2O- BaF_2 system in 0.1 MPa Ar, based on the results of DTA/TGA experiments. The S-shaped flattening of the liquidus is suggestive of incipient immiscibility.

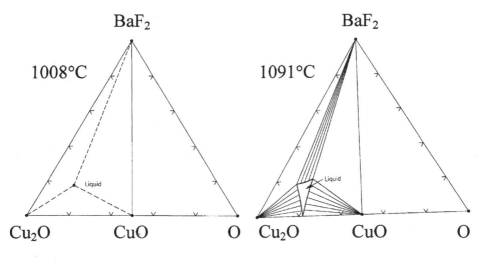

Fig. 7(a). $BaF_2 + Cu_2O + CuO \rightarrow$ Liquid

Fig. 7(b). $Cu_2O + CuO \rightarrow$ Liquid

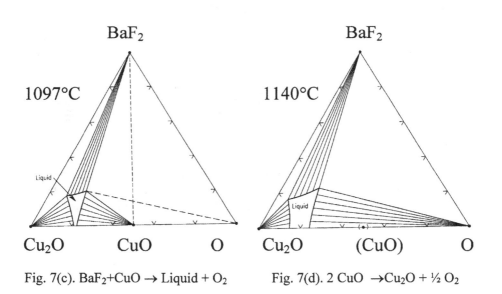

Fig. 7(c). $BaF_2 + CuO \rightarrow$ Liquid $+ O_2$

Fig. 7(d). $2\,CuO \rightarrow Cu_2O + \frac{1}{2}\,O_2$

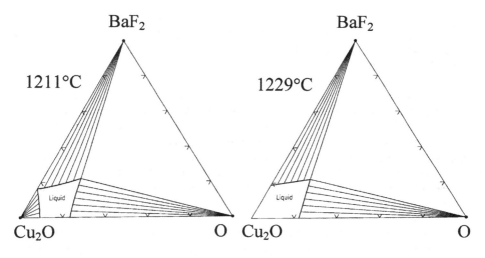

Fig. 7(e) BaF$_2$+Cu$_2$O → Liquid

Fig. 7(f) Cu$_2$O → Liquid

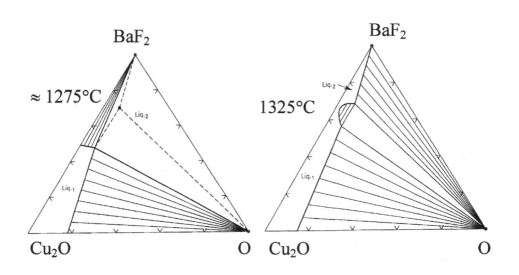

Fig. 7(g) BaF$_2$ + Liq1 +O$_2$→ Liq.2

Fig. 7(h) BaF$_2$ → Liquid

liquid expansion to the lower left corner, i.e. the melting of Cu_2O. At a temperature we estimate to be near 1275 °C (Fig. 7(g)), it is proposed that a second liquid phase appears in the ternary system. Finally, at 1325 °C (Fig. 7(h)), the BaF_2 end member melts.

SUMMARY

Melting in the BaF_2-CuO_x subsystem of the Ba,Y,Cu//F,O reciprocal system provides an important basis for understanding the origin of the more complicated quaternary low melting liquids. The presence of BaF_2-CuO_x immiscible liquids may extend to the interior of the multi-component system, where it could have an important impact on the melting of the Ba,Y,Cu//F,O reciprocal system as a whole. Investigation of the phase equilibria of other Ba,Y,Cu//F,O subsystems is being continued.

[1]*Certain trade names and company products are mentioned in the text or identified in the illustrations in order to adequately specify the experimental procedure and equipment used. In no case does such identification imply recommendation or endorsement by National Institute of Standards and Technology.*

ACKOWLEDGEMENTS

This work was partially supported by the US Department of Energy. The authors thank N. Swanson and P. Schenck for assistance in preparing the illustrations.

REFERENCES

1. R. Feenstra, T. B. Lindemer, J. D. Budai, and M. D. Galloway, *J. Appl. Phys.* **69**, 6569 (1991).
2. P.M. Mankiewich, J.H. Scofield, W.J. Skocpol, R.E. Howard, A.H. Dayem, and E. Good, *Appl. Phys. Lett.* **51** (21) 1753 (1987).
3. A. Gupta, R. Jagannathan, E.I. Cooper, E.A. Giess, J.I. Landman, and B.W. Hussey, *Appl. Phys. Lett.* **52** (24), 2077 (1988).
4. S.-W. Chan, B.G. Bagley, L.H. Greene, M. Giroud, W.L. Feldmann, K.R. Jenkin, II, and B.J. Wilkins, *Appl. Phys. Lett.* **53** (15) 1443 (1988).
5. P.C. McIntyre, M. J. Cima, and M.F. Ng, *J. Appl. Phys.* **68** 4183 (1990).
6. P.C. McIntyre, M. J. Cima, and A. Roshko, *J. Appl. Phys.* **77** (10) 5263 (1995).
7. P.C. McIntyre, M. J. Cima, J.A. Smith, Jr., R.B. Hallock, M.P. Siegal, and J.M. Phillips, *J. Appl. Phys.* **71** 1868 (1992).
8. R. Feenstra, US Patent 5,972,847 (1999).
9. V.F. Solovyov, H.J. Wiesmann, M. Suenaga, and R. Feenstra, *Physica C* **309**,

269 (1998).

10. L. Wu, Y. Zhu, V.F. Solovyov, H.J. Weismann, A.R. Moodenbaugh, R.L. Sabatini, and M. Suenaga, *J. Mater. Res.* **16**, 2869, 2001.
11. W. Wong-Ng, L.P. Cook, Materials for High-Temperature Superconductor Technologies, Materials Research Society Symposium Proceedings, **689**, eds. M.P. Paranthaman, M.W. Rupich, K. Salama, J. Mannhart, and T. Hasegawa; Nov. 26-29, 2001, Boston, MA; Materials Research Society, Warrendale, PA 15086, pp. 337 (2002).

KEYWORD AND AUTHOR INDEX